STYLES OF SCIENTIFIC THOUGHT

Science and Its Conceptual Foundations
David L. Hull, Editor

STYLES OF SCIENTIFIC THOUGHT

The German Genetics Community
1900–1933

JONATHAN HARWOOD

The University of Chicago Press
Chicago and London

Jonathan Harwood is senior lecturer in the Centre for the History of Science, Technology & Medicine at the University of Manchester, UK.

The University of Chicago Press, Chicago 60637
The University of Chicago Press, Ltd., London
© 1993 by The University of Chicago
All rights reserved. Published 1993
Printed in the United States of America

01 00 99 98 97 96 95 94 93   5 4 3 2 1

ISBN (cloth): 0-226-31881-8
ISBN (paper): 0-226-31882-6

Library of Congress Cataloging-in-Publication Data

Harwood, Jonathan.
   Styles of scientific thought : the German genetics community, 1900–1933 / Jonathan Harwood.
      p.   cm. — (Science and its conceptual foundations)
   Includes bibliographical references and index.
   1. Genetics—Germany—History.   2. Genetics—Research—Germany—Methodology—History.   3. Research—Methodology.   I. Title. II. Series.
QH428.2.G3H37   1993
575.1'0943'09041—dc20   92-15321
                                                                                                     CIP

∞ The paper used in this publication meets the minimum requirements of the American National Standard for Information Sciences—Permanence of Paper for Printed Library Materials, ANSI Z39.48-1984.

For Tobs, Munkie, and Rolandus

Contents

	List of Figures	ix
	List of Tables	xi
	Preface	xiii
1	Getting Started: The Argument, Method, and Context	1

Part I: The Peculiarities of German Genetics

2	The Genetics of Development	49
3	Genetics and the Evolutionary Process	99
4	Demarcating the Discipline: Germany versus the United States	138
5	Shifting Focus	181

Part II: Styles of Thought within the German Genetics Community

6	Mapping the German Genetics Community	195
7	Imputing Styles of Thought	227
8	Mandarins Confront Modernization	274
9	The Politics of Nuclear-Cytoplasmic Relations	315
	Conclusion	351
	Archives Consulted	365
	Interviews	367
	Bibliography	369
	Index	415

Figures

1.1	Richard Goldschmidt, Erwin Baur, and Hans Nachtsheim, ca. 1920s	37
1.2	Hans Nachtsheim in 1935	40
2.1	Richard Goldschmidt, ca. 1930	53
2.2	The animal physiology course at Göttingen, 1921–22	56
2.3	Patterns on the front wing of the flour moth, *Ephestia*	58
2.4	Karl Henke in 1937	59
2.5	Otto Renner about 1940	68
2.6	Fritz von Wettstein about 1940	71
2.7	Carl Correns about 1920	74
2.8	Peter Michaelis in 1952	76
2.9	Friedrich Oehlkers about 1950	78
2.10	Julius Schwemmle at Erlangen, 1935	81
2.11	Ernst Caspari about 1960	88
3.1	N. W. Timoféeff-Ressovsky, ca. 1940	112
5.1	Group portrait in the Kaiser-Wilhelm Institute for Biology during the 1920s, including Karl Belar	186
6.1	Alfred Kühn's assistants at the Kaiser-Wilhelm Institute for Biology from 1937: Ernst Plagge, Erich Becker, Hans Piepho, and Viktor Schwartz	198
6.2	Elisabeth Schiemann at Berlin Agricultural College, 1927	201
6.3	Hermann Kuckuck, Rudolf Schick, and Hans Stubbe	203
7.1	Alfred Kühn and Erwin Baur	230
8.1	Structural Analyses of the German Professoriate	298
8.2	The Relationship between Party Politics and Attitudes toward Modernization among German Academics	305
8.3	Curt Stern at the University of Rochester in 1938	309

Tables

4.1	Research Interests among German and American Geneticists, ca. 1920–1935	140
5.1	Content Analysis of the Obituaries of German and American Geneticists	182
6.1	Organization of the Kaiser-Wilhelm Institutes for Biology and for Breeding Research	205
6.2	Funding of the Kaiser-Wilhelm Institutes for Biology and for Breeding Research, 1932–1935	215
6.3	Capital and Running Costs of the Kaiser-Wilhelm Institute for Breeding Research, 1928–29	216
7.1	Styles of Thought within the German Genetics Community	228
7.2	Comparing Styles of Thought: The Baur School versus Those of Kühn and von Wettstein	247
7.3	Comparing Styles of Thought: Geneticists outside the Schools of Baur, Kühn, and von Wettstein	248
7.4	Universities Where Comprehensives and Pragmatics Received Their Doctorates, Worked as Assistants, and Acquired Their First Posts	270
8.1	Social Class Background and Secondary Schooling of Pragmatics and Comprehensives	289
8.2	Social Class Background and Secondary Schooling of Members of the Schools of Erwin Baur, Alfred Kühn, and F. von Wettstein	290
8.3	Career Trajectories	311
9.1	Positions Adopted in the Debate over the Plasmon Theory	316

Preface

From anthropology we have learned that members of different cultures often display distinctive modes of thought. Even within the same society, sociologists tell us, individuals from different social groups—be they social classes or those living in rural versus urban parts of the country—tend to think differently, not only about the way in which their society is organized, but also about fundamental metaphysical issues (e.g., conceptions of time or the role of chance in everyday life). One form of knowledge, however, is routinely excluded from this culturally relativist scheme: that of the natural sciences. We commonly think of science as quintessentially cosmopolitan, the very model of a classless and *inter*national community bound by shared norms of evidence and argument

And yet, it has periodically been suggested that "national styles" of scientific thought have persisted in the twentieth century (Forman, Heilbron, and Weart 1975; Reingold 1978). Could it be that scientists' choice of research problem, their methodological preferences, or the grounds on which they choose between theories are affected, still today, by differences of nationality? The very suggestion is deeply offensive to many scientists and others, doubtless because it conjures up associations with the attacks upon "Jewish physics" in Nazi Germany and upon "bourgeois genetics" in the Soviet Union from the mid-1930s. To abandon the search for styles because of these unwelcome associations, however, would be to throw out the baby with the bathwater. Unlike the Nazis, no one today is seriously suggesting that styles of scientific thought might be biological in origin. Nor can one attribute the damage done to science by either Lysenkoites or "Aryan physicists" simply to their belief in class- or ethnic-based forms of science. The intellectual diversity so common within science—often entailing the clash of theories, of methods, or of underlying metaphysical assumptions—has sometimes correlated with membership of particular social groups. But to recognize this historical fact is hardly to pave the way for totalitarianism. Since those party members in charge of the Soviet and German scientific communities were rarely competent to judge the intellectual merit of competing scientific claims, the only sensible policy for science would have been a pluralist one. Thus the crucial mistake common to Stalinists and Nazis alike was to put all their money on a single horse. And where no alternative sources of support for science were available, as in the Soviet Union, the damage was especially severe.

Apart from the unpleasant historical associations with the concept of styles of scientific thought, there remains the question of their theoretical plausibility. As long as one thinks of scientific method as a set of procedures driven by a single internal logic, the notion of styles of science seems paradoxical. We commonly associate the term *style*, after all, with activities which can be performed in *alternative* ways, above all with creative artistic work. During the 1950s and 1960s, however, the emergence of an antipositivist tradition in the philosophy of science began to undermine unitary conceptions of science:

> A historian of magnetism will have to ask himself the following logical question: "How far have these changes in idiom [from 'magnetic effluvia' to 'corpuscles' and 'fields'] been responses to brute experience, reflecting justifiable inferences from experience or observation? How far have they reflected the changing theoretical affiliations of magnetism? Or have these idioms sometimes altered rather for aesthetic reasons, as matters of intellectual fashion...." I suspect myself that only a small proportion of changes in magnetic theory have had any direct empirical justification.... If pursued beyond a certain point, the study of scientific ideas may thus bring one up against questions of style.... The question is at any rate worth investigating [Toulmin 1961, 97].

My own long-standing interest in styles of thought probably owes something to the fact that, although born and educated in the United States, I have spent twenty years on the other side of the Atlantic, primarily in Britain but also in Germany. The experience of being an outsider (as much in one's mother country as abroad) lends a certain persuasiveness to the concept of style which no abstract argument, however ingenious, can quite match. That said, the original stimulus for this book came from two academic works. One of these was Paul Forman's striking claim that the emergence of an acausal quantum mechanics in Germany during the 1920s owed much to a cultural climate characterized by hostility to materialism, atomism, and causality (Forman 1971). Although Forman's argument eventually came in for criticism (Hendry 1980), it has had considerable influence upon historians of twentieth-century German science, and it prompted me to wonder whether particular biological theories might also get a rough reception in Germany after the First World War. The other seminal work was Fritz Ringer's study of the "mandarin" ideology among German professors, 1890–1933 (Ringer 1969). Although based on a study of German academics in the humanities and social sciences, Ringer's analysis was sufficiently general in scope to invite extension to the natural sciences, especially since the mandarin ideology had much in common with the cultural climate so central to Forman's analysis. Ringer's principal thesis, however, was that certain intellectual controversies within these disciplines were attributable to a fundamental political rift within the mandarinate: between those sympathetic to the modernization of German society and those who equated such change with social decay. And this conflict within the professoriate—a phenomenon which Forman had explored only briefly—seemed to offer a promising backdrop for the study of controversy within the scientific community.

The next step was to find an appropriate area of early-twentieth-century science. In view of my original training, some area of the biological sciences was an obvious choice. I eventually settled on genetics, partly because there was a substantial secondary literature which had already mapped some of the early-twentieth-century cognitive terrain. And the fact that this literature was focused primarily upon developments in the United States and Britain would prove useful for comparative purposes. Most important of all was the fact that an American geneticist, T. H. Morgan, and his students had proposed a theory in 1915 which seemed tailor-made for my purposes. The manner in which it portrayed the structure and process of inheritance was diametrically opposed to the "hostile cultural climate" which Forman had described. Morgan's theory portrayed heredity in *materialist* terms by locating it in the chromosomes; in comparing the genes within a chromosome to beads on a string, it was an *atomistic* theory; and its explanation of recombination in terms of breakage and exchange was thoroughly *mechanistic*. Furthermore, it was of American provenance at a time when, for traditionally minded German academics, the United States epitomized the evils of modernity. Such a theory, I reckoned, ought to have aroused the ire of conservative mandarins while appealing to progressive ones. In the event, I found something rather different. It was not the ontological premises of the theory which disturbed most German geneticists but its narrow explanatory scope. Morgan's theory simply made no attempt to explain how chromosomal heredity was involved in the biological processes of most interest to the Germans: development and evolution. Eventually, the Germans' preoccupation with breadth—be it the breadth of a theory's scope or the breadth of a professor's interests—emerged as a central theme of the book, linking theory choice in genetics with social change in Germany society.

Although this book is, to my knowledge, the first to focus upon the history of genetics in Germany, it makes no attempt to chart that history fully.[1] For one thing, I discuss primarily the interwar period and have very little to say about the early development of Mendelism prior to 1914. Nor have I sought to cover the interwar period comprehensively; instead the focus is upon those areas of research which distinguish German genetics most sharply from its counterpart in the United States. By framing the problem in this way, I will understandably disappoint certain groups of readers. Most of the older German geneticists whom I interviewed, for example, would have preferred me to write a more comprehensive history, highlighting hitherto neglected German contributions to the body of theory constituting modern genetics. Some historians of genetics, on the other hand, will wish that I had inverted this emphasis, analyzing why geneticists in Germany often seemed—at least to American eyes—preoccupied with peripheral or insoluble problems, and thus did not make the impact upon modern genetics that one might have expected of them, say, in 1915. Although the book does provide an answer to this question, it also shows why the question is misconceived; what was peripheral to the Americans was absolutely central to the Germans. Other readers, needless to say, will wish that I had included human genetics and had followed the German genetics community beyond 1933, exploring the connections between genetics and racial hygiene. This book, however, is not primarily an analysis of the genetics community under National Socialism, a fact which the title ought to convey. Although I chose 1933 as the cutoff date, it is certainly not my view—nor that of most historians of science under National Socialism—that 1933 represented a break with the past in all respects. Continuity, especially in research programs, is much in evidence, and that is why I have elected to follow some of these into the 1940s. Nor is this book concerned with the development of human genetics in Germany. This theme has already been treated, at least tangentially, in some of the books on the German eugenics movement which have begun to appear over the last decade (Weingart, Kroll, and Bayertz 1988; Weindling 1989), and it is likely that other authors will soon do so more directly.

To disappointed readers my plea for leniency is twofold. Be patient; there is plenty of room for more books on the history of genetics in

1. That I have omitted Austria and various other parts of Europe from this study might seem arbitrary, since the German-speaking scholarly community extended across national boundaries at this time. Focusing upon Germany, however, made sense because almost all of the most important German-speaking geneticists of Austrian, Swiss, or Czech nationality made their careers at German institutions.

Germany, and perhaps mine will prompt others to fill in the gaps. Bear in mind, too, that this book is conceived as a contribution not only to the history of genetics, but to the historiography of science more generally. I hope to persuade others that "style of thought" is a useful analytical (rather than merely descriptive) concept in the history of science; that certain features of early-twentieth-century German scientific thought make sense only as the cognitive consequences of the modernization of German society; and at the most general level, that the history of ideas cannot reasonably be severed from a sociology of intellectuals. By now even those readers who share my interest in the historical sociology of knowledge are probably concluding that this book is rather "Germanic" in scope. They are right, of course. And by being so ambitious, the book becomes vulnerable to criticism by general historians and others in areas remote from the history of genetics. Accordingly, the wisdom of such an enterprise has sometimes seemed doubtful, not only because the project has taken more than a decade to complete, but because its deficiencies—both empirical and theoretical—are all too evident. Nonetheless, the rationale for books of this kind is a familiar one: better an interesting and ambitious theory which is wrong (well . . . partly wrong) than a cautious one which is irreproachable.

My debts are many and varied. To begin with, the book could never have been researched and written without the emotional and material support of Heide Ingenohl Harwood. Our progeny, too, deserve thanks for their patience, notwithstanding the occasional jibe ("When's the best-seller due?" etc.). Mothers-in-law are conventionally objects of ridicule; mine is different. From the start, Renate Ingenohl has not only encouraged my interest in the Germany of her youth but helped me with the rather more mundane mysteries of old German script.

I am also grateful to numerous people for having taken time to answer my questions. Apart from those who granted me interviews (listed separately), I thank Alfred Barthelmess, Adolf Butenandt, Kurt Gottschaldt, Friedrich Gruber, Rudolf Hagemann, Viktor Hamburger, Anton Lang, F. Markgraf, Renate Mattick (née Ehrensberger), Ursula Nürnberg, Hans Piepho, Gerta von Rauch, Viktor Schwartz, Hans-Jürgen Troll, Helga Völkl (née Michaelis), and K. G. Zimmer. Ernst Caspari (1909–1988) and Hans Stubbe (1902–1989) were particularly generous in this respect, and I much regret that it is now too late to repay my debt to them.

Many thanks are due as well to those who helped me to obtain relatively inaccessible materials: Albrecht Egelhaaf, Gisela Eyser (née

Nachtsheim), Ulrich Grossbach, Hermann Hartwig, W. Haupt, Maria Henke, J. Holtfreter, Brigitte Hoppe, Amélie Koehler, H. Kuckuck, Gertrud Linnert, Georg Melchers, Georg Michaelis, H. Mohr, W. Odenbach, Eeva Pätau, H. Piepho, E. Pratje, Hans Querner, Brigitte von Rosenstiel-Baur, B. Schwemmle, Helga Stromeyer-Baur, Millard Susman, Friedrich Vogel, G. Wagenitz, and Diter von Wettstein. I am particularly indebted to the late Gunter Mann (Medizinhistorisches Institut, Universität Mainz), who encouraged this project from the beginning and made his institute's library and seminar series available to me.

Among the many archivists and librarians whose knowledge has saved me time and effort, I am particularly grateful to Dr. Marion Kazemi and her colleagues at the Archiv zur Geschichte der Max-Planck Gesellschaft, Christiane Groeben at the Archive of the Naples Zoological Station, Beth Carroll-Horrocks at the American Philosophical Society Library, and the librarians at the Wissenschaftskolleg zu Berlin.

For comments on various draft chapters I am indebted to David Bloor, Frederick Churchill, David Edgerton, Peter Halfpenny, Jonathan Hodge, Peter Lundgreen, Ernst Mayr, Robert Olby, Nils Roll-Hansen, and Norton Wise, and especially J. V. Pickstone. I am also grateful to David Hull and two anonymous reviewers, who read and commented on the complete manuscript for the University of Chicago Press. At the start of the project I was fortunate to be able to draw upon Paul Weindling's advice and extensive knowledge of archival sources, and Adela Baer allowed me to consult her unpublished manuscript on the relations between geneticists and the racial hygiene movement in Germany. On sabbatical at the Technische Universität Berlin in 1982–83, I learned a great deal about the history of German science from Herbert Mehrtens and Christian Sund. I am grateful, too, to Reglindis and Walter Lantzsch (Halle) and Doré Vogel (Frankfurt) for their *Gastfreundschaft,* and to Firooza Kraft and Yvonne Aspinall for helping to prepare the manuscript. Lastly, many thanks to James Secord for some excellent advice: he recommended that I send the manuscript to Susan Abrams at Chicago.

For research funding I thank the Deutscher Akademischer Austauschdienst, the Royal Society, the joint Economic and Social Research Council/Deutsche Forschungsgemeinschaft "Research Exchange Scheme for Social Scientists," and the University of Manchester. Much of the manuscript was drafted in 1987–88 while I was a fellow at the Wissenschaftskolleg zu Berlin; I am grateful to Wolf Lepenies and his colleagues for creating a stimulating work environment there.

Early versions of sections of chapters 1 to 4 and 9 have previously appeared as journal articles. For permission to include this material I thank the editors and publishers of *Medizinhistorisches Journal, Freiburger Universitätsblätter, Annals of Science,* and *Isis.* Figure 8.1 first appeared (in German) in my paper "Mandarine oder Aussenseiter? Selbstverständnisse deutscher Naturwissenschaftler, 1900–1933," in Christophe Charle, Edwin Keiner, and Jürgen Schriewer (eds.), *Sozialer Raum und akademische Kulturen: Studien zur europäischen Hochschul- und Wissenschaftsgeschichte im 19. und 20. Jahrhundert* (Frankfurt a.M.: Peter Lang, 1992).

Altrincham, Cheshire
August 1992

1 Getting Started: The Argument, Method, and Context

> The rapid growth of communications, particularly after World War I, made of genetics a single entity all over the world, and there was very little national specialization. . . . I should say that genetics in one country is virtually indistinguishable from genetics in any other country.[1]

The author of this statement, the American geneticist L. C. Dunn, was well placed to make such a judgment. During his career, which stretched from World War I to the 1960s, he maintained close contact with geneticists in many countries. Most historians of science, furthermore, would probably agree with his general claim. The distinctive traditions of science which John Theodore Merz, Pierre Duhem, and others described for nineteenth-century Britain, France, and Germany were shaped by national variations in the organization of science—such as the comparatively amateur character of British science and the centralization of French science—and were sustained by the relative isolation of those scientific communities from one another (Merz 1896–1914; Duhem 1954, chap. 4; cf. Crosland 1977; Geison 1978, chap. 2). By the turn of the century, however, the professionalization of science throughout the industrialized world had substantially narrowed the range of institutional forms in which research and teaching were conducted. And whatever residual national differences might have remained were constantly undermined by the growth of *international* institutions from the end of the nineteenth century: societies, congresses, prizes, collaborative projects, and so forth.

While national differences may well have declined during the latter nineteenth century, a major aim of this book is to demonstrate that they most certainly did not disappear in the twentieth century. To judge from recent conferences, interest in national traditions is growing, especially among historians of science in the United States.[2] That this subject should be of particular interest to American historians may not be accidental. Into the 1960s, as Charles Rosenberg has pointed out,

1. Dunn, "Reminiscences," 935.
2. Sessions at the 1986, 1988, and 1990 meetings of the History of Science Society were devoted to "national styles of science," as was a conference sponsored by the Dibner Institute at Boston in 1988.

the study of American science attracted relatively little attention from American historians of science (Rosenberg 1983). By the 1980s, however, this theme had moved from the periphery to the center, reflecting a series of more general historiographical trends which were prompted, at least in part, by the experiences of the late 1960s. Like historians of science elsewhere, more Americans began to work on twentieth-century science; they began to talk to general historians; and more of them became interested in the impact of scientific knowledge—whether as ideology or as technology—upon society. Each of these trends made a focus upon the peculiarities of American science not only plausible but important.

As welcome as these trends are, the study of national traditions also has its pitfalls. Like any other weakly established discipline, history of science is perpetually obliged to demonstrate its importance—within the academic community, among school science and history teachers, to potential funding bodies of various kinds, and so forth. However resolutely we may reject "whiggism," there is no doubt that for many sponsors the importance of history of science is basically celebratory. For them the historian's task is to chronicle major scientists' achievements, paying implicit tribute to the institutions which employed them, the patrons who financed their work, or indeed the nation whose cultural significance they enhanced. The danger thus posed is obvious: The historians' analytical aims are at risk of being compromised by the sponsor's power. One of the purposes of a recent collection of essays, for example, was to identify the distinctive character of early-twentieth-century biology in the United States.[3] Although individually each essay makes a valuable contribution to our understanding of the conceptual and institutional changes which transformed the life sciences toward the end of the last century, collectively they fail to show that there was anything distinctive about American biology at that time. The reason is quite simple; although the editors (and no doubt the contributors) fully recognize that demonstrating a national tradition necessarily requires comparative analysis,[4] none of the essays—nor the book as a whole—is comparative. The volume, it must be said, was commissioned by the American Society of Zoologists in order to commemorate its centenary.

Obviously the lack of cross-national comparative work cannot be

---

3. Editors' introduction to Rainger, Benson, and Maienschein 1988.
4. Ibid.

chalked up solely to sponsorship. In some cases cross-national research of high quality is actively promoted by national sponsors; the German Historical Institute in London has been notably successful in this respect. Nor is cross-national work necessarily immune to celebratory or polemical purposes; Duhem's comparison of French and British physics is a case in point (Duhem 1954). Furthermore, the dearth of cross-national analyses is hardly peculiar to the history of biology, nor to the history of science. The problem is more fundamental: An explicitly comparative perspective is rare in many areas of history. Some authors have recently drawn attention to this deficiency in British urban history (Pickstone 1989), and I have encountered it in the historical literature on the organization of universities. Useful studies of American and German universities are available, but there have been exceedingly few attempts at systematic structural comparison of the two.[5] There are, of course, good reasons why history is so seldom comparative; it is hard enough getting to know *one* period, nation, or institution well, let alone two or more. And the problems are compounded for cross-national comparison where several languages may be involved. Perhaps the persistent appeal of "localism" also derives from the concern to reconstruct a particular historical context in all of its rich detail. Historians of this persuasion may well fear that the costs of a comparative perspective are too great, for comparative analysis requires much more than simply juxtaposing two or more case studies.[6] Insofar as it employs a set of relatively abstract analytical categories in order to render two unique historical contexts commensurable, the comparative method necessarily erodes particularity.

But just as traditional narrative political history has gradually been supplemented over the last generation or two by the growth of social and economic history, so also have historians of science—prompted by political concerns and/or by sociology—begun to make greater use of comparative analysis. As long as the history of scientific ideas was domi-

---

5. Veysey's (1965) *Emergence of the American University* is useful, and I refer to several similar analyses of the German university in chapter 4. Works which look superficially comparative usually turn out to do little more than juxtapose national differences (e.g., Jarausch [1983], *The Transformation of Higher Learning*). Flexner's (1930) *Universities: American, English, German* is broad in scope but fails to develop an analytical framework and apply it to each of the three systems. To my knowledge, the only attempts to systematic comparison so far are Ben-David and Zloczower (1962), "Universities and academic systems," and Ben-David (1968–69), "The universities and the growth of science."

6. For a discussion of this point with reference to the history of physiology and of medicine, see Pickstone (1990), "A profession of discovery."

nated by an internalist perspective, there was little point in comparing scientific development in different countries. One simply traced the growth of a theory or discipline at a transnational level, except where normal development was stunted in a particular country due to unusual conditions (genetics under Stalinism, physics under the Nazis, Darwinism in France, etc.). But since the 1950s knowledge growth has increasingly come to be seen as a contingent process requiring *decisions* at every point. Instead of this process being envisaged as an essentially linear trunk which only branched when logic gave way to irrationality, knowledge growth is coming to be seen as a *multi*-stranded affair in which a variety of rational research strategies are at work, with different strands being propelled by different aims in different contexts and no single strand enjoying privileged epistemological status. As a result we now have comparative studies of schools or generational shifts within a given discipline as well as many studies of controversy.[7] For reasons already discussed, cross-*national* studies still constitute only a small proportion of this newer comparative work,[8] but they are indispensable in preventing that version of historiographical myopia in which we mistake one particular route (e.g., in the development of a discipline) for the only rational one.

As in other areas of the history of science, cross-national studies in the history of genetics are practically nonexistent (Sapp 1987 is a recent exception). Worse, the existing literature has been dominated by English-speaking authors. Three of the four general surveys of the history of genetics since 1900, for example, have been written by Americans (Carlson 1966, Dunn 1965b, and Sturtevant 1965a; the exception is Barthelmess 1952). In consequence, we now know a good deal about those areas of genetics in which British and American biologists made important contributions: the elaboration of Mendelian chromosome theory, mathematical population genetics and its precursors, and the emergence of molecular biology.[9] But there has been very little work so far on the history of developmental genetics, human genetics, or evo-

---

7. For a guide to the literature, see Shapin 1982.
8. Among the most interesting cross-national studies of knowledge formation are Forman 1971, Danziger 1979, and Kohler 1982.
9. Book-length studies on the chromosome theory include Allen (1978), *Thomas Hunt Morgan*, and Carlson (1981), *Genes, Radiation, and Society*. On population genetics see Provine (1971), *Origins of Theoretical Population Genetics*, Mackenzie, (1981), *Statistics in Britain*, and Provine (1986), *Sewall Wright*. On molecular biology see Olby (1974), *The Path to the Double Helix*.

lutionary genetics, areas where continental European contributions have been important.[10] As I will show in the next three chapters, the existing historiography of genetics is misleading in several respects, precisely because it is so heavily based on the development of the field in the United States. Comparative analysis offers the best corrective; with it we can distinguish those developments which are genuinely international from those which are local and distinctive. Until more work has been done on other countries, of course, it would be foolish to claim that the path which genetics took in Germany was any more typical of that in other countries than was its path in the United States. All that can be safely said at the moment is that from what little we know about the development of genetics in the Soviet Union, France, and Germany, it is quite possible that historians will one day regard genetics' American trajectory—however triumphal—as highly atypical.

1.1 An Overview of the Argument

In part 1 I contrast the development of genetics in Germany with that in the United States in order to throw into relief "the peculiarities of the Germans."[11] Sections 3 to 5 of this chapter provide orientation, outlining the state of biological theory at the turn of the century, the institutional obstacles and opportunities with which advocates of the new genetics were confronted, and the early development of Mendelism in Germany. Chapter 2 considers the German response to Morgan's chromosome theory during the 1920s, focusing on the recurring German complaint that the theory neglected the process of development. One area of research in which this German concern was clearly visible was cytoplasmic inheritance. Given the difficulties during the interwar period in imagining how atomistic nuclear genes could account for the

---

10. On French contributions to developmental genetics see Sapp 1987 and Burian, Gayon, and Zallen 1988. On Soviet contributions to evolutionary genetics see Provine 1981 and the work of Adams, discussed in chapter 3. Of the various studies of the British and American eugenics movements which appeared during the 1970s and 1980s, only Kevles 1986 and Ludmerer 1972 devote more than passing attention to human genetics. Apart from Allen's earlier study of Richard Goldschmidt (1974), studies of German genetics have only begun to appear since the mid-1980s (e.g., Saha 1984 and Richmond 1986).

11. Readers unfamiliar with elementary genetics, embryology, or evolutionary theory need only skim chapters 2 and 3, as well as sections 1 and 2 of chapter 9, in order to get their general drift. The remaining chapters should be accessible to those from a variety of backgrounds.

temporally and spatially ordered events which constitute development, many German geneticists were interested in the role which heritable cytoplasmic structures might play in regulating nuclear genes' activities. While cytoplasmic inheritance did not attract much interest among American geneticists until the 1950s, other areas of developmental genetics did, notably the Nobel Prize–winning work of George Beadle and his collaborators from the mid-1930s. Nevertheless, Beadle's work does not undermine the contrast between American and German genetics. When one compares his work with that of his main German competitor (Alfred Kühn), it is evident that the two research programs, while nominally similar, had very different goals. Although eminently successful in its own terms, Beadle's approach illuminated the relatively simple and earliest stages of gene action, while Kühn focused upon the later and more complex processes whereby such gene products modify the matrix of developmental reactions.

Chapter 3 shifts to the other major problem of interest to German geneticists: the role of genes in evolution. Ernst Mayr and others have argued that most geneticists during the interwar period had little appreciation of the complexities of the evolutionary process. Essential for the evolutionary synthesis of the 1930s and 1940s, therefore, were broadly educated biologists who could bridge the intellectual gulf which separated the concerns of geneticists from those of field biologists. In this chapter I argue that this gulf was far narrower in Germany. While many geneticists in the United States were relatively uninterested in evolutionary issues or held overly simplified mutationist or selectionist views, a variety of German geneticists sought to develop more sophisticated theories of selection which could integrate genetic theory with the complexities of geographical distribution and the fossil record.

In chapter 4 I try to make sense of these distinctive German concerns, concluding that the boundaries which demarcated genetics from its disciplinary neighbors in Germany were different from those in the United States. "Genetics" designated a much more inclusive terrain in Germany. The reason for this, I argue, is that the structure and development of higher education in Germany made it very difficult to specialize. Mendelian genetics emerged at a time when German higher educational institutions—unlike their American counterparts—were no longer growing, such that new specialties had to be accommodated within departments devoted to older established disciplines rather than becoming independently established. Had the German university been structured along American lines, it might nonetheless have been possible to grant new specialties a substantial degree of autonomy within existing departments. But at several levels the organization of the German university actively discouraged specialization. The best career strategy for a young geneticist in search of a chair (necessarily in botany or zoology) was to cultivate a broad profile, working on genetic aspects of the major biological problems: development and evolution.

In chapter 5, however, the limitations of this kind of explanation become apparent. For one thing, the differences in values and attitudes between American and German geneticists were by no means confined to their conceptions of genetics. The Germans also seemed to place greater emphasis on breadth of scientific knowledge and general cultivation (e.g., in the arts) than the Americans. Fritz Ringer's work and comparable studies of the United States suggest that any attempt to explain national differences in academics' self-understanding is likely to require comparative social history on an intimidatingly large scale, rather than simply a sociology of research and teaching institutions. Furthermore, it is also clear that black-or-white generalizations about German versus American genetics will not do. The German genetics community contained an interesting minority whose approach to genetics (and attitudes toward breadth and cultivation) was remarkably "American," while a few American geneticists seem to have resembled the German mainstream. Thus the differences *between* these countries were also present *within* them. I conclude that the simplest way to study the genesis of such differences is to focus on one country. In part 2, therefore, the argument shifts away from national differences toward intra-German ones, attempting to identify those social processes outside the university which could account for the differences in outlook between two factions within the genetics community.

Chapter 6 sketches the institutional topography of the German genetics community, focusing upon the two most important schools of the interwar period. With its concentration upon developmental genetics, Alfred Kühn's school exemplifies the dominant tradition within the genetics community, while Erwin Baur's school, with its emphasis upon transmission genetics, illustrates the minority approach. These two schools also differed in the organization of their respective institutes, their sources of funding, and their responses to political pressure after 1933. Having characterized the institutional contrasts between dominant and minority traditions, we turn in chapter 7 to their cognitive differences. I survey the attitudes of approximately fifty geneticists on a range of matters, scientific and nonscientific, arguing that these attitudes are patterned. A "comprehensive" style of thought can be imputed to Kühn and other members of the dominant tradition, while Baur and others in the minority faction display a "pragmatic" style. Apart from their differing conceptions of the key problems in genetics, comprehensive thinkers were more concerned to master a wide range of biological knowledge (in order to combat the evils of specialization), more interested in the humanities and fine arts, and less likely to be party-politically active than were pragmatic thinkers.

Chapter 8 offers an explanation for the genesis of these styles. Ringer's work offers a useful starting point, since the ideas characteristic of his "mandarinate" correspond closely to the comprehensive style of thought. Central to both is the concept of *Bildung,* a form of cultivation and judgment much admired by the emergent educated middle class from the late eighteenth century, whose possession served to legitimate the social status enjoyed by this stratum during the nineteenth century. As industrialization began to erode their influence toward the end of the century, the importance of *Bildung* was forcefully restated, especially by academics who insisted that it conferred the wisdom and perspective without which Wilhelmian society would disintegrate under the strains imposed by what they regarded as excessive modernization. A growing minority within the professoriate, however, was both more sanguine about modernization and skeptical of *Bildung,* not least because its members came from newer industrial and commercial strata. This, I argue, is the origin of the pragmatic style.

While the foregoing chapters outline a sociological account of problem choice within the genetics community, chapter 9 addresses the question of theory choice. One of the prominent features of the mandarin ideology, according to Ringer, was its holism, manifest in both the mandarins' scholarship and their support for a harmonious social order. If this is correct, then in those areas of genetic controversy where holistic theories vied with atomistic ones, one might expect comprehensive and pragmatic geneticists to choose different theories. The debate during the 1930s over a particular theory of cytoplasmic inheritance (the plasmon theory) provides a way to test this hypothesis, since the available data on cytoplasmic inheritance left room for various interpretations. I show that comprehensive thinkers did, in fact, tend to endorse the holistic plasmon theory, while pragmatics either adopted a competing atomistic theory or rejected cytoplasmic inheritance altogether. The metaphors used by most participants in the controversy to describe nuclear-cytoplasmic relations suggest that the cell was widely perceived as a political microcosm, such that geneticists' differing conceptions of cellular order strongly resembled their differing views of the political order.

1.2 Styles of Thought

The concept of style was devised in order to classify cultural patterns observed in the study of the fine arts (Shapiro 1953; Gombrich 1968). Styles can be said to exist when various sectors of a society's cultural production (e.g., its pottery, myth, music, etc.) embody particular recurring elements *and* when those elements are distinctive; that is, they differ from one society to another. This can be represented so:

society x	society y
POTTERY	pottery
MYTH	myth
MUSIC	music

Notice that unless one identifies recurring elements in *several* cultural sectors, there is little point in using the term *style*.[12] One might just as well say that the pottery of society x and that of society y were different. The concept of style, however, implies not merely that the cultures of two societies are different, but that those differences are *patterned*. Thus the pottery of x differs from that of y *in the same way* (whatever "same" might mean) as the music of x differs from that of y. It is such patterned differences that allow us to distinguish form from content by using the term *style*. In the simple diagram above, therefore, one can claim that the style of the two cultures is distinct only when the horizontal rows display the same thing being done differently, while the vertical columns show different things being done in similar fashion. Thus imputations of style must be comparative, both within and across groups.

In the 1920s the sociologist Karl Mannheim extended the concept of style from art history to various sectors of a group's articulated thought in an attempt to identify styles of thought (Mannheim 1953; see also Scheler 1980). Adapting Mannheim's approach in turn, we can say that styles of *scientific* thought exist when particular ontological and/

---

12. To insist on recurring elements in *several* cultural sectors is, of course, overly stringent. There is good reason to deploy "style" even within a *single* cultural domain as long as a given contrast recurs in multiple instances. For example, the ways in which Stan Getz and Sonny Rollins interpret a given melody are radically different, yet characteristic.

or epistemological assumptions recur in a variety of scientific domains *and* those assumptions differ from one group to the next. For example:

group x	*group y*
BIOLOGY	biology
PHYSICS	physics
CHEMISTRY	chemistry

Of particular interest for those concerned to place the development of science within the context of general history are those cases where the assumptions underlying various scientific domains also recur in other cultural sectors:

group x	*group y*
SCIENCES	sciences
POLITICAL THOUGHT	political thought
POETRY	poetry

But the historian's job is not finished once cultural patterns of this kind have been described; we still need to explain why they have emerged. A century ago most explanations would have been racist: Different patterns of thought were generally ascribed to biological differences between the groups in question. The work of cultural anthropologists, however, has largely eliminated such explanations from serious consideration. Far more credible are various psychological theories which regard certain perceptual or cognitive tendencies as part of the personality. By so doing, one can then ascribe individual differences of thought or perception to the same kinds of variation in early experience which are thought to account for personality differences.[13] While some perceptual and cognitive styles undoubtedly vary from individual to individual, styles of thought also vary from one *group* to another. How these should be explained depends on the nature of the group. If the group is the product of self-selection (e.g., a club whose members meet to pursue shared interests), psychological explanations of its style of thought are very plausible. On the other hand, if membership of the

---

13. For a social psychological account of certain perceptual characteristics, see Witkin et al. 1962. Along similar lines, Sulloway has argued that whether scientists defend orthodox or radical positions in controversies is strongly correlated with the structure of the families in which they grew up (Frank J. Sulloway, "Orthodoxy and Innovation in Science: The Influence of Birth Order in a Multivariate Context," paper delivered at a meeting of the American Association for the Advancement of Science, New Orleans, Louisiana, February 1990).

group is ascribed (e.g., clan, nation, or, to a large extent, social class and religious group), then the explanation of its style of thought is necessarily sociological.[14] The problem which Mannheim addressed was why a style of thought with specific features should be associated with a particular social group at a particular time and place. His answer was that a style of thought has a purposive character; one cannot hope to understand why a style emerges and persists unless one looks at what its "carrier group" is trying to *do*. And if one wants to understand what ties together the various features of a style, one cannot hope to do so merely by examining thought patterns in isolation. Instead one has to analyze those patterns in action, looking at how they are deployed by particular social groups in particular circumstances in order to achieve particular aims.

Like most sociologists of knowledge until recently, Mannheim exempted scientific thought from the scope of sociological analysis. Since the 1970s, however, a substantial literature on the sociology of scientific knowledge has made the search for styles of science altogether plausible.[15] What this literature demonstrates is that the cognitive processes whereby scientific knowledge is constructed—for example, observation, classification, theory formation, and theory testing—are *routinely* shaped by the social circumstances of the scientists concerned.[16] A variety of empirical studies have shown that these processes are sometimes performed in quite distinctive, though empirically grounded, ways by different groups within the scientific community. The most striking instances are cases of controversy where the contending parties reach contradictory conclusions, despite altogether plausible arguments on both sides, because they start from contrasting premises or are pursuing different goals.

14. Without quite detailed biographical information about the historical figures with which one is concerned, it is difficult to rule out psychological explanations in principle, but there is also no need to do so. Psychological explanations can usefully complement sociological ones—for example, by accounting for individual variation within groups which, sociologically speaking, are uniform.
15. For guides to this literature see Shapin 1982 and Golinski 1990.
16. Sociologists working in this tradition remain divided over the extent to which "the real world" constrains cognition. The position which I adopt in this book—based on my own experience at the laboratory bench, in the archive, and waiting for the bus—is that the structure of the material world restricts the range of possible interpretations available to us, without specifying a unique one. That socially generated expectations cannot *dictate* how we perceive or interpret is evident whenever we are surprised by anomaly.

Although there are now both theoretical and empirical grounds for expecting a diversity of forms of scientific knowledge under particular circumstances, the concept of style itself has attracted remarkably little attention from sociologists of science. While sociologists have devoted much effort to identifying the fundamental ontological and epistemological assumptions upon which particular lines of scientific inquiry have been constructed, they have rarely been interested in looking at the extent to which these assumptions pervade *other* areas of scientists' thought. Thus stylistic patterns which might have emerged from a comparative analysis of a research group's scientific work with its religious, artistic, or political beliefs have gone undetected. Nor have scientists' social class or nationality commanded as much attention as has their membership of various sectors within the scientific community.[17] There are probably several reasons for such neglect. For one thing, as studies in the sociology of scientific knowledge got under way during the 1970s, its priorities lay elsewhere. The most urgent task at that time was to replace the rationalist emphasis in many philosophies of science with a demonstration of the contingent or "negotiated" character of science. For some sociologists, then as now, the mere demonstration of contingency sufficed. To ask *why* scientists pursued one line of inquiry rather than another was of little interest, since human action was seen to defy systematic causal explanation (Collins 1985; Latour and Woolgar 1979). For those sociologists bent on identifying causes, however, the next step was to show that the cognitive judgments made by scientists were generally collective (rather than idiosyncratic) in character. Thus sociological explanations of some kind were most likely, and it remained only to identify which social structural features of the society in question were responsible. But demonstrating cognitive contingency in a convincing manner was so demanding that no one wanted to spend more time than necessary on an exhaustive trawl through all potentially relevant social structures. The usual solution, therefore, was to attribute differing cognitive patterns to different social locations within the scientific community itself—laboratories, specialties, disciplines—thus avoiding the complications associated with scientists' class membership, nationality, and so forth (e.g., Mulkay 1972). The result, as one observer noted, was a certain asymmetry: Cognitive structures

17. Whitley (1984), for example, has demonstrated how various cognitive features of scientific knowledge (e.g., the scope of problems tackled, the extent to which results are theoretically integrated) are attributable to the structure of scientific institutions and their relations to patrons. But the cognitive patterns thus revealed are the result of comparative analysis *among* the sciences, rather than between them and other forms of culture.

received rather more analytical attention than did social structures (Rosenberg 1980).

It is quite likely, moreover, that the science of *today*—the object of study preferred by most sociologists—is unsuited for stylistic analysis. It is not that national styles of scholarly inquiry no longer exist; those with experience of academic communities abroad will know otherwise (e.g., Galtung 1981). It is rather that the study of contemporary styles is methodologically awkward. The search for cognitive patterns which are manifest in many sectors of thought poses problems of access. How do we establish what the scientist who is speaking into our tape recorder thinks about art, politics, or philosophy? Occasionally this information is volunteered or available in published form. But to solicit it is to jeopardize the interview as well as future access to the scientist's laboratory: "irrelevant" questions about personal matters can only arouse suspicion. In this respect it is simpler to work on the science of the past. Dead men may tell no tales, but they do leave records. Above all, in the days before cheap telephone calls, scientists' correspondence provided an invaluable source of their views on matters nonscientific. And since older figures are more likely than younger ones to publish their reflections upon such matters (or at least to draft an autobiography), there are advantages in waiting until they have had every opportunity to do so.

Methodological considerations of this kind may help to explain why the growth of interest in national styles of science over the last few years has been rather more evident among historians than among sociologists of science. Nevertheless, few historians of science have deployed the concept with much rigor. In most cases the term *style* is used casually, without definition and in passing. Scientists working in different contexts (e.g., institutions or nations) who approach the same problem in different ways are said to display different styles of science (e.g., Reingold 1978; Geison 1978; Nicolson 1989; Fruton 1990). Often no attempt is made to establish whether those stylistic elements identified within a given body of scientific work are also present in other disciplines or in other cultural sectors, so that it is unclear what is gained by the use of the term.[18] By contrast, even though Stephen Brush does not use style as an analytical concept, his wide-ranging study of homologous themes in nineteenth-century thermodynamics, philosophy, and

---

18. E.g., Fruton 1990. Similarly, Gerald Geison's discussion of "national styles" in nineteenth-century physiology makes no attempt to identify comparable stylistic features in other disciplines (1978: 15–23 and 338–55).

literature is comparative in just the way which stylistic analysis requires (Brush 1967). Although Brush describes these cultural patterns, however, he does not explain them.

Another historian of science whose work is of interest for those concerned with styles—though he rarely uses the term—is Gerald Holton (1973, 1978, 1986). Like Kuhn and Lakatos, Holton stresses the importance for the research process of scientists' nonrational commitment to "themata": those fundamental ontological assumptions (e.g., atomism, the continuum, a preference for simplicity) which channel the ways in which the scientist incorporates observations into laws and theories.[19] From my point of view, Holton's themata are one example of the kinds of elements which, providing they recur in various cultural sectors, define a style of thought. Nevertheless, Holton's historiographical concerns are very different from those proposed here. To begin with, he is not interested in the extent to which the themata central to a scientist's work might also be present in—thus perhaps derived from—the wider culture. Themata, he emphasizes, are not to be confused with metaphysics or worldviews (Holton 1978, 23; cf. Holton 1986, ix–x, and 1973, 11–40). And granting that we still know little about the *origin* of themata, Holton is more inclined to turn to psychology for explanation than to sociology: "To a much larger degree than . . . worldviews, thematic decisions seem to come not only from the scientist's social surrounding or 'community,' but even more from the individual" (Holton 1978, 23–24). This concern to insulate scientific themata from the wider culture, as well as from sociological explanation, is noticeable in the framework with which Holton analyzes scientific change. Of the nine "components" which he believes shape scientific inquiry, "cultural, ideological or political elements" and "themata" are treated as separate and equivalent categories. And although he does not deny that "sociological setting"—by which he means the structure of scientific institutions and their links with the state and economy—also affects the path of scientific inquiry, Holton again regards it as a component separate from, and equivalent to, themata. By making room for all three of these components (plus six others), Holton's framework appears admirably eclectic. But by treating them as *equivalent* components, he discourages us from considering whether some of them might be more equal than others. In so doing, he dodges the important theoretical question: the possibility that themata might

---

19. Cf. the discussion of "disciplinary matrix" in the postscript to T. Kuhn 1970 and of the "hard core" of research programs in Lakatos 1970.

frequently incorporate 'ideological components,' or that such incorporation might be fostered by the 'sociological setting.'

Holton's reluctance to pursue the sociological implications of his work was not shared by Ludwik Fleck.[20] Although largely unaware of the literature in the sociology of knowledge which was flourishing in his own time, Fleck was critical of those sociologists who exempted science from sociological analysis. Making abundant use of his own scientific training, he examined the ways in which perception and classification in science were shaped by what he called "styles of thought." Each style was borne by a "thought-collective"; distinctive styles could be found in different disciplines and occasionally within the same discipline. And within any given style, one could find individual variations, due to the fact that each member of that scientific thought-collective was simultaneously a member of many other collectives, both in and outside science. The scientist was thus a conduit by which meanings from the wider culture could influence styles of scientific thought.

Despite its sociological orientation, however, Fleck's concept of style is idealist in ways which can and should be avoided. He often writes that a style "dictates" and "coerces" how and what the scientist sees and thinks. The same tendency is evident in Holton's analysis. Themata are said to "filter" experience, to "guide" the work of the scientist, to "constrain" inquiry, and to have "power" over the scientific imagination (e.g., Holton 1986, x, 18–19, 52–53; 1978, 22). But ideas of themselves do not coerce. Wittgenstein, among others, has shown how our inventiveness allows us to evade the constraints seemingly imposed by any rule. It is rather human beings who coerce, using ideas to justify that coercion. In stylistic analysis, therefore, it is important not to reify styles, as if they were things which possessed the power to shape scientists' thought and action. Instead we ought to see styles as *indicators* that thought is patterned, that thought is not simply a hodgepodge of unrelated attitudes. And once alerted to the existence of that pattern, we can begin to ask where it came from.

Mine is hardly the first attempt to discuss styles of scientific thought in a sociologically informed manner. William Coleman, Donald MacKenzie, and Kenneth Caneva have previously adapted Mannheim's ideas in order to illuminate debates within twentieth-century British genetics and early-nineteenth-century German physics (Coleman 1970; MacKenzie and Barnes 1975; Caneva 1978). Nor is Mannheim's the only sociological approach to styles of thought. Mary Douglas's "grid-

---

20. For a discussion of the literature by and about Fleck, see Harwood 1986.

group" theory offers an alternative which has been profitably applied to nineteenth-century geometry, geology, and physics by David Bloor (1978), Martin Rudwick (1982), and Kenneth Caneva (1981), respectively. Unfortunately this earlier work has not had the historiographic impact which it deserves, perhaps because its arguments were not elaborated at book length or because much of it was published in places relatively inaccessible to historians. Moreover, it is my impression that enthusiasm for the grid-group theory has largely disappeared among historians of science over the last decade, probably because the theory faces fairly serious difficulties.

This is not the place to discuss grid and group at any length, and I would urge those interested in styles of thought to read that literature and decide for themselves.[21] But for those already familiar with the theory, my misgivings can be summarized by saying that the theory is both overly restrictive and ahistorical. Like any theory, grid-group draws our attention to a small number of variables deemed important. In this case four basic kinds of cosmology are seen to be characteristic of four basic kinds of social structure. While it is quite likely that some kinds of cognitive pattern (e.g., the ways in which scientists respond to anomalous findings; cf. Bloor 1978) and some aspects of social structure (e.g., certain relationships within and between small face-to-face groups) can be subsumed within the grid-group typology, it is difficult to imagine how such a simple typology could do justice to the diversity of social structures. For example, as Douglas has conceded, the theory does not allow for the fact that power is unequally distributed in most groups.[22] Thus powerful individuals in a particular social setting (defined in terms of grid and group) would probably welcome the cosmology "appropriate" to that setting, while powerless individuals in that setting would reject it. The other problem is that grid-group theory implies a *formal* homology between cognitive and social structures. According to one reading of the theory, for example, the members of any small threatened group will tend to be especially dismissive of anomalies. But this treats human reactions to ideas (or observations) in an ahistorical way; will *all* anomalies, regardless of their specific meanings, be rejected alike? Surely an individual's response to a given idea can only be fully understood by reference to the historically specific meanings which that idea has acquired in the context in question. Since

---

21. A good place to begin is Douglas, 1982 *Sociology of Perception*.
22. This point has been made by Law and Sharma 1977.

1945 the swastika has meant something radically different to us than it did before 1920. Thus grid-group theory has difficulty incorporating the context-dependence of meaning.[23] In setting out typologies of thought and social structure which are supposed to hold for all times and places, Douglas's theory strives for universal scope. But the cost of achieving such scope is that the typologies in question cover only a limited range of cognitive and social patterns. Ironically, the general approach which Mannheim devised in order to solve a very specific historical problem (the emergence of a conservative political philosophy in early-nineteenth-century Germany) promises to be more broadly applicable. In any event, I hope that this book will not only alert a wider audience to the virtues of earlier work on styles of scientific thought, but also make a convincing case on its own for the analytical value of that concept.

Enough of methodology for the moment; what about the story? If we want to understand why genetics developed so differently in Germany and the United States after 1900, we must consider both the state of biological theory at the turn of the century and the institutional landscape within which the new field's practitioners sought to find a niche.

1.3 The Revolution in Morphology

Let us begin by looking at the theoretical issues which preoccupied late-nineteenth-century biologists. To refer to "biologists" in this context, however, is convenient but misleading, since the life sciences in the nineteenth century are best seen as a cluster of loosely related research traditions rather than a coherent discipline in the modern sense. Nonetheless, historians are agreed that these fields underwent three major shifts of problem and method during the nineteenth century (cf. Coleman 1971). The first of these occurred around 1800 with the appearance of the term *biology*. In the eighteenth century the dominant tradition in the life sciences was "natural history": a descriptive activity which sought to document and classify the diversity of natural forms—plant, animal, and mineral—and their characteristic geographical distributions. In contrast, advocates of the new biology around the

---

23. Douglas's reply will be that grid-group analysis assumes meaning uniformity across all four social sectors, but this is rather more likely to obtain in structurally simple societies—for which she designed the theory—than in the highly differentiated ones in which historians of science are interested.

turn of the century (several of them admirers of Newton) called for the study of the laws which governed the living world. Early-nineteenth-century morphologists sought laws which related the diversity of animal forms to the requirements of the environments which they occupied. Others employed the microscope in search of the structural principles of plants and animals. Embryologists sought patterns in the developmental paths taken by different groups of vertebrates. Plant geographers sought, not merely to catalog geographical distributions, but also to explain why characteristic combinations of plant species were to be found in particular regions. And the experimental study of life processes (physiology), largely restricted in the eighteenth century to human subjects, was rapidly extended to animals and plants in general. A final strand of the new biology led to what is commonly regarded as a second revolution in the life sciences: A few adherents of the new perspective ventured to challenge the eighteenth-century assumption of a static living world, devoid of history since the Creation. Early in the nineteenth century Lamarck argued that natural laws, acting gradually over the eons, had transformed simple organisms into increasingly complex ones. Although initially controversial and speculative, the concept of laws of transformation was taken up by others and developed in quite different ways; by the 1870s evolution by "descent with modifications," as Darwin called it, had been widely accepted.

Nevertheless, while the general idea of evolution found acceptance relatively quickly, the particular mechanism which Darwin had invented—natural selection—did not. From the 1870s to the First World War the question of evolutionary mechanism was intensely debated, and natural selection was but one among several competing theories of evolution (Bowler 1983). And it is this long and inconclusive dispute which spawned the third major reorientation in the life sciences during the final decades of the century, sometimes called the "revolt from morphology" (Allen 1975). Many younger life scientists demanded greater rigor in the methods with which the evolutionary debate was conducted, and some of them opted out of the debate altogether, choosing to address different problems. For enthusiastic Darwinians after 1860, constructing phylogenetic trees solely upon the evidence from comparative anatomy required a certain amount of guesswork. Where intermediate forms were lacking among fossil and living species, relationships of descent had to be postulated. Haeckel's "biogenetic law," however, suggested that the study of embryos offered a way around this problem. The law encouraged the study of embryological development in detail, since the sequence of developmental stages was believed to mirror the historical sequence of evolutionary forms.²⁴ Thus the reasons why one developmental stage led to another were to be found, not in the structural requirements of the developing embryo, but in that organism's evolutionary history. Some physiologists were not impressed, arguing that phylogeny did not provide a truly causal account of embryological development (Querner 1980a; 1975). From the 1880s Wilhelm Roux and others presented a major new research program in embryology which challenged the reigning phylogenetic tradition. By calling it "developmental mechanics" (*Entwicklungsmechanik*), Roux not only distinguished his approach from the reigning conception in embryology (*Entwicklungsgeschichte*),²⁵ but also signaled his interest in the causal forces which shaped development.²⁶ This new embryology would no longer be subordinated to evolutionary ends, nor would its method be solely descriptive. Experiment was now the key to understanding the causes of development.

Above all, Roux's younger admirers defined the new approach to morphology in terms of *method*. William Bateson described their concerns thus: "Disgusted with the superficiality of 'naturalists,' [the younger reformers] sit down in the laboratory to the solution of the problem, hoping that the closer they look, the more truly will they see. For the living things out of doors, they care little. Such work to them is all vague."²⁷ In their introduction to the first issue of the *Zentralblatt für allgemeine und experimentelle Biologie* [*Journal for General and Experimental Biology*], the editors announced that "experimental biology is the research area . . . which is not satisfied with description."²⁸ In the 1890s H. F. Osborn excluded "all . . . purely descriptive work" from his new biological journal (Rainger 1988, 224). Courses at American universities were increasingly named according to the nature of the structures

---

24. That such recapitulation could occur was explained by positing a mechanism which took the evolutionary changes experienced by an adult and added them on to the sequence of embryonic stages undergone by that organism's progeny (cf. Gould 1977).

25. Literally, "the history of development." *Entwicklung* was ambiguous, since it could refer to either ontogeny or phylogeny, but the ambiguity was convenient for Haeckel et al., since it served to elide the two processes.

26. Others had suggested *Entwicklungsphysiologie,* with its implication of process and function rather than structure, or *experimentelle Morphologie,* which would have emphasized the continuity between the new tradition and its descriptive parent (Barfurth 1910; 1924). On Roux's work see Churchill 1973.

27. Bateson, *Materials for the Study of Variation* (1894), cited by Sturtevant 1965b, 200.

28. "Einführung," *Zentrallblatt für allgemeine und experimentelle Biologie* 1 (1910): 1–4, cited in Sund 1984.

or processes of interest rather than the group of organisms studied: for example, histology, cytology, or physiology rather than herpetology, ichthyology, or ornithology (Benson 1988, 71–72). August Weismann's grand theories—linking heredity, development, and evolution—were "pure speculation" as far as the experimentalist Wilhelm Johannsen was concerned.[29] The radicalism of Roux's program was not lost on his opponents. For their part, naturalists like C. Hart Merriam complained that the new laboratory-oriented generation had "an exaggerated notion of the supreme importance of their methods" (Merriam 1893, 353). Roux was using the term *mechanics,* Oscar Hertwig objected, "as a banner for an approach to biology which looks down its nose [*mit Geringschätzung herabblickt*] at the heretofore dominant approach to biology, its results and its aims."[30]

What, then, was the essence of this methodological revolution? For Garland Allen, the *Entwicklungsmechanikers'* most important contribution to biology at the turn of the century was the use of experiment to test hypotheses (Allen 1975, 38; cf. Allen 1979b). Experiment, E. B. Wilson observed, was "fast becoming the characteristic feature of latter-day biology" (Wilson 1901, 18). In Theodor Boveri's opinion (Horder and Weindling 1985, 214), one experiment was better than a hundred pages of Gegenbaur (the comparative anatomist). And critics like Oscar Hertwig singled out Roux's conception of experiment as "artificial, impatient, unreliable."[31] Experiment could only interfere with normal processes; only observation could bring understanding of normal development.[32] Various historians have argued that, among younger American biologists, this methodological shift was rather more gradual and diffuse. Wilson, Conklin, Morgan, and Harrison championed the value of experiment, to be sure, but as an addition to descriptive and comparative work.[33] What distinguished the new (and influential) biology program at Johns Hopkins University from the 1880s was less

29. The 1909 edition of Johannsen's *Elemente* is peppered with derisive references to the "Weismannian philosophy" (322), with its "fantastic speculations" (480; cf. 3, 319, 117, 335–36).

30. Oscar Hertwig, *Zeit- und Stretifragen der Biologie, Heft 2: Mechanik und Biologie* (Jena, 1897), 84, cited in Sund 1984.

31. Cited in Querner 1980a.

32. Since Hertwig was himself a cytologist—who took organisms out of their natural habitats, sectioned them, and stained them before examining them under the microscope—the methodological point at issue here was *how,* not *whether,* it was legitimate to intervene in nature (in a similar vein, see Maienschein 1981).

33. Werdinger 1980; Maienschein 1981 and 1983; Wilson emphasizes this point in his "Aims and Methods" (1901).

experiment per se than an emphasis upon *laboratory* research (Benson 1988). What characterized the new biology in the United States was the preference for "more obviously productive ways of working, more modern techniques, less speculative theories" (Rainger, Benson, and Maienschein 1988, 9). In Germany, too, there are similar signs that the new biology of the late nineteenth century remained methodologically eclectic. For example, the first project which Hans Spemann was assigned as a student during the 1890s was a descriptive one because his *Doktorvater* (Boveri) regarded skill in observation, description, and drawing as essential in experimental embryology (Horder and Weindling 1985, 184–85). To be sure, the editor of the new German genetics journal tried to keep the number of paleontological articles in it to a minimum and was disappointed at the initial dearth of experimental manuscripts[34] (cf. chap. 3). But the title which he chose for the journal—*Zeitschrift für induktive Abstammungs- und Vererbungslehre*—suggests that it sought to attract empirical contributions rather than explicitly experimental ones.

Whatever methodological reforms the younger morphologists may have been seeking, it is clear that the late-nineteenth-century revolution was not solely about method. The new generation also embraced different problems. Above all, a number of *Entwicklungsmechaniker* in both Europe and the United States insisted that embryologists should reject their traditional underlaborer status within the phylogenetic research program and devote themselves to the causal analysis of development as an important problem in its own right (e.g., Allen 1975, 25; 1979b; Werdinger 1980). Probably the most outspoken defender of this position was Hans Driesch:

What do the phylogeneticists want then? They cannot even do consistent biological research! . . . We know what we can do and what we cannot do for the time being. Our opponents think that *they* know what we don't even want to know. With their comparisons they deal with questions that by their very nature we have not dealt with and which in fact cannot be approached. . . . We are well aware of the problem of "Transformation" but we consider it for the time being an impregnable fortress and we address ourselves to "Developmental Physiology" as here we see the possibility of obtaining results. . . .[35]

During the 1890s Ross Harrison, too, thought that the study of evolution was too speculative to be scientific (Maienschein 1983, 111). However much E. B. Wilson may himself have valued evolutionary

34. Erwin Baur to Carl Correns, 21.8.10, Baur letters, Sammlung Darmstädter.
35. Driesch 1899, cited in Monroy and Groeben 1985, 37.

work, by the 1880s he felt that interest in big phylogenetic questions was clearly declining, partly in reaction to the "inflated speculation" of earlier morphologists.[36] Naturalists like Merriam and morphologists of a more traditional hue like Haeckel were disappointed: The new biology was obsessed with development and had little to say about evolution.[37]

Nevertheless, some members of the younger generation continued to wrestle with the problems of evolution, albeit from a new methodological angle. In the 1880s, for example, the plant physiologist Hugo deVries began to carry out a series of controlled crosses between plants differing in a single trait in order, he hoped, to provide the theory of natural selection with a less speculative theory of inheritance than Weismann's or Darwin's (Heimans 1978; Meijer 1985). William Bateson agreed with the *Entwicklungsmechaniker* that descriptive embryology

---

36. Wilson 1901, 17. Although continuing to work on evolutionary embryology, E. G. Conklin conceded that some phylogenetic work had been so speculative "that it is no wonder that the whole 'phylogeny business' has come into disrepute" (cited in Benson 1979, 314). The experimental embryologist Julius Schaxel (1887–1943) also felt that phylogeny had been in decline from the 1880s due to its speculative character (Schaxel 1919a [1922], 53–55). On Schaxel see Uschmann 1959.

37. Merriam 1893. Jane Maienschein has emphasized that the young American disciples of *Entwicklungsmechanik* "were not forsaking the earlier concern with evolutionary . . . questions; on the contrary, they thought they saw a new way to explore both evolutionary and anatomical concerns by studying the embryo with a new emphasis" (1981, 97). Similarly, Garland Allen observes that E. B. Wilson maintained a lifelong interest in evolution ("E. B. Wilson" in *Dictionary of Scientific Biography*). On the other hand, Maienschein shows that the actual *research* conducted by Wilson and the other American experimental embryologists was primarily concerned with unraveling the process of differentiation. Conversely, other biologists of the period continued to work on older problems (or with older methods), though interested in newer ones. For example, although essentially a descriptive morphologist interested in the embryology of invertebrates, the German zoologist Karl Heider (1856–1935) was far more charitable toward *Entwicklungsmechanik* than was Oscar Hertwig. And as early as 1905 he regarded as promising the combination of cytology and breeding analysis which had given rise to the new Mendelian chromosome theory. Like the younger generation of experimentalists, he came to doubt whether phylogenetic questions could ever be answered definitively, but unlike "extremists" such as Driesch, he never rejected the value of evolutionary inquiry (Kühn 1935; Ulrich 1960). Keith Benson has described W. K. Brooks's work in similar terms, emphasizing that although Brooks's own work remained largely descriptive, he acknowledged the value of the experimental embryology which was being conducted at the Marine Biological Laboratory during the 1890s (Benson 1979, 300–302). Perhaps, therefore, Wilson, Heider, and Brooks (and no doubt others) are best treated as transitional figures in the morphological revolution, in the sense that they came to terms with the new trends—albeit in different ways—*without* abandoning their interest in, and belief in the legitimacy of, the older phylogenetically oriented morphology.

could not shed light on phylogeny, but he turned his attention in the 1890s instead to what he saw as the key to understanding the mechanisms of evolution: the study of variation, both in the field and through the controlled breeding of domestic animals.[38] At about the same time, W. F. R. Weldon—Bateson's contemporary at Cambridge—also rejected descriptive embryology in favor of new methods for demonstrating the mechanism of evolution: the statistical analysis of the distribution of phenotypes in natural populations.[39] In the United States as well, a variety of biologists in the 1890s sought to devise experiments which would shed light on the nature of variation and heredity. Some of these were students of evolution—such as C. B. Davenport, George Shull, D. T. MacDougal—while others were employed as plant breeders. It was from these two groups, as the geneticist W. E. Castle recalled, that the new Mendelism was to recruit many of its first converts.[40]

1.4 The Response to Specialization

While historians continue to differ over the precise character of the morphological revolution,[41] there is little doubt that it provoked debate over appointments and the content of curricula. In the United States naturalists attacked what they saw as excessive specialization. Cell biology and embryology had "usurped too much of the attention" of staff in American universities (Benson 1988, 73). Students of the life sciences needed a "broad and comprehensive grasp of the phenomena of living things," C. Hart Merriam insisted, but "modern biologists" were too one-sided (1893, 353–54). Others argued that ecology was the modern successor to nineteenth-century natural history and could offer the integrative perspective necessary to counteract the dangers of specialization (Worster 1979, 203). The British physiologist Michael Foster put the point most graphically:

Anatomists, zoologists, physiologists, have thus from being brothers closely bound together become, through the very progress of their respective sciences,

---

38. See Bateson to the electors to the Linacre professorship, June 1890, reprinted in B. Bateson 1928, 30–36.
39. On Weldon's work see Norton 1973.
40. Castle 1950, 60–61. On MacDougal's attempt to study evolution experimentally, see Kingsland 1991. On the evolutionary interests of those biologists employed as plant breeders at agricultural experiment stations, see Kimmelman 1987, chap. 3.
41. For a discussion of these questions in the context of late-nineteenth-century American biology, see the *Journal of the History of Biology* 14 (1981). The debate continues in Rainger, Benson, and Maienschein 1988.

24   Getting Started

more and more estranged from each other. Instead of working hand in hand to build together the common tower of biology, each has been constructing his own chambers, not only without reference to, but in more or less complete ignorance of, what the others are doing. And now they are so far apart, that even when they wish to call to each other, they can rarely be understood.

[The naturalist]—a whole-minded inquirer into the nature of living beings—is for the most part a thing of the past. He has well nigh disappeared through that process of differentiation of which I have spoken. He has . . . been cut up into little bits, and while the bits have flourished and grown great, the whole has vanished from sight. . . . If you attempt to put the pieces together, you will find that they do not piece into a whole; gaps are left where fragments have fallen away [Foster 1899, 216, 218–19].

Twenty years later William Morton Wheeler echoed the same theme, identifying natural history as "the perennial root-stock or stolon of biological science." In its modern form (ecology) it provided the best foundation for a course in biology. Genetics—"so promising, so self-conscious, but, alas, so constricted"—was merely a "bud" from this root-stock (Wheeler 1923, 67).

But there seems to have been little that naturalists could do to retard the process. Although older biological programs at American universities were often dominated by traditional natural historians, concerned with broad evolutionary questions, those newer programs established in the 1890s—not least those established at the land-grant universities—were more likely to give pride of place to experimental work.[42] On his trip to the United States in 1907, William Bateson was overwhelmed by the support for the new biology there.[43] Systematics, however, fared particularly badly, as Merriam noted with regret (Merriam 1893). The new journal proposed by H. F. Osborn in the 1890s explicitly excluded systematics, and systematists were kept out of the new discipline-based professional societies which emerged from the late 1880s, such as the American Society of Zoologists (see Rainger 1988,

---

42. Maienschein 1988, 174; Kimmelman 1987, 23–25. According to one of his colleagues, T. H. Morgan left Columbia University in 1927, even though he was in line for the chair of the department of zoology there, because there were a number of "morphologists" there, whereas at the California Institute of Technology he would have a free hand to build up a new department from scratch (Garland Allen's interview with A. H. Sturtevant, file 8.8, box 8, Sturtevant papers).

43. "These have been wonderful days. . . . I am everyone's prey, being torn to pieces by my admirers. No exaggeration, I assure you! . . . All went splendidly [at one of his talks]. I had a record attendance, and great enthusiasm." "After the years of snubbing it is rather pleasant to get appreciation even though in an overdose." W. Bateson to Beatrice Bateson, 23.8.07 and 24.8.07, reprinted in B. Bateson 1928, 109–110.

224; Appel 1988, 97, 104). At the University of Chicago, C. O. Whitman's department concentrated upon studies of development and inheritance, and he made no attempt to hire broadly educated natural historians interested in evolution. There was no longer any point in an individual biologist's attempting to cover the whole field, he believed. The only proper way to organize a university department was to hire a wide range of specialists who could collectively cover the field and who were prepared to cooperate in their research (Maienschein 1988; cf. Werdinger 1980, 78–84). At least one organization, however, tried to restrain the lurch toward specialization. From the 1890s the American Society of Naturalists sought to unify professional biologists—hoping in the process to reverse its own declining membership—by providing a forum in which members of the newly expanding discipline-based societies might meet to discuss problems of common interest.[44] The strategy seems to have failed. Societies were reluctant to compromise their independence through affiliation to the ASN, and the conference symposia on "evolution in all its aspects"—with which the ASN had hoped to attract members from all of the biological societies—were soon dominated by papers on genetics.[45] By 1910, one historian of American biology has argued,

the experimental biologists had established themselves in American educational and scientific institutions and had overshadowed the old-time natural historians. By and large the experimentalists now held the best university appointments, dominated the leading scientific societies, published heavily in the major journals in the field, and controlled the research institutions.[46]

44. Appel 1988. Paul and Kimmelman (1988) observe that, unlike several other American biological journals, the society's journal, *The American Naturalist*, was slow to publish papers by the new Mendelians.

45. For the domination of the evolution symposia by geneticists between 1908 and 1910, see Appel 1988, 105. Genetics continued to dominate these programs into the early 1920s, aided perhaps by the substantial number of geneticists on the ASN's executive committee (cf. programs of ASN meetings and "Records of the Executive Committee, 1917–1937," in vols. 1 and 2, respectively, both in the ASN papers). The minutes of the executive committee for January 24 (1922?) suggest that the reason why genetics no longer dominated ASN meetings after the early 1920s was that geneticists had found another home. In 1921 they formed the "Joint Genetics Sections of the Botanical Society of America and the American Society of Zoologists" and began to meet separately ("History and Organization," *Records of the Genetics Society of America* 2 [1933], 5–8).

46. Cravens 1978, 33; cf. 18–33. Despite being newly elected as chair of the zoology section of the American Association for the Advancement of Science, one of these natural historians acknowledged the shift of power, describing himself as "one of [zoology's] tired old bisons from the taxonomic menagerie instead of . . . one of its fresh young bulls from the Mendelian byre" (Wheeler 1921).

26   Getting Started

In Germany, however, the trend toward specialization met with considerable resistance. During the 1850s new sections had been gradually established in the German equivalent of the British or American Associations for the Advancement of Science (*Gesellschaft deutscher Naturforscher und Ärtze*) (Pfetsch and Zloczower 1973, 20–23). During the 1860s and 1870s this process accelerated sharply, increasing the number of sections by threefold. From about midcentury the standard pattern in discipline formation had been for the members of a new specialty (*Fachgebiet*) to meet initially within a section of the *Gesellschaft*, subsequently acquiring their own section, and eventually forming their own independent professional society (von Gizycki and Pfetsch 1975). By the 1880s record numbers of members (4,000) were attending the annual meetings, but the number of professional societies "seceding" was also higher than ever before. Increasing resentment in the sections about the constraints which their membership of the *Gesellschaft* imposed led many to consider opting out in order to form their own societies. By 1890 there were even occasional complaints that the *Gesellschaft* had become redundant. Senior members of the *Gesellschaft* reacted with alarm and became increasingly preoccupied with the organization's unity. Some keynote speakers—such as the physicists Ernst Mach and Hermann Helmholtz or the experimental zoologist Richard Hertwig—expressed confidence that various centripetal tendencies would ensure a future commonality of aim and method, despite specialization. But more often speakers warned of the dangers of "fragmentation" (*Zersplitterung*). Greater emphasis on the plenary sessions as well as more review papers would be necessary to cure what the animal morphologist Karl Heider called "this sickness of our time."[47]

The *Gesellschaft* was by no means the only arena in which the trend toward specialization, as well as the reaction against it, was visible. Although the natural sciences at German universities were traditionally part of the "philosophical faculty," in 1863 Tübingen became the first university to regroup its scientists and mathematicians in a separate science faculty (followed by Strassburg in 1873, Heidelberg in 1890, and

---

47. Cited in Schipperges 1976, 85; among those voicing similar worries were the zoologist Carl Chun and the physiologist Max Rubner (ibid., 57–59, 64–69, 73–89). It would be interesting to compare these attitudes toward specialization with those common within the American Association for the Advancement of Science at this time. From one observer's account, it would appear that the AAAS embraced specialization quite happily. "Specialization," as he put it, "must necessarily accompany progress" (McMurrich 1974, 43).

Freiburg in 1910). For some—scientists among them—this was a regrettable signal that the old ideal of the unity of knowledge in the arts and sciences was being abandoned.[48] Within faculties, too, resistance mounted to the establishment of chairs in new specialties. Conservative critics complained that specialization was undermining the coherence of the curriculum and placing a premium upon the quantity rather than the quality of research.[49]

Perhaps the most systematic critique of fragmentation in the biological sciences was published in 1919 by the experimental embryologist Julius Schaxel (1919a). "Biology," as Schaxel called it, was in crisis. The field had become a theoretical patchwork: a "labyrinth of opinions" and a playground for "undisciplined speculation." Although specialists might not notice it, there were no fundamental principles yet capable of connecting these diverse perspectives. Distracted by rapid progress since the mid-nineteenth century, life scientists had not stopped to reflect upon either their premises or their aims. Creating theoretical coherence in this area would require both historical and critical /philosophical analysis. The history of the life sciences was still very undeveloped, but it offered in principle a means of identifying their changing premises and aims. What was required, however, was neither a mere chronicle of what biologists did in each era, nor history written with polemical intent (e.g., in order to attack what the historian perceived as biological dogmas). The kind of history necessary would have to trace the changing conceptual foundations of the various biological traditions. Once such a history had uncovered the field's conceptual underpinnings, philosophical analysis could proceed: clarifying concepts, identifying contradictions, and creating a unified and truly *general* "biology." Schaxel's book was designed to carry out just this program.

One of the six research traditions which Schaxel surveyed in the historical section of his book was the study of inheritance (*Vererbungslehre*) (Schaxel 1919a [2d ed.], chap. 4). Here he found a characteristic feature of modern biology: Theoretical development had lagged behind the accumulation of facts. For example, the central problem in the study of development, Schaxel argued, was that of determination. It em-

---

48. Rudolf Lehmann, appendix to Paulsen 1921, vol. 2, 710. The chemist August Hofmann was strongly opposed to the division of philosophical faculties (Jarausch 1982, 177–78), and at Göttingen this issue was discussed on and off for more than a decade—due in part to scientists' reluctance to secede—before the decision to create a science faculty was finally taken (Dahms 1987b, 170–72).

49. Riese 1977, 124–26, 133; Weindling 1989, 335; Weindling, 1991, 152.

braced the phenomena of both *Vererbung* ("heredity"; i.e., the organism's capacity to develop through a regular sequence of stages) and *Entwicklung* ("development") itself. Although the two phenomena were analytically separable, in actual developmental processes they were inextricably connected. But late-nineteenth-century cytologists as well as more recent students of heredity (Mendelians) had abandoned the analysis of development and were focusing solely upon heredity; they were concerned only with the localization and structure of the hereditary material, plus the mechanics of the process whereby the (already developed) traits of one generation were transmitted to the next generation. In so doing, they had created a theoretical gap between the study of heredity and of development which made solving the problem of determination yet more difficult. A similar gap had opened up between the study of heredity and evolutionary theory. Here the assumption that Mendelian factors were stable flatly contradicted Lamarckian theories, but also seemed to set limits upon what natural selection could achieve. The problem was aggravated by the Mendelians' reluctance to theorize about the *nature* of their factors or how they worked (a "machine without a mechanism," as Schaxel put it). So far these factors were little more than symbols designed to make sense of the statistical analysis of breeding results. For Mendelians to do no more than criticize those evolutionary theories which were inconsistent with the stability of the Mendelian gene was unsatisfactory; perhaps the study of mutagenesis would help to bridge the gap between these two traditions.

It is difficult, of course, to judge the impact of a book upon the scientific (or any other) community. But Schaxel's appears to have been read fairly widely. The first edition was sold out the year it was published, and Schaxel claimed that it was regarded as "stimulating" even by those biologists who were critical of it (Schaxel 1919a [2d ed.], preface). The second edition received a favorable review in *Die Naturwissenschaften*, and eminent biologists such as Alfred Kühn, Hans Spemann, and Otto Renner were impressed by Schaxel's writings on this issue.[50] Since Schaxel regarded his book essentially as a diagnosis of modern biology's theoretical incoherence, the quasi-philosophical task of reconstruction remained. For the latter purpose he founded in 1919 a monograph series, *Abhandlungen zur theoretischen Biologie* (*Treatises*

50. The review was by Kurt Lewin in *Die Naturwissenschaften* 12 (1924), 461–62; cf. Otto Renner to Alfred Kühn, 18.11.44, Kühn papers, Heidelberg. In Britain, J. H. Woodger cited Schaxel with approval (1930, 5).

on Theoretical Biology). By bringing together work by those biologists and philosophers who shared his concern with "the unsatisfactory diversity of biological theorizing," Schaxel hoped that the series would point the way "out of the crisis of contemporary biology."[51] This undertaking found support from a number of well-known biologists, and by 1928 the series had published twenty-seven volumes, among others those by Hans Driesch, Victor Franz (professor of zoology at Jena), F. Alverdes (professor of zoology at Marburg), Ernst Küster (professor of botany), Paul Jensen (professor of physiology at Göttingen), Valentin Haecker (a geneticist discussed below), and Hans Przibram (professor of biology at Vienna), to name only the better known.[52] It would seem, therefore, that Schaxel's concerns resonated within the biological community, not least among well-known experimentalists.

A variety of institutional solutions to the perceived problem of specialization were proposed in Germany. Some argued that the one-sidedness of academics' research was already adequately compensated for by the obligation to prepare lectures which necessarily drew connections between specialized research areas and provided the student with an overview (Nauck 1937, 96–98; Paulsen 1906, 197). Others, such as Max Planck, regarded it as the prime function of the academies of science to curb the evils of specialization by providing a forum for interdisciplinary exchange (Heilbron 1986, 60–61; Riese 1977, 250).

In the *Gesellschaft deutscher Naturforscher und Ärtze* the crisis induced by rapid section formation led to discussion of structural reforms which could enable the *Gesellschaft* to preserve its unity and serve as a "bulwark against sterile specialization."[53] In 1891 new statutes were passed, making it much more difficult to create additional sections. Their effect was dramatic: The number of sections remained essentially constant from 1891 until the First World War (von Gizycki and Pfetsch 1975). These reforms alone, of course, were hardly likely to discourage German scientists from specializing, nor could they block the institutionaliza-

---

51. The first phrase comes from a statement of the series' aims, set out on the inside cover of the first volume (Schaxel 1919b). The second is from p. 34 of that article.

52. Additional monographs were promised (though not all were delivered, at least by 1928) by, among others, the biologist Paul Kammerer, the zoologist Alfred Kühn, and the geneticist Hans Nachtsheim. Among those who indicated an interest in contributing were the physiologists E. Abderhalden, A. Bethe, and A. von Tschermak; the biologists S. Tschulock and A. Thienemann; the zoologists M. Rauther, Hans Spemann, and Wilhelm Roux; and the anatomist A. Fischel (cf. inside cover of vol. 1 of the series).

53. Richard Hertwig used the phrase at the 1891 meeting; cited in Schipperges 1976, 74.

tion of new research areas outside the *Gesellschaft*. What is interesting, nevertheless, is the contrast between the *Gesellschaft*'s impact and that of the American Society of Naturalists. While both organizations tried to slow the pace of specialization, the German one seems to have been more effective.

In German universities those specialties which emerged toward the end of the century found it more difficult to gain a foothold. Various advocates of the new experimental biology, for example, complained that *Entwicklungsmechanik* was finding a much warmer welcome in American universities (e.g., von Uexküll 1913, 800; Barfurth 1910, xxxii; Uschmann 1959, 214). If some of the new German funding for science could be directed toward *Entwicklungsmechanik,* Hans Driesch observed wryly, "then the German founders of the field will no longer need to say that their seed only really germinated on American soil."[54] In the event, Driesch's hopes were realized. In 1914 the Kaiser-Wilhelm Institute for Biology was established in order that Germany could catch up with the United States in the newer branches of experimental biology.[55] Of its five departments, one was earmarked for experimental embryology. Nevertheless, by 1918 the first head of this department (Hans Spemann) did not feel the battle had yet been won, noting "the resistance which [experimental biology] still often encounters from older zoologists."[56] He might have been referring to Jena, where Julius Schaxel ran into considerable opposition over his attempts to found an institute for developmental physiology. The professor of zoology there suggested that if Schaxel did not like the provision for experimental work in the existing Zoological Institute, he should move to the KWI for Biology.[57] But the KWI itself was not altogether secure. When Spemann left the institute to take up the chair of zoology at Freiburg in 1919, his department at the KWI for Biology was closed

---

54. Driesch, *Die Biologie als selbständige Grundwissenschaft und das System der Biologie* (Leipzig, 1911), iii, cited in Sund 1984. When the new Kaiser-Wilhelm Society was founded in 1911, it was by no means a foregone conclusion that *Entwicklungsmechanik* would be one of its beneficiaries. In that year one prospective donor to the planned KWI for Biology was worried that the views of two powerful critics of *Entwicklungsmechanik* would prevail (Wendel 1975, 175).

55. Cf. Theodor Boveri's plan for the institute, dated 25.9.12, in the KWIB file, Emil Fischer papers.

56 Spemann to Eurer Exzellenz (A. Harnack), 13.4.18, no. 3 Harnack file, box 19, Emil Fischer papers.

57 Uschmann 1959, 214–18. With funding from the Carl-Zeiss Foundation, Schaxel eventually succeeded in establishing at Jena the *Anstalt für experimentelle Biologie* in 1918.

for lack of funds (cf. chap. 4).⁵⁸ During the 1920s, accordingly, Roux advised some young *Entwicklungsmechaniker* to apply for positions at American universities.⁵⁹ One of those who emigrated was Emil Witschi. Two weeks before leaving Basel for a professorship of zoology and experimental embryology at the University of Iowa in 1927, Witschi told Richard Goldschmidt, "I formed the very definite impression that [compared with the situation in Germany] the approach to science which you and I share finds greater sympathy and a more certain future in the United States."⁶⁰

As every academic knows, of course, establishing a new field is hard work in *any* country. Most representatives of older disciplines are likely to regard the newcomer as little more than a competitor for scarce resources. Moreover, in bidding for such resources in one's own country, there is an understandable tendency to exaggerate the speed with which the new field has developed elsewhere. Nevertheless, the limited evidence so far available seems to bear out the German experimentalists' complaint; systematics and descriptive morphology seem to have remained well established longer in German universities than in American universities. This situation was particularly clear in the case of zoology. When Julian Huxley complained in the 1920s to Richard Goldschmidt that experimental biologists in Britain were in danger of being squashed between "the old fashioned comparative anatomists, who occupy all the chairs of zoology, and the scientific medical people and physiologists who are stealing all the experimental problems," Goldschmidt replied that "the situation is more or less the same in all countries with the exception of America."⁶¹ The lack of a compulsory retirement age at German universities meant that more traditional zoologists hung on into the 1920s. The systematist Ernst Ehlers (1835–1925), for example, only retired as professor of zoology at Göttingen in 1919 (aged eighty-four). The professor of zoology at Königsberg, forced to retire in 1921 because of age, had a reputation as an opponent of the newer approaches in zoology. Eugen Korschelt (1858–

58. This "reduced yet further the already small number of experimental embryological laboratories" in Germany (Julius Schaxel to Richard Goldschmidt, 17.4.21, Goldschmidt papers).
59. Erwin Bauer (sic) to R. G. Harrison, 16.9.23, Harrison papers.
60. E. Witschi to R. Goldschmidt, 23.8.27, Goldschmidt papers.
61. Huxley to Goldschmidt, 6.1.22, and Goldschmidt's reply, 26.1.22, both in Huxley papers. Erwin Baur was hopeful in 1912 that Goldschmidt might get a chair of zoology "so that we get at least *one* zoological institute in Germany where experimental work on heredity is encouraged" (Baur to Goldschmidt, 14.2.12, Goldschmidt papers [emphasis in original]).

1946), professor of zoology at Marburg until 1928, devoted most of his research to descriptive embryology and comparative morphology. F. E. Schulze (1840–1921), a morphologist of very traditional cast, remained in the chair of zoology at Berlin until his death. And since the life sciences were represented in the prestigious Prussian Academy of Sciences by older men working in traditional areas, the academy's research projects during the 1930s included compiling enormous catalogs of plant and animal taxa.[62] Leadership of the German Zoological Society also remained with traditionally minded figures into the 1920s. In 1922, however, a colleague warned Goldschmidt that the experimentalists among the membership had organized a putsch (*Staatsaktion*) designed to elect Goldschmidt president at the next meeting. The "conspiracy," he wrote, would work only if the "revolutionaries" acted in concert.[63]

There are probably several reasons for zoology's persistent traditionalism. One contemporary attributed it to an informal division of labor: Since their colleagues in the local medical faculty were often working on problems of animal anatomy or physiology, many German zoologists continued to work in systematics. In consequence Roux felt that the new experimental embryology had better prospects of being established in anatomical than in zoological institutes. That medical faculties at several universities in the late 1920s still offered better provision for experimental zoology was one reason why both Ernst Caspari and Hans Kalmus originally trained in medicine.[64] Moreover, since the number of chairs of zoology grew very slowly between 1880 and 1914 (in contrast with the situation in the United States; see chap. 4), it took a long time to displace the older zoological traditions.

Experimental work in botany fared rather better, partly because additional chairs were established for the new plant physiology at a num-

---

62. Kühn 1926; on Königsberg see Richard Hertwig to Richard Goldschmidt, 5.4.21 and 8.2.21, Goldschmidt papers; on Korschelt see Querner 1980b; on Schulze see Tembrock 1958–59; on the Prussian Academy see Schlicker 1975, 46, 151–53, 282, and 341–42.

63. Jakob Seiler to Goldschmidt, 18.10.22, Goldschmidt papers. Success seems to have come a few years later; Karl Belar described the 1925 meeting of the society—when Alfred Kühn and Karl von Frisch got most of the votes in the presidential election—as a "revolution" (Belar to Goldschmidt, 10.6.25, Goldschmidt papers).

64. The contemporary in question was Reinke (1904); on Roux's strategy see Nyhart 1986 and interviews with Caspari and Kalmus. Disappointed that the professor of zoology at the University of Vienna about 1905 taught little but traditional morphology, Karl von Frisch conducted his first research in a physiological institute (von Frisch 1967, 31–33).

ber of German universities during the 1870s. And over the next few decades, according to one contemporary, "modern" botanists succeeded in displacing systematic botanists, even at those universities which had only one chair of botany. By the early twentieth century, systematists were found only at universities which had two botanical chairs or where there was an associate professorship (*Extraordinariat*) in addition to the chair.[65] Nevertheless, this was not a trivial number: In 1910 one-third of the universities had two botanical chairs, and associate professorships of botany existed at half of the universities.[66] Some experimentalists resented the continuing influence of traditional botany: "When will they finally realize here in Europe that halfway decent *experimental* plots [Versuchsgärten] are more necessary than the ridiculously large botanical gardens [*Sammlungsgärten*] that we still have everywhere."[67]

Although the evidence assembled above is patchy, it suggests that biology of a rather traditional kind was alive and well in Germany between the world wars. (It also indicates how badly we need a full-scale comparative study of the growth and reception of experimental biology in the United States and Germany between 1890 and 1920.) As we shall see in chapter 4, this fact was to constitute an important institutional obstacle for the development of genetics after the First World War. In the years *prior* to the war, however, the new "Mendelism" found an enthusiastic response in Germany.

1.5 The Early Days of Mendelism in Germany

During the 1920s it was often said—even by Germans—that the United States was the leading country for genetics. The Mendelian chromosome theory developed by T. H. Morgan and his students was widely acknowledged as a major breakthrough, and no geneticist can

---

65. Reinke 1904. When discussions were under way as to which specialties should be included in the planned KWI for Biology, many referees pressed for zoological areas on the grounds that experimental botany was much better developed in Germany (Wendel 1975, 170–85, passim).

66. Von Ferber 1956, 209. As in zoology, the influence of traditional perspectives was prolonged by the lack of a compulsory retirement age. At Berlin, for example, Adolf Engler (professor of systematic botany and director of the Botanical Museum) only retired in 1921, aged seventy-seven, while his colleague Simon Schwendener (the professor responsible for plant anatomy and physiology) retired in 1909, aged eighty (see the entries for Engler and Schwendener by Frans Stafleu and A. P. M. Sanders, respectively, in the *Dictionary of Scientific Biography*).

67. Erwin Baur to Carl Steinbrinck, 31.12.09, Steinbrinck papers.

have been very surprised in 1934 that Morgan was the first geneticist to receive a Nobel Prize. I suspect, however, that the image which most historians of biology have of German genetics during this period is one of backwardness. After all, little work was done there on *Drosophila* or on transmission genetics more generally, and several German geneticists—among them Carl Correns—directed their attention to *non*-Mendelian phenomena. In addition, the early work by Wilhelm Weinberg found little resonance in Germany, with the result that by the 1930s mathematical population genetics was far weaker there than in Britain or the United States.[68] Furthermore, the German geneticist best known to Anglo-American historians is probably Richard Goldschmidt, iconoclast extraordinaire, whose very considerable talents were deployed against both the chromosome theory and natural selection— in short, a two-time loser. And the forces which hindered the institutionalization of genetics as a discipline (which we will examine in chapter 4) seem to sustain the image of Germany as a backwater. Nevertheless, this view is deficient in two respects. First, it bears saying that a discipline's institutional success is only weakly correlated with its intellectual vigor. As Kohler has shown, for example, German biochemistry at this time was outstanding, despite its institutional weakness (Kohler 1982). Similarly, in chapters 2 and 3 I will show that although the general thrust of interwar genetic research in Germany was *different* from that in the United States, there were no grounds for dismissing the former as inferior or misguided. But the image of Germany as an interwar backwater may be even more misleading if it is implicitly assumed to hold for the *pre-1914* period as well. Quite the contrary; as we shall see in this section, several observers at this time believed that genetics was developing more rapidly in Germany than in the United States.

As a more-or-less coherent body of knowledge, "genetics" is conventionally regarded as beginning with the rediscovery of Mendel's work in 1900 (e.g., Olby 1985). In the early 1860s the Moravian monk Gregor Mendel had conducted a series of controlled crosses between contrasting forms of the pea (among other plants), noting the frequency with which particular traits (e.g., seed color or shape) appeared among the offspring. In 1865 he published an account of this work, formulating several general principles which he believed could account for these results—among them, the postulate that a small number of "factors" were responsible for each trait—but his paper attracted little attention

---

68. On Weinberg's work see Provine 1971, 134–36.

until "rediscovered" in 1900 by three men, all of whom had been carrying out similar crosses in plants during the 1890s: the Dutch plant physiologist Hugo deVries; the Austrian plant breeder Erich von Tschermak; and the German plant physiologist Carl Correns. Within a few years, Mendel's work had attracted considerable interest among experimentalists interested in the nature of heredity as well as among practical plant breeders looking for greater predictability and control. Over the next decade the quantitative analysis of breeding results was used in order to explore how widely Mendel's laws were applicable among plant and animal species, and the laws themselves were gradually modified and supplemented in order to account for more complex patterns of inheritance. The cellular location and physical nature of Mendelian "factors"—dubbed "genes" by the Danish Mendelian Wilhelm Johannsen in 1909—however, remained unclear.

German interest in Mendelism grew rapidly. In 1903, simultaneously with Walter Sutton in New York, Theodor Boveri drew attention to the apparent fit between Mendelian principles and the behavior of chromosomes. Correns began teaching the new "science of heredity" (*Vererbungswissenschaft*) in 1901, and by 1910 Erwin Baur, Valentin Haecker, and Richard Goldschmidt had followed suit.[69] A variety of textbooks soon appeared: by Baur, Goldschmidt, Haecker, and Ludwig Plate.[70] These quickly sold out, and by 1914 second editions of most of them had been published. In 1908 Baur founded the world's first journal devoted to the new field: the *Zeitschrift für induktive Abstammungs- und Vererbungslehre* (*ZIAV*). By 1916 *ZIAV* had published papers by a number of foreign geneticists, including the Americans Raymond Pearl, George Shull, T. H. Morgan, W. E. Castle, A. H. Sturtevant, and R. A. Emerson, perhaps because a comparable American journal (*Genetics*) was not founded until 1916.[71] In higher education establishments several men with research interests in genetics had acquired chairs of botany or zoology by the First World War: Haecker at Halle (1900), Correns at Münster (1909), Plate at Jena (1909), Baur at the Berlin Agricultural College (1911), and Hans Winkler at Hamburg (1912). The year 1914 saw two major institutional innovations which the

69. Correns papers, folders 168–78; Schiemann 1934, 66; R. Haecker 1965; Goldschmidt 1960, 75.

70. Baur, *Einführung* (1911); Goldschmidt, *Einführung* (1911); Haecker, *Allgemeine Vererbungslehre* (1911); Plate, *Vererbungslehre* (1913). Among the most important foreign textbooks available in German translation before the war was Johannsen's *Elemente der exakten Erblichkeitslehre* (1909).

71. On the founding of *Genetics* see Kimmelman 1983.

American geneticist George Shull reckoned would soon make Berlin "a great center for genetic research."[72] At the Berlin Agricultural College, Baur became Germany's first professor of genetics, and in the suburb of Dahlem the Kaiser-Wilhelm Institute for Biology was created with Correns as director.[73] Its department of plant genetics (headed by Correns) employed two younger postdoctoral assistants (*Assistenten*) and two gardeners, while Goldschmidt's department of animal genetics employed one assistant and one technician.[74] Once these new institutions were in prospect, Baur was keen to host the fifth International Congress of Genetics, projected for 1916, and Bateson heartily agreed. "You have now splendid work going on and in several places, too." A conference in Berlin "would certainly be in every way the best possible thing for genetics," at any event preferable to one in San Francisco, since "the Americans are rather late in the field anyhow."[75]

At the time of writing, Bateson could not have anticipated that the publication of a Mendelian chromosome theory two years later by Morgan and his students at Columbia University (Morgan et al. 1915) would force him—or at least others—to revise this judgment. Nonetheless, Bateson's remarks are a useful reminder that prior to the First World War the United States was by no means the only center of genetics. The appearance of the Morgan school's book, however, represented a landmark in the development of genetic theory; using an ingenious combination of cytological and breeding analysis, they produced a variety of evidence for the chromosomal location of Mendelian factors. (For a summary of this work, see Allen 1978.) In his biography of Morgan, Garland Allen suggests that although the Mendelian chromosome theory gained acceptance between 1915 and 1930 in the United States, Norway, Sweden, Denmark, and the Soviet Union, it was less well received in Germany and quite poorly received in Britain

---

72. George Shull to C. B. Davenport, 16.10.13, Davenport papers. Shull was keen to bring both Correns and Baur to the United States for lecture tours (Shull to E. B. Babcock, 28.6.14, papers of the Genetics Department of the University of California / Berkeley).

73. On Baur's career see chap. 7; on the KWI for Biology see Burchardt 1975 and Wendel 1975.

74 "Abgeänderter Entwurf des Haushaltsplanes . . . für das Geschaftsjahr 1918/19," folder: "KWI für Biologie 1913–1918," carton 1, Emil Fischer papers. In many respects Max Hartmann's department of protozoology was also in effect a department of genetics, since many of the younger assistants whom he employed after 1918—Karl Belar, Curt Stern, Joachim Hämmerling, Victor Jollos—worked primarily on genetic problems.

75. These three quotes are from Bateson to Erwin Baur, 31.1.13, 25.2.13, and 9.3.13, respectively (Bateson papers).

Figure 1.1 (left to right) Richard Goldschmidt, Erwin Baur, and Hans Nachtsheim, ca. 1920s (courtesy: Archiv zur Geschichte der Max-Planck-Gesellschaft).

38   Getting Started

and France (Allen 1978, 278–83). Since it was made over a decade ago, this summary has stood up quite well to subsequent research.[76] Its characterization of the theory's reception in Germany, however, while sustaining the image of interwar Germany as a backwater, is misleading. Let us now look in detail at this reception.

The Morgan school's work on *Drosophila* was first reviewed in *ZIAV* (the German genetics journal) in 1911, followed by a favorable review of Sturtevant's chromosomal maps in 1913.[77] Two years later Sturtevant published in *ZIAV* an extended account of his mapping work and an argument for the chromosome theory (Sturtevant 1915). While review abstracts on British and American research continued to appear in *ZIAV* during the war, the number of original articles declined sharply, and no further report of the Morgan school's work appeared until 1919, when Hans Nachtsheim devoted a review paper to Morgan's chromosome theory (Nachtsheim 1919). The decisive step in transmitting the theory to a broader German audience followed in 1921 with Nachtsheim's translation of Morgan's *Physical Basis of Heredity* (Morgan 1921). Reviews of this book provide a quick, preliminary way to assess the German reception of the chromosome theory. Of the twenty reviews of this book which I have located in German, Swiss, and Austrian journals, the vast majority offered either straightforward exposition of the book's contents or strong praise.[78] Only two voiced criticism, one

---

76. In view of the reservations about the scope and significance of the chromosome theory which Johannsen expressed during the early 1920s (cf. chap. 2), I am not sure which geneticists Allen might have had in mind when characterizing the Danish reception as favorable.

77. The review was by M. Daiber, *ZIAV* 10 (1913): 293–94.

78. The reviews in question are: J. Seiler, *Zoologischer Bericht* 1 (1922): 197–98. H. Stieve, *Archiv für Entwicklungsmechanik* 51 (1922): 332–33. Agnes Bluhm, *Archiv für Rassen- und Gesellschaftsbiologie* 15 (1923): 197–200. J. Seiler, *Archiv für Zellforschung* 17 (1923): 382–84. O Koehler, *Berichte über die gesamte Physiologie und experimentelle Pharmakologie* 17 (1923): 283–85. Blumm, *Zeitschrift für ärztliche Fortbildung* 19 (1922). Anonymous, *Unsere Welt* 14 (1922): 226. H. W. Siemens, *Deutsche Medizinische Wochenschrift* (1922): 105. Schicke, *Journal für Psychologie und Neurologie* 27 (1922). H. Löwenthal, *Sozialistische Monatshefte* (1923): 436–37. P. Meyer, *Entomologische Zeitschrift* 36 (1922/23): 105–6. E. Lehmann, *Zeitschrift für Botanik* 14 (1922): 467–70. W. Schleip, *Botanisches Zentralblatt* 1 (1922): 240–42. Anonymous, *Anatomischer Anzeiger* 55 (1922): 190. C. Hart, *Medizinische Klinik* no. 16 (1922): 515. E. Hedinger, *Schweizerische Medizinische Wochenschrift* 52 (1922): 356. F. Lenz, *Münchener Medizinische Wochenschrift* 68 (1922): 1564. G. Just, *Zeitschrift für Sexualwissenschaft* 9 (1922): 211. E. Bresslau, *Frankfurter Zeitung*, Literaturblatt, 17.2.22. Anonymous, *Zeitschrift für Pflanzenkrankheiten* 32 (1922).

of them by a neo-Lamarckian.[79] Similarly, of the three reviews of Morgan's *Theory of the Gene* which I have found, all were favorable.[80] Of four reviews of various other books or papers by Morgan, only one— again by a neo-Lamarckian—was critical.[81] To judge from book reviews, therefore, one would have to say that the Morgan school's work was well received in German-speaking circles during the 1920s.

Of course, book reviews provide a rather crude and indirect way of judging an audience's views. A more meaningful indicator of a theory's impact requires that we examine the extent to which the theory was actually incorporated into research programs. In the case of Morgan's chromosome theory, one finds three types of response among German biologists. Some individuals actively championed the theory and advanced it. One of these was Hans Nachtsheim (1890–1979), whose principal contribution was to propagate Morgan's work through the aforementioned publications and subsequent review papers in the 1920s.[82] Another was Curt Stern (1902–1981), one of interwar Germany's very few *Drosophila* geneticists, whose cytogenetic analysis of Y-chromosome mutants is generally recognized as an important part of the evidence for the linearity of genes' arrangement in chromosomes (Stern 1928). The cytogeneticist Karl Belar (1895–1931) was an important defender of the theory during the mid-1920s (see below). (Stern, Nachtsheim, and Belar all spent substantial periods in Morgan's laboratory during the 1920s.)[83] Finally, Erwin Baur (1876–

---

79. The neo-Lamarckian was H. Stieve; the other was Bluhm. As we shall see in the next chapter, it is significant that one of Bluhm's complaints in her moderately critical review was that Morgan had neglected cytoplasmic inheritance.

80. The reviews are: G. Heberer, *Zoologischer Bericht* 11 (1927): 332–33. Anonymous, *Zeitschrift für Pflanzenzüchtung* 11 (1928): 375. Simon, *Botanisches Zentralblatt* 8 (1926): 440.

81. The critical review was by Ludwig Plate, *Archiv für Rassen- und Gesellschaftsbiologie* 19 (1927): 70. The favorable ones were: P. Hertwig [review of Morgan, *Evolution and Genetics*, 1925], *Berichte über die gesamte Physiologie /Abteil B: Biologie* 38 (1927): 29–30. W. Reinig [review of Morgan, "Recent Results Relating to Chromosomes and Genetics," *Quarterly Review of Biology* 1 (1926): 186–246], *Zoologischer Bericht* 15 (1928): 49. F. Alverdes [review of Morgan/Bridges/Sturtevant, "The Genetics of *Drosophila*," *Bibliotheca Genetica* 2 (1925): 1–262], *Zoologischer Bericht* 7 (1925/26): 558–59.

82. Nachtsheim, "Ergebnisse und Fortschritte" (1922) and "Einige Ergebnisse" (1924).

83. Folder 997 ("Curt Stern, 1924–1927"), box 60, subser. 3, ser. 1; folder 1254 ("H. Nachtsheim, 1925–1927"), box 75, subser. 4, ser. 1; both in International Education Board papers. On Belar see the obituaries by Curt Stern and Hans Bauer.

Figure 1.2 Hans Nachtsheim in 1935 (courtesy: Archiv zur Geschichte der Max-Planck-Gesellschaft).

1933) was one of the first German geneticists to take up the Morgan group's chromosome-mapping techniques and apply them to the snapdragon, *Antirrhinum* (cf. chap. 6).

The second category, who were critical of the chromosome theory, were primarily zoologists and anatomists. Some of these figures need not detain us; although they published in important journals, they seem to have been entirely ignored by the genetics community, which suggests that they posed little threat to the young discipline (e.g., Dembowski 1925; Herz 1923; Jensen 1919). Two neo-Lamarckian anatomists, however, were too important to be ignored. Both Rudolf Fick (1866–1939) and Hermann Stieve (1886–1952) had published cytological analyses of chromosome behavior prior to the First World War, disputing the evidence for crossing-over and rejecting claims for the chromosomes' individuality and continuity.[84] At the German anatomical congress in 1922 Stieve unleashed a sharply worded attack upon "the vast majority of geneticists" (naming Morgan) for interpreting their experiments without any regard for actual facts. Goldschmidt and Correns decided not to respond, on the grounds that a response would draw further attention to Stieve, but one sympathetic anatomist warned them not to forget that Stieve had close friends among "leading people": "The unfortunate thing is that such plants [as Stieve] always flourish on suitable soil and that in this case one sees once again just how favorable the anatomical soil is."[85] The same year Stieve published a lengthy critique of the chromosome theory in an anatomical journal (Stieve 1923), and in 1925 Fick followed suit in a rather more prominent place: *Die Naturwissenschaften* (Fick 1925). This time geneticists counterattacked. The cytologist Karl Belar rebutted Fick's arguments in *Die Naturwissenschaften*, and although Fick was permitted a reply, neither anatomist seems to have pursued the debate further.[86]

---

84. For a summary of Fick's work see Stieve 1939; on Stieve's see Romeis 1953.

85. Herman Braus to Richard Goldschmidt, 22.12.22, Goldschmidt papers. When Braus had tried to prevent the most intemperate of Stieve's remarks from being published in the congress proceedings, other anatomists had rounded on him (Braus to Goldschmidt, 10.12.22, Goldschmidt papers).

86. Belar's rebuttal appeared as "Chromosomen und Vererbung" (1925). Earlier the same year, Belar had published criticisms of Fick's work in a book review. Goldschmidt later praised Belar's "bravery" in taking on such a "dangerous" critic (Goldschmidt 1956, 146–47), presumably because in 1925 Fick was professor of anatomy in Berlin and a member of the Prussian Academy of Sciences who managed to block successive attempts to get Goldschmidt elected to the academy between 1924 and 1928 (Schlicker 1975, 239–40). Belar, on the other hand, was an untenured postdoctoral fellow at the KWI

42   Getting Started

It should be emphasized that only one member of this second category could be described as a geneticist (even if we define this term very broadly to denote biologists whose primary research concerned the study of inheritance). That geneticist was Valentin Haecker (1864–1927), who acknowledged the importance of the chromosome theory but remained dissatisfied with several of its features throughout his life. Many of his criticisms were also voiced by Bateson: the cytological evidence for crossing-over was inadequate; vastly different kinds of organisms sometimes possessed the same number of chromosomes; sex chromosomes were less likely to be the *cause* of sex differences than merely *indicators* of the real cause; and the evidence for purity of the gametes was unconvincing.[87] Richard Goldschmidt's opposition to the chromosome theory attracted much attention after he emigrated to the United States in 1936. Nevertheless, it is important to recognize that he did not adopt this position until the mid-1930s. Existing historical accounts overlook the fact that Goldschmidt was sympathetic to the chromosome theory as early as 1913 and remained a supporter of almost all aspects of the theory for the next twenty years.[88]

Some critics of the chromosome theory registered ontological objections. Philosophers of biology such as Ludwig Bertalanffy and Adolf Meyer-Abich, along with a few biologists, rejected the theory as the embodiment of atomistic and mechanistic science: It represented genes as independent units which could be reshuffled at random through

---

for Biology with a reputation as a sharp and fearless critic who had publicly drawn attention to several senior biologists' errors (Ernst Caspari, "Karl Belar," *Dictionary of Scientific Biography*).

87. See Haecker's *Allgemeine Vererbungslehre* (1921); "Einfach-mendelnde Merkmale" (1922); *Pluripotenzerscheinungen* (1925), chap. 9 and pp. 141, 165–66, 196; and "Aufgaben und Ergebnisse," 96–97. Haecker's work is further discussed in chapters 2 and 3. On Bateson's criticisms of the chromosome theory, see Coleman 1970.

88. In 1913 Goldschmidt wrote to Bateson that "I am familiar with your skepticism toward the chromosome theory, but I believe that the day is near when you will have to endorse the theory" (Goldschmidt to Bateson, 27.1.13, Bateson papers). In the second edition of his introductory textbook, published the same year, Goldschmidt treated the theory—including the idea of the chromosome as a linear array of minute particles—as very promising (*Einführung* [1913], 300–301, 312–13, 315, 323). In the third edition (1919) he felt that "the *Drosophila* work of the Morgan school has placed the [chromosome theory] on a fully secure basis" (preface). His support for the theory is also abundantly clear, both in the fourth edition (1923) of this textbook (e.g., 180, 288, 300–301, 308–14, 441) and four years later in *Physiologische Theorie* (e.g., 1–2, 5, 235–36). These facts seem to have been overlooked by both Carlson (1966, 77–79, 124–30) and Allen (1974).

chromosomal breakage and re-formation.[89] Other biologists were reluctant to explain life processes in terms of material objects. Referring to the chromosome theory, one of them remarked that "such crudely materialist conceptions could probably only be developed in America."[90]

It is the third category of responses to the chromosome theory, however, which best characterizes the majority of the German genetics community. Members of this category were neither so unqualified in their praise of the theory as the first, nor so hostile or philosophically motivated as the second. These geneticists had no doubts about the general validity of Morgan's theory in the 1920s, but they were decidedly unhappy with the theory's narrow explanatory scope. No adequate theory of inheritance, they insisted, could leave out the role of genes in the developmental process. Furthermore, many of them also chose to study genetic dimensions of the evolutionary process. As we shall see in the next two chapters, it was this concern to integrate Mendelism with a wider range of theoretical issues which distinguished German geneticists most sharply from their American counterparts.

Thus almost all of the chromosome theory's *critics* were to be found, not among geneticists, but in older established disciplines. Frequently this skepticism was directed, however, not merely at the chromosome theory, but at genetics as a whole. For it was within traditional areas of zoology, anatomy, and botany that more general reservations about the use of experiment were most commonly expressed. The "seventh heaven of genetics," one aging botanist remarked, was far removed from down-to-earth botany. Why, another asked, should a perfectly good plant physiologist like Correns want to bother with Mendelism? As one of Baur's first students recalled, genetics was then regarded by outsiders as "a curiosity of little significance."[91]

---

89. Bertalanffy 1927, 216, 241; Bertalanffy 1928, 36–43; Meyer-Abich 1935, 37. Hamburger (1980a, 306) reckoned that Spemann found the genes' atomism difficult to accept.

90. Kappert 1978, 46. J. Hämmerling felt that one of the reasons why the chromosome theory was not developed by Europeans, following its formulation by Sutton and Boveri, was that Americans had fewer inhibitions about materialist explanations in biology ("Mein wissenschaftlicher Lebenslauf," Stern papers).

91. The view of genetics as a seventh heaven was expressed (probably about the mid-1920s) by F. Oltmanns (1860–1945), professor of botany at Freiburg (Marquardt interview). The student of Baur's was Gerta von Ubisch (von Ubisch 1956, 421). Both Max Hartmann (1934a, 258) and Baur himself (quoted in von Rauch 1944, 258) complained of similar views. The remark about Correns was made by Wilhelm Pfeffer (1845–1920); see Melchers 1987a, 380. In his *Antrittsrede* to the Prussian Academy in 1915, Correns

44   Getting Started

Moreover, given the contemporary concern among senior German scientists with the dangers of "fragmentation," genetics was vulnerable to the charge of being a "narrow specialism." Whenever the opportunity arose, accordingly, geneticists sought to convey a more favorable image of their field. As late as 1924, for example, in a paper nominally concerned with "some issues in modern genetics," Goldschmidt went out of his way to emphasize that genetics was not just another specialty within biology.[92] The history of biology since the publication of *The Origin of Species,* he argued, had so far passed through three phases: the reconstruction of phylogeny using primarily descriptive methods, the analysis of development via experiment, and finally the study of the structure and transmission of the hereditary substance, ushered in by the rediscovery of Mendelism. At the moment, the biological sciences were entering a new phase in which genetics promised to *unite* the concerns and methods of the three preceding phases by illuminating the old fundamental issues of evolution and development. A few years later Fritz von Wettstein argued along similar lines: In addition to its significance for agriculture and medicine (i.e., eugenics), genetics—more than any other field—*integrated* botany and zoology by focusing upon the common elements of life. Furthermore, genetics' methodology made it the area of biology which was closest to the exact sciences (F. von Wettstein 1930b). And again a decade later, Max Hartmann took advantage of the publication of Dobzhansky's *Genetics and the Origin of Species* to get back at genetics' critics. At first, he conceded, Mendelians had steered clear of evolutionary questions because it

---

referred to "the indifference and resistance" which the new genetics had encountered from many quarters (Correns 1915).

The academy itself was one outpost of resistance. The application submitted in 1918 by the academy's biologists, nominating Rudolf Fick for membership, devoted a paragraph to Fick's "thorough and largely justified criticisms" of genetic theory ("betr. Personalien der Ordentlichen Mitglieder 1917–," Sign: II–III, 37, papers of the Prussian Academy). Significantly, the application submitted in 1924, nominating the geneticist Richard Goldschmidt for membership, stressed his breadth as a zoologist. It devoted one long paragraph to his publications in morphology and one short one to his genetic work. The strategy failed; Fick mobilized the opposition, and Goldschmidt was never elected ("Personalia, Mitglieder," Sign: II–III, 41, papers of the Prussian Academy).

92. Goldschmidt 1924b. That this was an important way of legitimating a new field at this time can be seen from the fact that German protozoologists (*Zellforscher*) pursued the same strategy. According to Jacobs, theirs was a "basic science . . . concerned with the organizational principles and elementary properties of life," which stood "above the secondary fields of embryology, genetics and evolution studies" (Jacobs 1989, 235–36).

seemed that they could not be solved until the causes of variation were better understood. But as these causes began to be studied in the 1920s,

it became evident that geneticists were not at all such narrow laboratory- and garden-specialists as they had been regarded by other biologists. Rather it turned out that they had not only maintained but deepened their knowledge of the diversity of form in nature and that they had not at all lost sight of the problem of evolution—*the* issue in biology—in the course of detailed laboratory research.[93]

As the next two chapters will demonstrate, German geneticists' sensitivity about "narrowness" produced more than mere rhetoric; their research agenda during the 1920s went some way toward fulfilling genetics' promise as a unifying discipline. Of course, the suspicion directed at "narrow specialties" was not confined to Germany; naturalists in the United States shared similar misgivings about the new experimental biology. But as we shall see in chapter 4, the greater institutional power enjoyed by traditional elements within the life sciences in Germany gave them far more influence upon the development of genetics as a discipline than was possible in the United States.

93. Hartmann 1939 (introduction to the German edition of Dobzhansky, *Genetics and the Origin of Species*).

Part I THE PECULIARITIES OF GERMAN GENETICS

2 The Genetics of Development

> ... one of the things a scientific community acquires with a paradigm is a criterion for choosing problems that ... can be assumed to have solutions. To a great extent these are the only problems that the community will admit as scientific or encourage its members to undertake. Other problems, including many that had previously been standard, are rejected ... as the concern of another discipline or sometimes as just too problematic to be worth the time [T. Kuhn 1970, 37].

The research strategy pursued in the United States by T. H. Morgan and his school after 1911 neatly illustrates this function of paradigms. At Columbia University and later at the California Institute of Technology, Morgan and his students elected to do "genetics" without reference to genes' role in either evolution or development. Although Morgan acknowledged that heredity was central to an understanding of both evolution and development, he felt that these issues were too complex for geneticists to handle in the 1920s. Far better to concentrate on more tractable problems such as the structure of the hereditary material and the processes of its transmission between generations, leaving development to "embryologists." Having worked as an embryologist himself until his conversion to the chromosome theory about 1910, Morgan seems to have been somewhat uneasy about abandoning the problem of development in this way. Nonetheless, it was expedient to do so, he argued in 1917, and one hoped that genetics and embryology could be reunited at a later date (Morgan 1917, 535, 539). Six years later Morgan noted that genetics still had "almost nothing to say" about development, but he asked those embryologists who had criticized this deficiency of the chromosome theory to accept "the limitations that the geneticist places on his own work" (Morgan 1923, 623, 627). Again in 1932 Morgan conceded that there was an "unfortunate gap" in geneticists' understanding of the physiological processes linking genes with traits (Morgan 1932, 285).

To Richard Goldschmidt, this gap was unfortunate indeed. In the United States during the First World War Goldschmidt complained that American geneticists were not interested in his ideas about a physiological (or developmental) approach to genetics. While the European reception of his book on the mechanism and physiology of sex determination was enthusiastic, only a few Americans understood it.[1] Part of this indifference, he implied, was due to ignorance. Especially in the United States a school of genetics had grown up whose members thought that biological knowledge beyond "pure Mendelism" simply was not necessary.[2] In private he felt that the development of physiological genetics in the United States had been "constantly blocked by Morgan":[3]

It is really too bad that Morgan and his students—who are after all unusually clever and experienced researchers—have got stuck in such a narrow [*engherzige*] interpretation of genetic phenomena and oppose at all costs any new idea, especially a physiological one, which might invigorate an otherwise somewhat boring Mendelism. I have discussed this issue at length with my dear friend Morgan, but he insists that a thing [phenotype] has been explained once one has mapped a corresponding Mendelian factor.[4]

While Goldschmidt's call for a physiological genetics fell largely on deaf ears in America during the 1920s, his views were widely shared in Germany (e.g., von Wettstein 1924, 178–79; V. Haecker 1918, 7). Friedrich Oehlkers, for example, argued that the field of genetics was undergoing a transition in which Mendelism—by which he meant transmission genetics—was giving way to developmental genetics (or *entwicklungsphysiologische Genetik,* as it was called in Germany) (Oehlkers 1927). Others paid tribute to both Theodor Boveri and Carl Correns as students of hereditary transmission who did not lose sight of the developmental genetic problem (Hartmann 1939, 11–13). Work during the mid-twenties by the botanists Hans Winkler and Fritz von Wettstein was often cited in Germany as opening up new possibilities for a physiological genetics (e.g., Michaelis 1933a, 1). Throughout the

1. Goldschmidt 1960, 158 and 317. The book in question was presumably *Mechanismus und Physiologie der Geschlechtsbestimmung* (1920).
2. Goldschmidt 1927, 2. The Swiss embryologist Emil Witschi tried to reassure Goldschmidt that American geneticists were not dismissive of his work, but agreed that some were narrowly educated, citing Calvin Bridges, "whose knowledge of zoology is confined to *Drosophila* and the female *Homo sapiens*" (Witschi to Goldschmidt, 23.8.27, Goldschmidt papers). Asked by the Rockefeller Foundation to compare the quality of the Norwegian Otto Lous Mohr with the Swede Gerd Bonnevie, Goldschmidt conceded that while the former was probably brilliant, he was "only a Drosophilist whereas Professor Bonnevie . . . is really a broad, first class biologist" (W. E. Tisdale to Warren Weaver, 28.8.34, folder: "KWIB, 1931–34," box 4, subser. 717D, ser. 1.1, RG 6.1, Rockefeller Foundation papers).
3. Diary of HMM, 13.12.33, Rockefeller Foundation papers.
4. Goldschmidt to Julian Huxley, 27.5.20, Huxley papers.

interwar period German geneticists acknowledged the United States as the stronghold of transmission genetics while perceiving themselves as leaders in developmental genetics.[5] The American geneticist L. C. Dunn seems to have agreed. In his report to the International Education Board on the state of genetics in various European countries, Dunn noted the Europeans' "somewhat different and broader conception of genetics" than that which was common in the United States. Geneticists in Germany and elsewhere, he observed, were often interested in embryological and physiological problems.[6] Let us look in more detail at this German work.

2.1 Developmental Genetics in Germany during the 1920s

Richard Goldschmidt's spirited defense of various unorthodox views, following his emigration to the United States in 1936, has guaranteed him a place in virtually all English-language surveys of the history of genetics. Since his work in developmental genetics has already been discussed in some detail by others (Allen 1974; Richmond 1986), I need only single out three features of his work which are important for my purposes. The first point is that his interest in this area emerged very early. Already in 1909 he wanted to understand the genetic basis for the *development* of sex differences in animals, rather than merely identifying the Mendelian factors which caused them. And by 1917 he had worked out the basic details of the theory which was published a decade later as *Physiologische Theorie der Vererbung* (physiological theory of inheritance) (Richmond 1986, 380–82). The second point is that it is wrong to dismiss Goldschmidt as historically insignificant because he "left few lasting contributions" (Carlson 1983). Although his work met

---

5. Lang to Pätau, 31.1.46, Pätau papers. In a wartime report to the German government on developments in genetics, von Wettstein pointed out that "transmission genetics, especially in Drosophila and corn, remains the focus especially of American research, even if the most exciting results are probably over with. . . . Developmental genetics is often attempted, but Germany remains the center of such work" (Report to Forschungsdienst, Fachsparte Landbauwissenschaft und allgemeine Biologie, Dahlem, 17.12.44, folder 200, Correns papers).

6. L. C. Dunn to C. B. Hutchinson, 26.1.28; in a similar vein see Dunn to Hutchinson, 2.11.27 (Dunn papers). Even as late as the 1960s a number of German-speaking geneticists still felt that in terms of textbooks American biology was very strong in molecular biology but weak in embryology and developmental genetics. See U. Clever to A. Kühn, 20.2.66 (Kühn papers, Heidelberg), Ernst Caspari to Springer Verlag, 2.6.66 and 20.6.66, and Caspari to M. Hazlett, 13.3.63 (Caspari papers).

with criticism and indifference in the United States following his emigration there in 1936, his earlier influence in Germany had been substantial. His work on the gypsy moth was widely read in Germany, and his books on physiological genetics attracted a great deal of attention there, even when his readers did not agree with his theories, because Goldschmidt's interest in development was widely shared by his German colleagues.[7] And if scientific knowledge grows through a clash of conflicting theories, one cannot adequately assess a thinker's importance solely in terms of the number of "lasting contributions" which he or she leaves. As Garland Allen put it, "[Goldschmidt's] theories, though often missing the mark, raised serious questions about the Mendelian-chromosome theory to which the silence of Morgan's group did not do full justice" [Allen 1974, 81]. Thirdly, if we are to account for Goldschmidt's poor reception in the United States, it is not enough to say that "his publications were strewn with errors" (Carlson 1983). Admittedly, his later work on *Drosophila* was sloppy and duly ignored, and his unorthodox theories from the early 1930s placed him on the periphery of the American community (Caspari 1980, 19; C. Stern 1958) And the conviction—verging on arrogance—with which Goldschmidt defended his speculative theories hardly helped his reputation in the United States (Allen 1974, 60–61). But German geneticists were also critical of Goldschmidt's fondness for speculation[8] *without* thereby dismissing the value of his work as a whole and *without* writing off the problem of development as insoluble. There is no avoiding the conclusion that Morgan et al. and the Germans were simply interested in different problems.

Another German geneticist interested in development was Valentin Haecker (1864–1927). After receiving his doctorate at Tübingen in 1889 (supervised by the orthogeneticist Theodor Eimer), Haecker spent the next decade at August Weismann's institute in Freiburg, as assistant, *Privatdozent*[9] (1892), and finally associate professor (*Extraordinarius*). Although he continued to publish in ornithology, most of his work became cytological, his principal aim being to demonstrate the chro-

---

7. Melchers interview; C. Stern 1967. Caspari was persuaded to work in developmental genetics because of Goldschmidt's *Physiologische Theorie* (1927). The book provided the first simple and consistent model relating gene structure and function and was attractive to students because it integrated genetics, embryology, and biochemistry (Caspari 1980).

8. For a few examples among many, see Karl Belar to Curt Stern, 22.12.29, and Hans Nachtsheim to Stern, 27.9.63, both in the Stern papers.

9. An unsalaried lecturer; see chapter 4 for a discussion of this type of position.

Figure 2.1 Richard Goldschmidt, ca. 1930 (courtesy: Archiv zur Geschichte der Max-Planck-Gesellschaft).

mosomal reduction-division which Weismann had predicted. In 1900 he was called to the chair of zoology at the *Technische Hochschule* Stuttgart, where he taught students of agriculture and veterinary medicine and became interested in Mendelism. In 1908 he joined the editorial board of Germany's newly founded journal of genetics (*ZIAV*), and the following year he became professor of zoology at the University of Halle, where he remained until his death.[10]

As we saw in the previous chapter, one of Haecker's objections to the Mendelian chromosome theory was its failure to bridge the gap between hereditary factors and traits. His solution for this deficiency was a research program called *Phänogenetik* (phenogenetics), which he outlined in detail (Haecker 1918). In essence, Haecker argued, *Phänogenetik* combined the aims of genetics with the descriptive analysis of development (*Entwicklungsgeschichte*). Unlike experimental embryology, which started with the fertilized egg and followed development forwards, *Phänogenetik* began with mature phenotypes and worked backwards, seeking to trace their causes into the germ cells. One began with the normal form and a variant of a well-defined trait and then followed their (differing) developmental stages back until one could identify a "branchpoint," prior to which their patterns of development were identical. Such branching, Haecker proposed, might be due (in the case of different colored variants) to the appearance of different kinds or degrees of pigment formation or to differing rates of division among pigment cells. The final stage of the analysis was to locate the cause of such phenotypic branching in various properties of the germ cells. The kinds of causes he had in mind were differences of chromosome number, differences in early cleavage, chromosomal relations in maturation divisions, and chemical differences in the germ cells.[11]

By 1925 Haecker believed that phenogenetics could claim a substantial number of adherents. Besides his own doctoral students, he cited seventeen biologists—most of them German-speaking, to judge from their names—who had adopted his research program during the previous decade, almost all of them using his term, *Phänogenetik* (Haecker 1925, 100). But although it had been said that his 1918 book contained a thousand potential dissertation topics, Haecker's approach to

10. See the biographical essays in Haecker's *Festschrift: Zoologischer Anzeiger*, vol. 174, nr. 1 (1965).

11. Quite how one was supposed to know what *kind* of chemical differences to look for, Haecker did not say, but he liked E. G. Conklin's suggestion that the left- and right-handed coiling of snails' shells might be due to an intra- or intermolecular asymmetry in the hereditary substance (Haecker 1923).

developmental genetics seems not to have survived his death in 1927.[12] Unlike Goldschmidt's, his method was essentially descriptive, little different from the late-nineteenth-century cytological work against which experimental embryologists and Mendelians had reacted. And despite the programmatic character of Haecker's claims, the program offered very little testable theory which a geneticist interested in development could have adopted.[13] Unlike the grand theory in Goldschmidt's *Physiologische Theorie*, Haecker's 1918 book was unremittingly empirical. Almost every chapter dealt with a different set of traits (mostly coloration and patterning) in a different group of animals, cataloguing what was known about its development. Given his seniority within the genetics community, however, Haecker's research program is significant as an early statement of the priority which the problem of development enjoyed among German geneticists.

Although N. W. Timoféeff-Ressovsky's contributions to molecular biology and population genetics are relatively well known (e.g., Mayr and Provine 1980; Olby 1974), his work on developmental genetics with his wife, Helena, during the 1920s is not. Although Russians by birth and citizenship, they were important members of Berlin's genetics community from their arrival in 1925 until N. W. was arrested by the Red Army and deported to a Soviet prison camp in 1945. Throughout that period they worked at the Kaiser-Wilhelm Institute for Brain Research, where N. W. became head of the department of genetics in 1931, the department becoming an independent institute in 1937 (Eichler 1982; Medvedev 1982). During their first ten years in Berlin they published about a dozen papers on the genetic and environmental factors which affected the phenotypic consequences of particular mutations in *Drosophila funebris*. On the basis of this research they introduced the now standard concepts of "penetrance" and "expressivity" to the genetics community.[14] Following Muller's report of X-ray mutage-

12. Freye 1965, 409. Presumably Freye is referring to Haecker's method; the terms *phänogenetisch* and *phänokritische Phase* (the stage during development when two closely related variants began to diverge) were certainly taken up by developmental geneticists later (e.g., Kühn 1937a, 440).

13. Richard Hertwig greatly preferred Goldschmidt's textbook on genetics to Haecker's; despite the fact that the latter drew upon a wide range of facts, he missed the forest for the trees ("*die grossen Grundzüge [gehen] verloren*") (Hertwig to Goldschmidt, 10.10.21, Goldschmidt papers).

14. *Penetrance* refers to the proportion of individuals within a population in which a given gene's effects are manifested; *expressivity* refers to the intensity with which a given gene's effects are manifested. For a review of their work see N. W. Timoféeff-Ressovsky 1934. Who originally coined these terms is not clear. The Timoféeff-Ressovskys' former

Figure 2.2 The animal physiology course at Göttingen, 1921–22. Alfred Kühn is standing, second from left; Karl Henke is standing, third from right (courtesy: Frau Maria Henke).

nesis in 1927, however, the Timofeeff-Ressovskys devoted less and less time to developmental genetics as the group became increasingly interested in mutation genetics and, as we shall see in the next chapter, population genetics.[15]

The most important center of developmental genetics during the 1920s, however, was in Göttingen. At the Institute of Zoology in 1924

---

colleague K. G. Zimmer credits N. W. with the terms (Zimmer 1982). According to Paul and Krimbas, the Timofeeff-Ressovkys took over these terms from the neurobiologist Oskar Vogt, who was interested in why the members of a pair of monozygotic twins suffering from a mental disease sometimes differed in the severity of their condition or even in whether they contracted the disease at all. Vogt brought the Timofeeff-Ressovskys to his Institute for Brain Research in Berlin because their related work on *Drosophila* seemed to offer a promising model system for studying gene manifestation (Diane Paul and Costas Krimbas, "N. W. Timofeeff-Ressovksy," *Scientific American*, February 1992, 86–92).

15. This shift of interest can be inferred from the titles of the papers published by the members of Timofeeff-Ressovsky's department between 1927 and 1938 (see *Tätigkeitsberichte der Kaiser-Wilhelm Gesellschaft,* published each May or June in *Die Naturwissenschaften*).

Alfred Kühn and Karl Henke set out to bridge what they saw as a gap between transmission genetics and embryology.[16] Goldschmidt's work during the twenties on the genetics of wing coloration in the gypsy moth (*Lymantria*) was promising, they felt, but his grand theory of developmental genetics needed more empirical grounding. The patterning of moth wings offered many experimental advantages: The pattern was comprised of relatively few elements; it displayed great phenotypic and genotypic variability; its development was rapid and easy to observe; the stages of its development were easy to manipulate experimentally; and moths were easily bred and developed rapidly, such that large numbers could be observed over many generations. Kühn and Henke opted for a species which offered most of these advantages: the flour moth, *Ephestia*.

Compared to the phenotypes best known to embryologists, *Ephestia*'s wing pattern was relatively simple. (Figure 2.3 illustrates the patterning on the upper side of *Ephestia*'s front wing, consisting of nine different marking systems.) The wing's overall appearance was affected by its color, patterning, and brightness. This appearance, in turn, was largely attributable to the pigmentation of the individual scales covering the wing's surface, but also to the size and form of the scales. For example, other things being equal, small scales look darker than large ones. Furthermore, these scales are arranged in layers, the top layer itself consisting of four different types of scales which differ in color and patterning.

In 1929 Kühn and Henke published their first monograph on the genetics of pattern formation (Kühn and Henke 1929). They showed that the relative intensity of the various marking systems on a wing was hereditary, since it varied from one strain to another. In two particular strains they could trace a number of phenotypic differences to several different pairs of alleles. The constituent elements of a given marking system, they found, were all affected by a single locus. Different marking systems (e.g., cross-banding, edge spots) were controlled by different loci. Yet other loci acted at a higher level, affecting the color of markings throughout the wing. The absence of scales in certain regions of the wing was also heritable, and the expression of this and other phenotypes was temperature-dependent. The segregation of scalelessness and the normal phenotype in various crosses suggested that three loci were involved.

Three years later the next installment of Kühn and Henke's results

16. For further details of Kühn and his school see chapters 7 and 6, respectively.

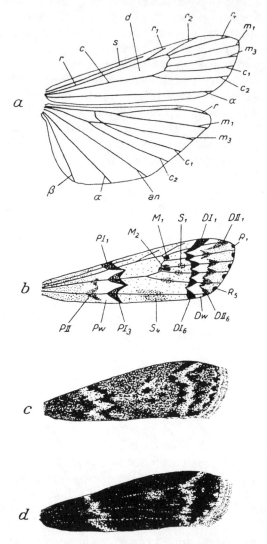

Figure 2.3 Patterns on the front wing of the flour moth, *Ephestia*. "a" shows the veins serving the wing. "b" is a schematic diagram of the nine marking systems on the top of the wing. "c" and "d" show the patterns for the wildtype moth and a dark mutant, respectively, with sixfold magnification (reproduced from Kühn and Henke [1936], p. 6, courtesy: Akademie der Wissenschaften zu Göttingen).

Figure 2.4　Karl Henke in 1937, newly appointed to the chair of zoology at Göttingen and delighted that a butterfly has just settled upon his hat (courtesy: Frau Maria Henke).

was published (Kühn and Henke 1932). They continued to identify genes involved in further marking systems and noted the pleiotropic effects of pigmentation genes on both speed of development and viability. Much of this monograph, however, was concerned with the characteristics not of marking systems, but of scales. Kühn and Henke observed that the characteristic color, size, and shape of a given type of scale were heritable and constant, wherever it was located on the wing and whether it occurred on the wing of a strain which was dark or light, weakly or strongly patterned. Thus variations in wing pattern were not reducible simply to the extent of scales' pigmentation. Instead the pattern was a higher-order phenomenon, due to *the distribution of different types* of scale across the wing.

The final report of their work was devoted to mutants which modified the development of pattern formation.[17] For example, in the mutant *Sy* the cross-bands were shoved closer together than normal. In order to get at the stage of development in which these mutated genes acted (or, more precisely, at the stage in which these actions became manifest), they employed what Goldschmidt was later to call the "phenocopy" method. To begin, they observed three phases during pupation. In the first of these the symmetry of the overall pattern was laid down, spreading from the underside of the wing to the leading and trailing edges, and thence to its upper surface. This spread could be interrupted at critical periods through raising the temperature; some of the resulting wing patterns of these "phenocopies" resembled those of particular mutants, from which Kühn and Henke inferred the times during development when these mutant genes acted. During the second phase the different types of scale were formed with their characteristic sizes and shapes, and in the third phase pigment was deposited in these preformed scales. Again Kühn and Henke found a particular kind of environmental manipulation which could inhibit pigment deposition, mimicking the wing pattern of one of their mutants.[18]

In 1933 Henke left Göttingen to join Goldschmidt's department at the Kaiser-Wilhelm Institute for Biology in Berlin. Four years later he became Kühn's successor as professor of zoology at Göttingen and began to work on mutants affecting the size and patterning of the *Drosophila* wing. Again he used heat-induced phenocopies to study the time at which mutant genes acted and sometimes their mode of action as well. Although Henke's work during the 1940s was "barely known" in

---

17. Kühn and Henke 1936. A summary of their findings appeared in Kühn 1936.
18. A preliminary account appeared in Kühn 1934a, 46.

the United States, Ernst Caspari judged Henke to be one of developmental biology's most original minds.[19]

2.2 Cytoplasmic Inheritance

One area of genetics which was dominated by German-speaking geneticists during the interwar period[20] illustrates the Germans' interest in development particularly clearly: the study of cytoplasmic inheritance. As Sapp has shown (1987), the origins of the interest in cytoplasmic inheritance can be found among late-nineteenth-century theories of development. According to Roux and Weismann, cell differentiation was due to the unequal distribution of nuclear elements at each cell division. Driesch's classic experiment of 1892, however, marked the beginning of an alternative research tradition which was to dominate embryology through the interwar period. Once Driesch had found that separation of the first two blastomeres of a sea urchin embryo yielded two *complete* organisms, embryologists concluded that the first nuclear division could not have been unequal and that the causes of differentiation in the early blastula must be sought elsewhere: in the distinctive cellular environments in different regions of the blastula and /or in the distribution of cytoplasmic elements within the egg.

Henceforth embryologists focused their attention upon the cytoplasm (Maienschein 1981; 1984). "Throughout my scientific life," E. G. Conklin declared, "I have been waging a fight for the recognition of the importance of the cytoplasm of the egg."[21] Some degree of "cytoplasmic preformation," Joseph Woodger believed, was necessary to

19. V. Hamburger to Gerard Pomerat, 15.2.50, folder 321, box 48, ser. 717, R7RG 2 GC 1950, Rockefeller Foundation papers; Caspari 1957.

20. Surveying the literature on cytoplasmic inheritance reviewed by Ernst Caspari (1948), one finds that fourteen of the twenty-two cited authors for the period covered (approximately 1920–1946) were German-speaking. Of the ten who published prior to 1930, eight were German-speaking. Both of the authors who published before 1920 were German. (In those very few instances of authors unknown to me, I categorized them as "German-speaking" on the basis of their names and of the journals in which they published.)

Since my own papers (Harwood 1984; 1985) were published in relatively inaccessible journals, Sapp's *Beyond the Gene* (1987) was an important step in bringing the research on cytoplasmic inheritance to the attention of historians. Unfortunately, however, despite German domination of this field before the 1940s, Sapp's coverage of German work is much thinner than his discussion of that in France and the United States. None of the available unpublished sources is used, and two major figures (F. Oehlkers and J. Schwemmle) receive little or no attention.

21. Cited in Plough 1954.

account for development; genes merely provided the materials for that process (Woodger 1930, 20). That the egg cytoplasm was perhaps more important than the genes drew support from observations that the most fundamental early characters of the embryo (e.g., the pattern of its symmetry) were those of the maternal parent, while the sperm's influence was only evident later and appeared to affect less fundamental traits.[22] On the other hand, almost all of the Mendelians assigned heredity solely to the nucleus. As Erwin Baur commented in 1910, "In recent years cytologists' interest has concentrated ever more on the cell *nucleus;* everything else has been neglected."[23] The sole exception, to my knowledge, was Carl Correns. In one of the earliest discussions of the new Mendelism, Correns expressed a view which was subsequently voiced in one form or another by many students of cytoplasmic inheritance over the next few decades, namely that Mendelian genes alone could not account for the ordered patterns of development:

> I propose to locate the Mendelian factors in the nucleus, in particular in the chromosomes, and to assume the existence outside the nucleus in the cytoplasm, of a mechanism which guides their action [*für ihre Entfaltung sorgt*]. The factors can now be mixed together at will like the colored bits of glass in a kaleidoscope, but they will exert their effects at the correct place.[24]

But how was the existence of *genetic* structures in the cytoplasm to be demonstrated? In general, demonstration of cytoplasmic inheritance required some means of situating the same nucleus in several different cytoplasms, such that the resulting phenotypes differed persistently over many generations. Among the earliest classical attempts of this kind were Boveri's experiments with "merogones": eggs or portions of eggs which no longer possess a nucleus. He found that sea urchin egg fragments lacking a nucleus, when fertilized with sperm from another species, seemed to develop normally. Thus egg cytoplasm rather than the egg nucleus seemed to bear the maternal genetic contribution. By 1914, however, Boveri had decided that these experiments were seriously flawed. Genuinely haploid merogone hybrids actually developed only up to gastrula stage, such that the relative genetic importance of

---

22. E.g., Schaxel 1916; Penners 1922. Even Theodor Boveri, who defended the view that the nucleus was the prime locus of heredity, acknowledged that some traits (e.g., location of the fertilized egg's axes) might be cytoplasmically inherited (Boveri 1904, 113).

23. Baur to William Bateson, 10.5.10, Bateson papers (emphasis in original).

24. Correns 1901, 87. The kaleidoscope metaphor was deployed subsequently by Loeb (1916) and by Woltereck (1931b, 285).

nucleus and cytoplasm for the mature organism could not be assessed (Boveri 1918).

Another difficulty, with the merogone experiments as well as many others which claimed to demonstrate cytoplasmic inheritance, was that some phenotypic effects which appeared to be associated with the cytoplasm were actually of nuclear origin. Experimental embryologists had observed various cases of such "predetermination" (*Nachwirkungen*)—for example, where the distribution of components in the egg cytoplasm could affect early features of zygote development such as the plane of cell division. Other cases ("pseudo-inheritance") were known in which environmental effects upon the egg cytoplasm became manifest in the following generation. When female moths were fed on wool dyed with Sudan Red, for example, their egg cytoplasm incorporated the dye, and the offspring's tissue retained a pink hue. Since such effects were diluted out over a few generations, however, they could not be regarded as true examples of cytoplasmic inheritance.

The most important evidence for cytoplasmic inheritance before the 1940s relied upon plant hybrids. Two plants of different strain, species, genus, or even subfamily were crossed with each other. Usually the results of such crosses are reciprocal: the characters of the offspring are the same whether species A is used to pollinate B or vice versa. Occasionally, however, the results are nonreciprocal. If one assumes that the chromosomal composition of each hybrid of a pair is identical, nonreciprocity cannot be attributed to nuclear elements. Where the pollen contributes a nucleus to the zygote but no cytoplasm, the cytoplasm of the zygote is entirely of maternal origin and must be responsible for nonreciprocity.[25] Finally, given the existence of predetermination, conclusive proof of cytoplasmic inheritance required that such nonreciprocity persist undiminished over many generations.

Correns and Baur were the first to demonstrate cytoplasmic inheritance through nonreciprocal crosses. From the very beginning of his Mendelian work, Correns was interested in the inheritance of variations in chlorophyll content in *Mirabilis jalapa*. One variant which arose spontaneously (named "albomaculata") was variegated: Some of its stems were green, others were white, and still others were mixed. When

---

25. Strictly speaking, of course, proof was conclusive only once the nucleus of such species hybrids no longer contained *chromosomes* of maternal derivation. This was achieved by repeatedly backcrossing the first-generation hybrid against the male parent until the hybrid's nucleus was, in all probability, exclusively of male derivation. Demonstrating cytoplasmic inheritance by this method also assumed that various complicating factors can be neglected (e.g., differential zygote survival, chromosomal irregularities).

Correns carried out controlled crosses using gametes from these three zones, he found that the phenotype always followed that of the egg donor. Pollen from the mutant behaved just like that from the original green strain. His explanation for these results was that the mutant's nucleus was normal, but its cytoplasm was labile, sometimes permitting and at other times inhibiting the formation of normal chloroplasts. It was not certain whether the cytoplasm's "sickness" was localized in its chloroplasts or in the unformed cytoplasm, but he inclined toward the latter possibility, perhaps because it fit better with his conception of an *Entfaltungsmechanismus* in the cytoplasm. As Correns pointed out, his hypothesis rested on the assumption that only the male nucleus—and not the male cytoplasm—was transferred through the pollen tube, but he presented indirect evidence which made this plausible (Correns 1908–09; 1909).

Baur was also interested in the genetics of color variation at this time. Some variants' pattern of inheritance was simple and Mendelian, but one variegated strain of *Pelargonium zonale*'s—whose green leaves had white edges—was not. Microscopic examination of the leaves revealed that their green core layer was covered with a colorless layer of cells which extended out beyond the core. Baur suspected that although derived from the same zygote, the two layers had separated early in development. Happily, one plant of this strain had developed an all-white stem and an all-green one, so he could conduct controlled crosses between gametes from these regions and normal green plants. As with Correns's mutant, the crosses proved nonreciprocal, and the phenotype was (for the most part) maternally inherited. In order to account for the distinct green and white layers in the mutant, however, Baur hypothesized that the mutant contained two kinds of chloroplasts, one of which was normal while the other failed to become green. The two types would be distributed randomly to the daughter cells during cell division, but an unequal distribution early in development ("vegetative segregation") would yield either all-green or all-white cells. But how did a zygote acquire two kinds of chloroplast? Since Baur's breeding results showed that occasionally the offspring's phenotype was influenced by its male parent, he suggested—contrary to Correns and the conventional wisdom—that chloroplasts might enter the egg via the pollen tube (Baur 1908–09).

In subsequent years new cases of cytoplasmically inherited variegation in other plants were reported, some of which accorded with Correns's model and others with Baur's. While Correns seems to have felt that his theory alone was sufficient, Baur reckoned that both theories were necessary to account for all forms of variegation.[26] Since the assumptions upon which each theory was based were vulnerable (an issue to which I return in chapter 9), both theories remained on the agenda into the 1930s, providing alternative conceptions of the nature of cytoplasmic inheritance.

German interest in cytoplasmic inheritance intensified after the First World War. The first meeting of the German Genetics Society in 1921, for example, opened with three botanical papers, all of them dealing with species hybrids (and two of them reporting nonreciprocity). The zoological section, similarly, opened with a general discussion of the relations between nucleus and cytoplasm and included a paper by Paula Hertwig on hybrid amphibian eggs.[27] During the 1920s the number of papers on species hybrids increased rapidly (e.g., Oehlkers 1929–30, 538; Geith 1924, 123). At the third meeting of the Genetics Society in 1923 there were papers on generic hybrids in mosses as well as reciprocal crosses in *Epilobium* and *Oenothera*.[28] What attracted most attention, however, was Hans Winkler's review of the evidence for cytoplasmic inheritance and his reflections on its significance (Winkler 1924). At that time the evidence for cytoplasmic inheritance was inconclusive, Winkler conceded, and most geneticists endorsed the principle of '*Kernmonopol*' (literally, "nuclear monopoly"): that the hereditary substance was located exclusively in the chromosomes. Nevertheless, like Loeb (1916) and Johannsen (1922; 1923) before him, he argued against the sufficiency of the chromosome theory on the grounds that known chromosomal genes dealt only with biologically trivial functions; after all, Mendelian analysis was necessarily restricted to the study of characters which, when mutated, still left the organism viable. Functionally and taxonomically important traits, therefore, had to be borne by hereditary elements which lay outside the chromosomes. Because there was no known mechanism for the precise distribution of cytoplasm to daughter cells, the postulated "plasmagenes" would have to exist in hundreds or thousands of copies in order to ensure that random "segregants" lacking genes for key functions were very rare. Although Winkler presented no new evidence and his argument for the existence of cytoplasmic inheritance was not widely accepted at the

26. On Baur's view see Oehlkers 1927, 80–82; for Correns' view see his 1909, 340, and his 1922.
27. Conference proceedings are in *ZIAV* 27 (1922).
28. Conference proceedings are in *ZIAV* 33 (1924): 237–355.

time, his paper was cited repeatedly over the next decade and seems in retrospect to have intensified geneticists' interest in the phenomenon.[29] By 1927 Lehmann and Schwemmle felt able to claim that nuclear-cytoplasmic relations were at the top of (German) geneticists' agenda (Lehmann and Schwemmle 1927, 68).

Some attempted to rejuvenate the merogone method. In the late twenties, for example, Richard Harder used microsurgery to construct a haploid fungus strain which contained the nucleus of one parent combined with the cytoplasm of both parents (Harder 1927; 1928; 1929). Unlike Boveri's sea urchin hybrids, these developed to maturity and displayed at least some phenotypes of that parent whose contribution had been solely cytoplasmic. Since he failed, however, to demonstrate that such cytoplasmic effects survived further crosses with appropriate strains, Harder's experiments could not be decisive. Although his work was often cited by advocates of cytoplasmic inheritance in the late twenties, Harder was reluctant to generalize from his experiments to other organisms or to grant the cytoplasm a major role in inheritance. By the 1930s his work attracted little attention; he published nothing further on cytoplasmic inheritance, and no one sought to refine his microsurgical techniques.

Despite the methodological difficulties of research on cytoplasmic inheritance, most German geneticists were persuaded by the mid-1930s that chloroplasts were independent genetic elements. The decisive experiments had been conducted by Otto Renner (1883–1960). After some early experiments with species hybrids in *Epilobium* which suggested cytoplasmic effects (Renner and Kupper 1921), Renner turned to the genus *Oenothera*.[30] In certain crosses he noticed that although both parent species were normally green, a sizable number of the hybrid

29. On Winkler's reception, see Renner 1950, 6; Michaelis 1937, 284. One reviewer predicted that Winkler's "programmatic remarks" would have considerable impact upon genetics' future development (Noack 1925).

30. *Oenothera*'s cytological peculiarities were ideally suited for demonstrating cytoplasmic inheritance. Diploid *Oenothera* species are complex heterozygotes whose chromosomal makeup consists of two "complexes" which are inherited under certain circumstances as integral packages. During meiosis, therefore, *Oenothera*'s chromosomes are distributed to the gametes, not randomly as in most plant species, but together as a complex. Instead of producing a wide variety of gametes whose chromosomal compositions reflect all of the statistically possible combinations of maternal and paternal chromosomes, therefore, *Oenothera* species produce only two: one complex deriving from the male parent of the plant in question, and the other from the female. This means that the chromosomal constitution of nonreciprocal hybrids can be known with precision, thus getting around some of the methodological problems mentioned in n. 25 above.

offspring displayed variegation. A series of reciprocal crosses among various species indicated that the chloroplasts of any given species would only become normally green when combined with certain chromosomal complexes. With all other complexes they remained pale and sickly. Renner explained the occasional variegation in such hybrids by hypothesizing, as had Baur earlier, that a small amount of cytoplasm was occasionally transferred through the pollen tube, creating either a few pale patches upon otherwise healthy green leaves, or a few green patches on otherwise pale leaves. As with the Baur-Correns debate, however, it was unclear whether such variegation was due to the presence of two kinds of chloroplast in the variegated plant or to only one kind which inhabited an unevenly mixed cytoplasm, thriving in one type but not in the other. Although he admitted to being unable to demonstrate it directly, Renner thought it more likely that paternal chloroplasts (rather than "unformed cytoplasm") were transferred via the pollen tube, followed by the uneven distribution of maternal and paternal chloroplasts during subsequent cell divisions (Renner 1922; 1924a; 1929). By the mid-1930s he had strengthened his theory by showing that the green sectors of a variegated species hybrid contained the chloroplasts characteristic of one species, while the pale sectors contained the chloroplasts of the other. And a large number of carefully constructed crosses showed that whether or not a particular kind of chloroplast became normally green was simply a function of which chromosomal complex—and not which unformed cytoplasm—was present.[31] Surveying reported cases of variegation in a variety of related genera, Renner identified in each instance the most likely tissue-developmental history which could account for the color pattern reported, often taking issue with other authors' interpretations where he had experience with the same plant. In nearly every case he found Baur's chloroplast-distribution theory to offer a better explanation than other theories, taking care to spell out wherever possible each theory's predictions so that others could check for themselves. Although their normal greening was affected by chromosomal genes, chloroplasts were self-reproducing genetic elements which could mutate independently of the nucleus, and each species of *Oenothera* had a distinctive kind of chloroplast. At a meeting of geneticists interested in cytoplasmic inheritance which took place in the KWI for Biology late in 1936,

31. Renner 1934; 1936. He acknowledged, nonetheless, that the unformed cytoplasm's composition might affect the rates of separation of different chloroplast types (1936, 264) or their rates of replication (1937, 649–52).

Figure 2.5 Otto Renner about 1940 (reproduced from Erwin Bünning [1944], p. vi; courtesy: Gustav Fischer Verlag, Jena).

Renner was evidently able to convince his colleagues: "Oehlkers and Schwemmle were there, too; they now conceive of the chloroplasts just as I do, and even von Wettstein has overcome his reservations."[32]

More controversial throughout the 1930s, however, was Fritz von Wettstein's "plasmon" theory of cytoplasmic inheritance. The theory's central claim was that the plasmon—a genetic structure in the cytoplasm which was *distinct from* the chloroplasts or other particulate elements—exerted a formative influence upon all traits in all organisms. Although the plasmon's genetic role was distinct from that of chromosomal genes, the two were equal in significance. However much various geneticists may have differed with von Wettstein over the structure and function of the plasmon (see chap. 9), his distinction between two kinds of cytoplasmic inheritance was quickly incorporated into a standard definition of heredity. The total hereditary material (*Idiotypus*) consisted of that which was located in the chromosomes (*Genom* or *Genotypus*), that in the chloroplasts and other plastids (*Plastidotypus* or *Plastom* or *Plastidom*), and that elsewhere in the cytoplasm (*Plasmon* or *Plasmotypus*).[33]

An assistant of Correns's at the Kaiser-Wilhelm Institute for Biology after the war, von Wettstein (1895–1945) was called to the chair of botany at Göttingen in 1925.[34] Never as well known outside Germany as Baur, Correns, or Goldschmidt, von Wettstein became one of the dominant figures in German botany from his call to Germany's largest botanical institute and garden at Munich in 1931, followed by his appointment as director of the KWI for Biology in 1934, until his early death in 1945. As editor of *Fortschritte der Botanik*, *ZIAV*, and *Biologisches Zentralblatt* from the mid-thirties, and as an important referee in appointments in botany, von Wettstein exercised considerable influence over developments in German botany and genetics during his brief career (H. Stubbe 1951; Renner 1959).

In 1927 von Wettstein published his first paper on cytoplasmic inheritance (von Wettstein 1927a). Obtaining viable hybrid offspring from reciprocal crosses between different species, genera, and even subfamilies of mosses, he was able to demonstrate differences—for example, in leaf shape, leaf width, and capsule form—which were

32. Renner to Ralph Cleland, 28.12.36, Cleland papers. That von Wettstein was, in fact, persuaded by then that chloroplasts were independent genetic elements is confirmed by his 1937.
33. E.g., Baur 1911 (1930 ed.); Jollos 1939; Correns 1928; Michaelis 1941.
34. For short biographies of von Wettstein, see Renner 1946; Kühn 1947; Melchers 1953.

attributable to maternal cytoplasm. Since such differences persisted unaltered into the second backcross generation, he postulated in the cytoplasm a genetic structure (the plasmon), which—along with chromosomal genes—shaped the development not just of the traits whose inheritance he had analyzed, but of all traits in all organisms. A year earlier T. H. Morgan had insisted that since most species hybrids show reciprocity, cytoplasmic inheritance could not be a general phenomenon (Morgan 1926a). In fact, some of the traits which von Wettstein studied in hybrid crosses did display reciprocity, but he argued that this did not rule out a plasmon. The plasmon-gene combination for various traits could produce either nonreciprocity or reciprocity, depending on whether the plasmon or the associated genes predominated in each particular instance.

Over the next few years it became clear that von Wettstein foresaw a key role for the plasmon in regulating gene action. The plasmon, he suggested, may be "the carrier of the developmental reactions . . . merely their rate and direction are steered by genes, on the one hand, and external conditions on the other" (1928b, 47). It made little sense to regard cytoplasm "merely as substrate . . . with which the nucleus works." It was more plausible to see genes and plasmon as "opponents" (*Gegenspieler*) in the "game" of development, acting upon and reacting to each other (von Wettstein 1934a, 36). In a subsequent review paper (1937) he refined his conception of the plasmon's structure and function and distinguished it sharply from Winkler's discussion of "plasmagenes." The plasmon and chloroplasts had different structures; while the former was a "a unitary mass" (*eine einheitliche vererbte Masse*), the latter were in some respects similar to nuclear genes: Both were independent "particles" (*Einzelelemente*) (von Wettstein 1937, 351). This meant that whereas each gene acted in a specific and independent manner, so that the organism's overall phenotype was a kind of mosaic, the plasmon's action was general, affecting the whole organism uniformly. Normal development required a mechanism which could regulate gene action. Although the condition of the cytoplasm itself was constantly in flux, the plasmon was the genetic structure which made regular change possible: "the moving belt [*das laufende Band*] upon which individual genes could exert their adjustments in an ordered sequence" (1937, 362). Just as the plasmon imposed coherence upon the activities of uncoordinated genes, so also did it provide continuity between generations via the egg cytoplasm.

Responding to criticism that his results could have been due to predetermination and would gradually decline, von Wettstein showed in

Figure 2.6 Fritz von Wettstein about 1940 (courtesy: Archiv zur Geschichte der Max-Planck-Gesellschaft).

specially constructed polyploid strains that even when the plasmon was exposed to multiple foreign genomes, its phenotypic effect remained undiminished after seven years. And by 1930 his fourth backcross generation still showed the effect of maternal cytoplasm (von Wettstein 1930a).

The concept of the plasmon as a regulatory mechanism clearly derived from Correns's 1901 paper cited above. From that time onward, Correns was far more interested in the physiology of gene expression than in transmission genetics (von Wettstein 1939b). And from the late 1920s until his death in 1933, Correns was certainly the most eminent advocate of the plasmon theory. In the same year that von Wettstein coined the term *plasmon,* Correns presented a review of the literature on non-Mendelian inheritance at the Fifth International Congress of Genetics (Correns 1928). Declaring at the outset that the chromosomes were virtually certain to be the bearers of Mendelian heredity, he posed what he took to be the central question surrounding non-Mendelian inheritance: What was the role of the cytoplasm?

Everyone agrees, even the keenest defenders of nuclear monopoly, that the cytoplasm is necessary for the development of each Mendelian trait. The question of whether there are limits to Mendelian inheritance is thus a question of whether the cytoplasm plays a role in trait-formation above and beyond that of *a substrate* which can at best influence the phenotype, rather the way nutrition can affect growth [1928, 132; emphasis in original].

Correns then discussed various explanations for non-Mendelian patterns of inheritance: pseudo-inheritance, predetermination, and plastids as genetic elements. Since these mechanisms, in his view, could not account for all non-Mendelian patterns, Correns devoted the second half of his paper to evidence for the plasmon. Like von Wettstein, he rejected Winkler's concept of plasmagenes and stressed the plasmon's role in development: "The fact of the free combination of genes demands their independence or at least the independence of the chromosomes in which they lie. But then it is difficult to understand how these genes themselves can determine the strict sequence in which they act" (1928, 163). He therefore posited in the cytoplasm a mechanism to guide gene action, repeating almost verbatim the phrases he had employed in 1901 (1928, 163). In the division of labor between genes and plasmon, Correns suggested, Mendelian genes' effect is usually, perhaps always, quantitative. One could imagine that nuclear genes accelerate, retard, or inhibit reactions in the cytoplasm, but it seemed unlikely that the reactions themselves were caused by genes. In contrast to the conventional view of cytoplasm modifying gene action, Correns proposed that genes merely modified cytoplasmic processes, so that a trait could develop without the participation of a gene but not without that of the plasmon. With this speculative model, Correns stressed, he did not wish to reduce Mendelism's significance (unlike Winkler), but rather to extend its scope so that it could better account for gene expression.

Correns's final discussion of gene-cytoplasm relations appeared posthumously in a monograph (1937) on non-Mendelian inheritance, the last section of which was devoted to evidence for the plasmon theory. The position which he outlined was little different from that published in 1928, except that he cited the views of the cytologist Victor Grégoire on the role of the cytoplasm. Since Correns was careful to dissociate himself from several of Grégoire's views, it is important to notice a quote from Grégoire which he allowed to go unchallenged:

> It is the cytoplasm which develops and differentiates itself, and it is here where all the capacities lie which determine the course of development and differentiation in all phases of ontogeny. The task of the chromosomes during ontogeny is to provide the cytoplasm with certain substances which the cytoplasm makes use of in carrying out its normal activities and in developing, step by step, in the direction in which its own nature tends [cited in Correns 1937, 128].

Although obviously committed to the plasmon theory, Correns did not spend much time articulating it. On those few occasions during the 1920s and 1930s when he discussed cytoplasmic inheritance, his empirical contributions far outweighed his theoretical ones.[35] It is clear, nonetheless, that his support for the theory was grounded in his longstanding interest in the problem of gene action in development.

Since species hybrids in animals are generally infertile, almost all of the evidence for cytoplasmic inheritance came from hybrids in plants. This fact naturally encouraged critics (see below) to marginalize the significance of cytoplasmic inheritance by suggesting that it was a phenomenon restricted to plants. The zoologist and geneticist Alfred Kühn was thus an important ally for the plasmon theorists. Kühn's only empirical contribution to cytoplasmic inheritance was a study of nonrecip-

35. The rarity of Correns's theoretical reflections on the plasmon, as well as their lack of development over substantial periods, is consistent with most accounts of his intellectual style. Virtually all of his obituarists agree that he was a very quiet and modest man, an extremely meticulous and self-critical gatherer of data—Hartmann called him "a veritable fanatic for experimental work"—whose writings on occasion lacked the impact they deserved because their theoretical significance was rarely spelled out (Bluhm 1933; Haberlandt 1933; Baur 1933; Stein 1950; Hartmann 1934b; Renner 1961b; Saha 1984).

Figure 2.7 Carl Correns about 1920 (courtesy: Archiv zur Geschichte der Max-Planck-Gesellschaft).

rocal hybrid crosses in a species of wasp (Kühn 1927). In crosses between a darker- and a lighter-pigmented strain, he found that the pigmentation of the offspring followed more closely that of the mother than that of the father, the effect persisting despite two generations of backcrossing. Although he did not rule out the possibility that this effect might decline in future under the influence of gene action, Kühn interpreted his results as evidence of a plasmon. Although he never worked on cytoplasmic inheritance again, Kühn's paper was important, not only as one of the few reported instances of the phenomenon in an animal species, but also because the cytoplasmic differences which he had discovered were between two varieties of the *same* species. This made it more difficult for critics to dismiss plasmon differences as a genetic peculiarity arising only in hybrids between more distantly related taxonomic groups where incompatibilities are often observed. Perhaps, therefore, plasmon differences were as normal and common as gene mutants? Empirically valuable though the paper was, however, Kühn's greatest significance in the debate was probably as a publicist. In both review papers and in his well-known textbooks of zoology, genetics, and embryology, Kühn presented the evidence for cytoplasmic inheritance (including the plasmon theory) to several generations of German students from 1928 until 1965.[36] As with Correns, Kühn's support for the plasmon theory accords well with the fact that his principal research interest was the genetics of development.

The youngest of the plasmon theorists was Peter Michaelis (1900–1975), a botanist who worked exclusively on cytoplasmic inheritance from the 1920s to the 1960s.[37] On receiving his Ph.D. at Munich in 1923, he went to Jena, where he began to work on cytoplasmic inheritance as an assistant in Otto Renner's institute. Four years later he took up an assistantship with Richard Harder at the *Technische Hochschule* in Stuttgart before moving in 1933 to Erwin Baur's KWI for Breeding Research, where he remained for the rest of his career. In the early 1920s Michaelis began to analyze species hybrids between *Epilobium hirsutum* and *Epilobium luteum*. The crosses were nonreciprocal with respect to such quantitative traits as fertility, leaf size, and leaf width, and persisted over several generations of backcrossing. Accordingly, his first paper on cytoplasmic inheritance was published in 1929 (Michaelis 1929), and in the next few years he endorsed the basic features of the plasmon theory. During the mid-1930s his group's reports that certain

36. Kühn 1937b; 1922 [1928], 252–53; 1955, 486; 1934e [1950], 97–100.
37. For a summary of Michaelis's work and career see W. Stubbe 1987.

Figure 2.8 Peter Michaelis, making a point in 1952 (courtesy: Prof. Georg Michaelis).

properties of the cytoplasm—its viscosity, permeability to various small molecules, and resistance to certain chemicals—were associated with a particular plasmon excited considerable interest within the cytoplasmic inheritance fraternity.[38] Here were perhaps the first indications of the plasmon's biochemical structure. By the late 1930s, however, his conception of the plasmon's structure was closer to Winkler's model than to von Wettstein's: For the remainder of his career he regarded the plasmon as a swarm of mutable, self-replicating units (e.g., Michaelis 1938, 456–57;1963).

The year 1940 marked a shift in Michaelis's methodology which allowed him to articulate the developmental features of his plasmon theory further. Discovering a particular strain of *E. hirsutum* whose plasmon inhibited the genomes of other *E. hirsutum* strains to varying extents, he began to study the interaction between genes and plasmon, an analysis which had not been possible with his interspecies hybrids (Michaelis 1940a). Noting that his data supported Correns's model, Michaelis argued that while genes determined to what degree inhi-

38. The work is summarized in Michaelis 1937.

bition occurred, the plasmon determined the kinds of inhibition (Michaelis 1940a, 220–21; 1942b, 455; Michaelis and von Dellingshausen 1942, 426). The plasmon was anything but a passive substrate. For Michaelis a gene's phenotypic outcome was assigned by the plasmon rather than intrinsic to the gene; the phenotypes which a gene's action affected could be redirected by the presence of particular biochemical reaction systems in the cytoplasm, which in turn were partly under plasmon control (Michaelis 1940b, 335–36; Michaelis and von Dellingshausen 1942, 375, 424–25).

Although colleagues complained from time to time that Michaelis's publications were unreadable, his work filled a serious gap which had been created by the fact that von Wettstein and his students largely abandoned experimental work on cytoplasmic inheritance after 1930, apparently in search of plants which offered greater experimental advantages than did mosses.[39] Furthermore, Michaelis was the only interwar student of cytoplasmic inheritance who had found a way to study the biochemistry of gene-plasmon interactions, and by 1949 the persistent nonreciprocity of his species hybrids over more than twenty-five generations of backcrossing constituted the best evidence for the presence in the cytoplasm of stable—thus genetic—structures.

On the other hand, *Epilobium* was not an ideal organism in which to study cytoplasmic inheritance. The fact that many traits in Michaelis's nonreciprocal species hybrids appeared to be functionally impaired invited the criticism that the *Epilobium* work merely demonstrated the inhibition of normal gene functioning by foreign cytoplasm (see chap. 9). Moreover, cytogenetic work with *Epilobium* was far more difficult than with, for example, *Oenothera*, where one could quickly construct genomes of known composition in species hybrids.[40] One could thus rule out various nuclear explanations for nonreciprocity more convincingly with *Oenothera* (one of the reasons why it remains today a preferred organism for plastid genetics). Given *Epilobium*'s limitations,

---

39. To my knowledge, none of von Wettstein's students at Göttingen (Wagenitz 1988) worked on cytoplasmic inheritance for their dissertations. Similarly, his 1937 review paper on cytoplasmic inheritance cites no work after 1930 on mosses, either by himself or by his students. The only one of his students cited in this paper was L. A. Schlösser (1906–1973), who chose to work on the tomato because of its technical advantages. Cytological work in mosses was difficult, and the quantities of material deriving from a given cross or generation were too small for biochemical analysis (Schlösser 1935). Judging from Caspari's 1948 review of the literature, however, this was the one and only paper on cytoplasmic inheritance (of any significance) which Schlösser ever published.

40. Oehlkers 1948, 38; W. Stubbe, personal communication.

Figure 2.9   Friedrich Oehlkers about 1950 (courtesy: Prof. H. Mohr).

1938 represented an important turning point in research on cytoplasmic inheritance: Two new works on other genera got around some of these technical difficulties, and both provided new evidence for the plasmon theory.

Friedrich Oehlkers (1890–1971) is perhaps best known for his work during the 1920s in Tübingen with Ralph Cleland which clarified the cytological basis of the peculiar patterns of inheritance in *Oenothera*.[41] As professor of botany at Freiburg from 1932, Oehlkers, along with his assistants, developed a research program on "the physiology of meiosis": assessing the effects of various environmental conditions upon crossing-over. Oehlkers had followed the literature on cytoplasmic inheritance with considerable sympathy from at least the mid-1920s, but his own empirical contribution first appeared in 1938 when he reported that species hybrids in the genus *Streptocarpus* differed in the development of their sexual characters.[42] When the hybrid's cytoplasm was derived from *S. Rexii*, its flowers were "masculinized": the anthers were normally fertile, but the stigma was less so, or even altogether absent. When the cytoplasm came from *S. Wendlandii*, on the other hand, the hybrid's flowers were "feminized": additional (fertile) stigmas grew out of the style, and some anthers were missing. In the latter hybrid, furthermore, Oehlkers found a new trait (slit petals) which segregated in Mendelian fashion, suggesting that *S. Rexii* contained a gene which was only expressed in *Wendlandii* cytoplasm. Finally, flower size was also affected in the hybrids: cytoplasm from *S. Rexii* increased it, while that from *S. Wendlandii* decreased it. These reciprocal differences appeared to be stable; no diminution of effect was observed after four generations of backcrossing or six generations of vegetative propagation (Oehlkers 1940). These findings were important, not simply because they extended the evidence for cytoplasmic inheritance to a new genus, but because the cytoplasm's effects were so different from those in *Epilobium*. Instead of inhibiting many traits, the cytoplasm in *Streptocarpus* hybrids seemed to affect only specific phenotypes, inhibiting the development of some of them but *enhancing* others. In keeping with von Wettstein's theory, Oehlkers used this work to develop a model of sex determination in which genes provided the potential for both male and female organs—the relative strength of these potentials, however, varying from one species to the next—while the

41. For a brief overview of Oehlkers's life and work, see Marquardt 1974.
42. For his summaries of the literature on cytoplasmic inheritance see Oehlkers 1927, followed by annual review papers in *Fortschritte der Botanik* from 1932. For his own work on cytoplasmic inheritance see his 1938b, 1940, and 1941.

plasmon regulated the expression of these genes. An imbalance between the "strength" of genes for maleness and femaleness in any given species would be compensated for by its plasmon (Oehlkers 1941).

The other breakthrough that year was the appearance of a magnum opus on species hybrids in *Oenothera* by Julius Schwemmle (1894–1979) and his students. As a student and assistant of Ernst Lehmann's at Tübingen from 1919 to 1929, Schwemmle had published primarily on species hybrids in *Epilobium*. On arriving at Erlangen as professor of botany in 1930, however, he concentrated exclusively on *Oenothera*.[43] While Renner focused on the causes of variegation in species hybrids of this genus, showing that leaf color was in part due to heritable characteristics of the chloroplasts themselves, Schwemmle looked for nonreciprocity in other traits. *Oenothera*'s biological peculiarities provided the ideal tool. Its complex heterozygosity, plus the transfer of chloroplasts from the male parent via the pollen tube, allowed him to construct strains of known chromosomal composition which contained the ("unformed") cytoplasm of one species with the chloroplasts from another. With this method he could establish whether a given phenotypic effect associated with the cytoplasm was due to chloroplasts or to the plasmon. This was an important analytical advance; since workers on cytoplasmic inheritance until this time had been unable to separate these two causes, they had generally assumed that those cytoplasmically inherited traits which seemed unrelated to the chloroplasts must be due to the plasmon. It came as a surprise, therefore, when Schwemmle showed that leaf shape—heretofore attributed to the plasmon—was in fact due to the *chloroplasts*. The impact of this discovery upon debate over the plasmon theory is difficult to judge. One contemporary felt that it undermined the theory's credibility: From now on, one had to recognize that "plasmon" was a catchall concept which might include chloroplasts, unless ruled out by careful experiments (Oehlkers 1952, 214–15). Von Wettstein's sharp distinction in 1937 between the structure of chloroplasts and that of the plasmon would thus have lost some of its weight. Nevertheless, Schwemmle also demonstrated that *some* of the nonreciprocal differences which he found in *Oenothera* (e.g., length of blossom) were attributable not to the chloroplasts but to the plasmon (i.e., the remainder of the cytoplasm). And while he had little or nothing to say about the plasmon's structure or function, Schwemmle's comments about the chloroplasts' function—that they differentially af-

---

43. For brief accounts of Schwemmle's life and work, see Arnold 1981 and Ziegler 1980.

Figure 2.10 Julius Schwemmle at Erlangen, 1935 (courtesy: Prof. Berthold Schwemmle).

fected the expression of leaf-form genes, located on either half of the chromosome complex—displayed the plasmon theorists' characteristic concern with the cytoplasm as a regulatory agent (Schwemmle et al. 1938, 649–50, 674).

By 1940, therefore, a substantial German literature pointed to the existence of cytoplasmic inheritance, especially in plants where the evidence for chloroplasts as genetic elements was particularly strong. To be sure, some German geneticists were not persuaded (cf. chap. 9). But the most interesting reaction to cytoplasmic inheritance came from the United States, where geneticists were almost uniformly skeptical or indifferent. In view of embryologists' persistent emphasis upon the importance of the cytoplasm in *development,* it is easier to understand why T. H. Morgan repeatedly dismissed the cytoplasm as a *genetic* element. In view of the limited evidence for cytoplasmic inheritance by the mid-1920s, his claim that apart from plastid inheritance "the cytoplasm may be ignored genetically" is perhaps understandable.[44] But American geneticists remained skeptical throughout the 1930s, despite the mounting evidence. In the mid-1930s Renner was still trying to convince E. M. East that chloroplasts were independent genetic elements.[45] The response to Oehlkers's paper on cytoplasmic inheritance at Columbia University in 1939 was uniformly critical. In private, however, several Americans admitted to Oehlkers that they had found various puzzling results—consistent with cytoplasmic inheritance—which they had decided not to publish.[46] Standard American genetics textbooks (unlike their German counterparts) took the same line. In 1939 Sturtevant and Beadle accepted the evidence for chloroplasts as genetic elements, but found all other evidence for cytoplasmic inheritance inconclusive.[47] In view of Sinnott and Dunn's interest in developmental questions, it is not surprising that their treatment of cytoplasmic inheritance was somewhat more sympathetic, but even they concluded that the cytoplasm's genetic role was "a relatively minor one and clearly subsidiary to the dominant genic mechanism" (Sinnott and Dunn 1939, 253).

From about 1940, however, cytoplasmic inheritance found a talented and energetic advocate in the United States: Tracy Sonneborn.

44. The quote is from Morgan 1926a, 491. See also his 1917 and 1923a, 626–27.
45. Otto Renner to Ralph Cleland, 19.1.36, Cleland papers. See also East 1934.
46. Interview with Cornelia Harte, 1.5.83, and Harte to me, 24.6.88.
47. Sturtevant and Beadle 1939, 327–33. During his lecture tour of the United States in 1938, von Wettstein had found Sturtevant skeptical of all but chloroplast-borne cytoplasmic inheritance (von Wettstein, travel diary, 1938, pp. 55–56).

And like the plasmon theorists, Sonneborn hoped that the study of cytoplasmic inheritance would contribute to "a deeper genetics," in which "the two major divisions of the cells—nucleus and cytoplasm—which have necessarily been torn asunder in the preliminary analysis of the last fifty years, will again be reunited in an integrated, interreactional [sic] conception of the genetic and developmental system of the cell" (Sonneborn 1950b, 311). By the late 1940s, however, the American reception of Sonneborn's work on cytoplasmic inheritance was still predominantly critical.[48] And while preparing a review paper on the subject for an American journal, Ernst Caspari (a German emigré to the United States since 1938) wrote to his *Doktorvater* that the German work on cytoplasmic inheritance was "almost unknown here . . . cytoplasmic inheritance has been thoroughly rejected by geneticists here."[49] Following the publication of Caspari's review, however, there are signs of growing American interest in cytoplasmic inheritance. Michaelis was invited to Cold Spring Harbor in 1951, and during the 1950s both he and Oehlkers were invited to summarize their work in *Advances in Genetics* (Michaelis 1954; Oehlkers 1964). Nevertheless, in the mid-1950s Sonneborn still found his colleague, H. J. Muller, unreceptive. As he complained to Alfred Kühn, "[Muller's group] is strongly 'classical' in bias and not easily taken in by claims of cytoplasmic factors."[50]

What can account for American geneticists' overwhelmingly critical response to the evidence for cytoplasmic inheritance? It cannot have been due to sheer ignorance. It is clear that the Americans in question were not only aware of the German literature but also had the linguistic competence to understand it.[51] Nor is there any reason to think that American geneticists had higher standards of proof than the Germans; as we will see in chapter 9, many of the critiques of cytoplasmic inheritance were *a priori* or otherwise uncompelling. Jan Sapp has rightly

---

48. Beale 1982, 549, and G. H. Beale to me, 13.1.83.
49. Caspari to Alfred Kühn, 2.11.46, Kühn papers (Heidelberg). The review paper was Caspari 1948.
50. Sonneborn to Kühn, 16.10.56, Kühn papers (Heidelberg).
51. For example, American geneticists cited the work of Renner, Correns, Oehlkers, and von Wettstein. To imply that the Americans dismissed cytoplasmic inheritance because the German literature was "almost unintelligible" (Sapp 1987, 86) does not explain very much. To be sure, even German colleagues complained that Schwemmle's and Michaelis's papers were poorly written and incomprehensible in places (Caspari to me, 5.7.83; J. Hämmerling to Curt Stern, 3.2.47, Stern papers; interview with J. Straub, 3.5.83; interview with H. Ross, 2.5.83). But I have never seen this objection leveled at any other German writers in this field.

pointed out (1987, 159, 230, 232) that work on cytoplasmic inheritance was much slower and cumbersome than the established techniques of transmission genetics in *Drosophila,* and this may, indeed, be one of the reasons why Morgan's school preferred to steer clear of cytoplasmic inheritance. But it cannot explain why German geneticists— as well as like-minded souls elsewhere, such as Sonneborn or Boris Ephrussi—chose to do otherwise. As we will see below and in chapters 3 and 4, there is no getting around the fact that most American geneticists, unlike their German counterparts, were simply not interested in the theoretical issues which the study of cytoplasmic inheritance promised to illuminate. Just as the study of cytoplasmic inheritance was frequently prompted by a theoretical concern with the problem of development, so also was skepticism toward cytoplasmic inheritance associated with the relatively simple view of gene action held by the Morgan school. Sturtevant and Beadle (1939) were confident: "There is no question that genes determine the nature of developmental reactions and the direction in which they will lead" (335). The question was simply *how* genes accomplished this. The cytoplasm was merely "a necessary medium for the growth of genes and for the production of their specific effects" (332). Similarly, Muller refused to alter the title of his 1926 paper from "The Gene as *the* Basis of Life" to "The Gene as *a* Basis of Life." During the evolution of the cell, he insisted, "the great bulk . . . of the protoplasm was, after all, only a by-product . . . of the action of the gene material; its 'function' (its survival-value) *lies only* in its fostering the genes, and the primary secrets common to all life lie further back, in the gene material itself" (Muller 1929 [1962], 195, emphasis mine; cf. Carlson 1981, 140). By taking the problem of development seriously, the proponents of the more radical forms of cytoplasmic inheritance, such as the plasmon theory, posed a challenge to those American geneticists who had chosen to focus narrowly upon transmission genetics.[52]

2.3 In Search of Simplicity: George Beadle's Approach to Physiological Genetics

Although the German genetics community was undoubtedly more developmentally inclined than its American counterpart during the 1920s, the contrast began to fade in the early 1930s. By 1933 Goldschmidt was admitting that the role of genes in development was be-

---

52. On American resistance to cytoplasmic inheritance, see also Sapp 1987.

ginning to attract American geneticists, even some in Morgan's group.[53] His textbook *Physiological Genetics* appeared in 1938 and was reprinted twice (Goldschmidt 1960, 322). American geneticists began to organize conference sessions devoted to the topic, as at the meeting of the American Society of Naturalists in 1933.[54] In a similar symposium at the Genetics Society of America's meeting in 1936, Curt Stern remarked on this shift of interest:

The study of the role of genes in ontogeny is a field for developmental physiology. Not that the geneticist will be excluded from the working out of this problem—he will become a genetic experimental embryologist. After a long journey which took him sometimes out of sight of his fellow biologists, he returns home again with some new concepts and tools [C. Stern 1936, 35].

At the Rockefeller Foundation, Warren Weaver was keen to promote physiological genetics. In the early thirties he began to fund six North American research projects in this area, and by the late thirties, physiological genetics dominated the Natural Sciences Division's funding for genetics (Kimmelman 1981). One of the geneticists who enjoyed the foundation's support was George Beadle, whose collaborative studies of gene action with Boris Ephrussi and Edward Tatum culminated in the "one gene—one enzyme" hypothesis.

As we shall see, the experiments which Beadle and Ephrussi conducted on the genetic determination of eye color in *Drosophila* are nearly identical to those done by Alfred Kühn's Rockefeller-funded group in Göttingen at about the same time.[55] At first sight, therefore, it looks as though the German-American contrast so evident during the twenties was being rapidly eroded during the thirties. One might thus be tempted to conclude that this contrast was both short-lived and inconsequential: It simply took the American genetics community a bit longer to become interested in physiological genetics. That temptation should be resisted. As I will show in this section, a comparison of Beadle's work with that conducted in Göttingen reveals that very similar experiments were embedded in research programs with radically different aims.

When George Beadle (1903–) joined Morgan's group at the Cali-

---

53. Diary of HMM, 13.12.33, folder: "Kaiser-Wilhelm Institute for Biology 1931–1937," Rockefeller Foundation papers. Embryologists also recognized this change of direction; see Harrison 1937, 372.
54. The conference proceedings appeared in the *American Naturalist* 68 (1934).
55. In 1935 Kühn had five Rockefeller-funded assistants working in his institute (Kühn to Reinhard Dohrn, 18.6.35, ASZN: Be, 1935, K; Naples Zoological Station).

fornia Institute of Technology in 1931 as a postdoctoral fellow, he hoped to learn something about the nature of the gene. Having trained in maize cytogenetics, he spent the first three years acquainting himself with *Drosophila* genetics and working on recombination with Theodosius Dobzhansky, Stirling Emerson, and A. H. Sturtevant. But interest in physiological genetics was growing at Cal Tech, stimulated in part by J. B. S. Haldane, who was a visiting professor there about 1933 (Kay 1986, 199–200). Having worked on the genetics of hemophilia and color blindness, Haldane talked with Beadle about the pioneering work of Archibald Garrod on the genetic determination of alcaptonuria, and presumably brought news of the latest work on the genetics of flower pigmentation being done at the John Innes Horticultural Institution by Muriel Wheldale Onslow and others. More important for Beadle was the arrival from Paris early in 1934 of a young embryologist funded by the Rockefeller Foundation: Boris Ephrussi. During the previous decade Ephrussi had been studying cellular differentiation in classical fashion (using sea urchin eggs), but he was also the first person in France to adopt the new technique of tissue culture to that end. In the early thirties he became interested in the mechanisms by which various lethal mutants altered development in mice, and decided that he ought to learn some genetics (Burian, Gayon, and Zallen 1988, 389–92). Beadle recalled numerous discussions in Pasadena with Ephrussi over the relations between genetics and embryology, and by the end of the year they had decided to collaborate on a project designed to bridge the "unfortunate gap" (Beadle 1974; 1966; Kay 1986).

They began work in Paris in May 1935. Faced with the fact that the organisms favored by embryologists were very awkward for genetic work, while very little was known about the development of the organisms preferred by geneticists, they decided nonetheless to work on *Drosophila*. Sturtevant had found an eye mutant (vermilion) whose eye color seemed responsive to substances (presumably pigments) produced in wild-type tissue. The original plan to study gene action in tissue culture proved overly optimistic, so Beadle and Ephrussi switched to transplantation between different *Drosophila* strains. When they transplanted embryonic eye buds out of the vermilion-eyed mutant larvae into the abdominal cavity of wild-type larvae, the eye buds developed the wild-type color. Since the transplant was not in contact with those host tissues which normally surround a developing eye bud, Beadle and Ephrussi inferred that a gene deficient in the vermilion mutant controlled the formation of some kind of diffusible substance which was necessary for the formation of the brown pigment missing in the mutant. Similar analyses of two dozen other eye mutants turned up one more (cinnabar) which behaved as if it possessed a gene defective in a later step of the series of reactions involved in the formation of eye pigment.

From 1936 the intensity of their collaboration declined as Beadle moved to Harvard and Ephrussi returned to Paris after a stay at Cal Tech. But in 1937 the next stage of their project began: They each engaged a biochemist to help in identifying the nature of the diffusible substances in the vermilion and cinnabar mutants. By 1940 Beadle and his biochemist (Edward L. Tatum) had succeeded in identifying the substance missing in the vermilion mutant as the tryptophane derivative kynurenine (Kay 1986, chap. 6).

Just why Beadle and Ephrussi decided to attempt transplantation experiments at all is at first rather puzzling. Although the technique had been widely used by embryologists from early in the century in order to shed light on the causes of differentiation, various *Drosophila* experts took the view that *Drosophila* larvae were too small to be operated upon and very difficult to anesthetize successfully.[56] In one of their first papers (1935), Beadle and Ephrussi stressed that it was Sturtevant's work on genetic mosaics in *Drosophila* which led them to try transplantation. When one turns to Sturtevant's paper, however, one finds that he had chosen to work on mosaics in order to *get around* the difficulties with transplantation (Sturtevant 1932, 304). Evidently the encouragement to try transplantation came from elsewhere. In the same paper they acknowledged that Ernst Caspari had previously conducted similar experiments with another insect, and they continued to cite his work in most of the papers which they published during the 1930s. Nevertheless, since they never quite clarify the nature of their debt, it is worth looking more closely at Caspari's work.[57]

In 1931 Ernst Caspari (1909–1988) began his dissertation research in Göttingen with Kühn (Eicher 1987; Grossbach 1988, 90–93). Aware that Dobzhansky and Sturtevant had shown during the twenties

---

56. Interview with Caspari; transplantation was "thought by some to be extremely difficult if not impossible" (Beadle 1974, 6–7).

57. In his Nobel acceptance speech in 1958 Beadle makes no reference to Caspari, despite devoting nearly a third of the speech to the genetics of eye color in *Drosophila* (Beadle 1977). In 1974 Beadle noted that the one gene—one enzyme hypothesis was "much encouraged by previous related work by Caspari and others," but he did not elaborate (Beadle 1974, 7).

Figure 2.11 Ernst Caspari about 1960 (courtesy: American Philosophical Society).

that certain tissues in genetically mosaic strains of *Drosophila* developed those phenotypes specified not by their own genotypes, but by *neighboring* tissue, Caspari noticed that no one had yet tried to investigate such effects using transplantation. *Ephestia*, the institute's preferred organism for developmental genetic work, was large enough to permit transplantation, and Kühn and Henke had isolated a promising mutant. It had red eyes rather than the wild type's black, and its phenotype was determined by a single recessive allele dubbed *a*.

Caspari began by checking whether the red-eyed mutant's other organs were also different in color. He could show not only that the mutant's gonads were colorless instead of violet-brown and that the mutant larvae's skin was pale green instead of reddish, but also that eye color and gonad color correlated perfectly in various crosses as well as in experiments where environmental factors were manipulated so as to alter eye color. The gene A/a was therefore pleiotropic. Caspari then transplanted gonads from normal black-eyed larvae into the body cavity of the developing red-eyed mutants and found that such mutants acquired both dark eyes and dark gonads. Since the transplant (genotype AA) and host eyes (genotype aa) had not been in physical contact, Caspari concluded that gene A controlled the production of a diffusible substance (present in the gonads) which altered pigmentation of the eye. Kühn published a preliminary report of these results in December 1932, and Caspari's dissertation appeared the following year (Kühn 1932; Caspari 1933). Thus many features of Beadle and Ephrussi's experimental design had been anticipated two years earlier by Caspari.

Grasping the importance of Caspari's work, Kühn arranged in 1932 to collaborate with Adolf Butenandt (1903–), then a *Privatdozent* and head of the Division for Organic and Biochemistry in the Institute of Chemistry at Göttingen, in order to identify the structure of the diffusible substance (or "hormone," as it was then called).[58] Soon two of Kühn's other assistants—the zoologist Ernst Plagge (1911–1945) and the biochemist Erich Becker (?–1942)—also began to work on the mutant.[59] By 1935 the Kühn group could show that the "hormone" was not specific to *Ephestia*, since organs from numerous other moth species could produce the same effect when transplanted into the red-

58. Kühn to Butenandt, 7.4.57 (Kühn papers, Heidelberg). On Butenandt's career see his 1981, vol. 2.
59. For brief biographical details on Plagge see G. Wagenitz 1988. I am grateful to Plagge's and Becker's contemporaries at Göttingen for additional information: Hans Querner to me, 28.10.87; Hans Piepho to me, 20.10.87 and 7.11.87; and Viktor Schwartz to me, 1.10.87.

eyed mutant, and by 1936 they had characterized another eye-color mutant analogous to cinnabar in *Drosophila* (Plagge 1936; Kühn, Caspari, and Plagge 1935).

Correspondence between Ephrussi and Beadle late in 1937 indicates that they were in fierce competition with Kühn's group.[60] In 1938 Beadle had to concede that Becker and Plagge had just beaten them into print with the demonstration that the cinnabar and vermilion mutants were homologous with two of Kühn's *Ephestia* mutants (Beadle, Anderson, and Maxwell 1938). To make matters worse, by 1938 the Nazi threat was beginning to make the Beadle-Ephrussi collaboration difficult. In October the evacuation of Ephrussi's institute began, and work stopped. A year later Ephrussi was mobilized and assigned to work in a different laboratory on a different problem. His laboratory in Paris was empty, and some of his former colleagues were already on the front. By December 1939 his salary had been cut by over 40 percent, and there was no opportunity for him to continue with genetic research.[61] Finally, at the AAAS meeting in 1939 Beadle heard the news that he and Tatum had been scooped yet again: Butenandt, Weidel, and Becker (1940) had just identified the "hormone," which was missing in both the *Drosophila* vermilion mutant and the *Ephestia a* mutant, as kynurenine. A very competitive man by nature, Beadle is reported to have been furious.[62]

The job of isolating pigment precursors, Beadle has recollected, was "slow and discouraging"; it was the "great difficulty of making progress in the identification of the cn+ hormone" which led him "in frustration" to abandon *Drosophila*.[63] Of course these tasks were presumably just as slow and difficult for Butenandt et al.; what made them especially frustrating for Beadle was the fact that he lost the race. Against this

60. When Kühn failed to provide Ephrussi with *Ephestia* mutants as agreed, Ephrussi's angry interpretation was, "In other words he is a s. of a b. . . ." (Ephrussi to Beadle, 18.10.37; see also Ephrussi to Beadle, 15.9.37 and 8.5.38). Ephrussi expressed the same opinion about the parentage of Georg Gottshewski, a *Drosophila* geneticist in Kühn's department from 1937, and warned Beadle not to tell him anything (Ephrussi to Beadle, 29.11.37 and 26.12.37; all letters in the Beadle papers).

61. Ephrussi to Beadle, 15.10.38, 23.10.39, and 7.12.39 (Beadle papers). In 1940 the Rockefeller Foundation discussed the possibilities of getting Ephrussi a job at various British and American universities, and early in 1941 Ephrussi fled to Lisbon, sailing for New York in May (see Ephrussi's Fellowship Recorder cards, Natural Sciences Division, Rockefeller Foundation papers).

62. Caspari to me, 17.9.87. On Beadle's competitive outlook, see Srb 1974.

63. The quotes are, respectively, from Beadle's Nobel speech (1977, 57) and from Beadle 1950a, 224.

backdrop, it is easier to understand why he has found it difficult to spell out the extent of his debt to Caspari's 1933 paper. In any event, Beadle soon had an idea for a new research strategy which would be "capable of adding *rapidly* to our understanding of the relation of genes and specific known chemical reactions."[64] With the new method, "it would obviously be *much less time-consuming* to discover additional [metabolic mutants]" (Beadle 1974, 8 [emphasis added]). Until then, like Kühn's group, he had opted to work with an organism which was genetically well characterized but with a phenotype (eye color) which, though *relatively* simple, still required laborious biochemical analysis. Why not choose instead a group of phenotypes whose biochemistry was already worked out? Then one would merely have to demonstrate that these were heritable. By starting with such simple phenotypes and working out their genetics rather than vice versa, as Beadle put it, "we could stick to our specialty, genetics, and build on work chemists had already done" (1963; cited in Kay 1986, 196).

Curt Stern has described this change of research strategy as "the courageous step of calling a halt to their successful exploration of the genetic-biochemical determination of the eye-colors of *Drosophila* and to begin work with a completely different kind of organism...." (1954, 229). In view of the outcome of his competition with Kühn's group, it is doubtful whether even Beadle would regard the *Drosophila* work as altogether successful. And to call the shift in strategy "courageous" overlooks Beadle's desperation in 1940 to find a method of studying gene action which would allow him to publish rapidly. Beadle's choice of strategy is thus better described as pragmatic.[65]

The story from 1940 is well known (Olby 1974; Kay 1986). Beadle and Tatum began to work with the bread mold *Neurospora*. Following mutagenesis with X-rays or ultraviolet light, they selected for mutants which were unable to grow on minimal medium by virtue of having lost the ability to synthesize a particular metabolite (e.g., an amino acid or vitamin). Appropriate crosses were then carried out in order to show

64. Beadle 1966, 29 (emphasis added). Similarly, Beadle had decided in the early 1930s to shift to *Drosophila* genetics because corn work was "just too slow" (Kay 1989b).

65. Beadle's former colleague, Norman Horowitz, has claimed that there were technical reasons for abandoning *Drosophila*. Once the "hormones" missing in the vermilion and cinnabar mutants had been characterized, he argues, the work on eye color would have drawn to a halt, since no other eye mutants in *Drosophila* involved diffusible substances (Horowitz 1979, 263). Although this argument seems plausible, Beadle himself has not used it in any of his retrospective accounts (Beadle 1966, 1974, 1977) when justifying his decision to switch to a new organism.

that the loss was attributable to a particular defective gene. The approach paid off almost immediately. By the AAAS meetings later the same year, Beadle and Tatum were already able to report the discovery of three such mutants (Beadle 1966, 29; Beadle and Tatum 1941). Over the next four years, as Lily Kay has shown, Beadle was extremely successful in attracting funding and building up a large research group with a clear-cut division of labor: Each member was responsible for isolating mutants relevant to a particular biosynthetic pathway (Srb 1974). By arguing that the *Neurospora* mutants could be used for bioassays of important nutritional factors—for example, the growth rate of a mutant which was unable to synthesize a particular vitamin would depend on the concentration of that vitamin in its growth medium—Beadle was able to attract research funding from pharmaceutical companies involved in the manufacture of vitamins and amino acids. Just as important, the nutritional angle enabled him to persuade the United States government to classify his research as war-relevant so that his staff could remain at Stanford working on *Neurospora*, rather than being drafted onto government research projects elsewhere (Kay 1989a). This gave Beadle an enormous advantage over Kühn's group, which fell apart under the pressures of Nazism and war. Caspari was forced to flee to Turkey in 1935; Becker was drafted into the military and died in 1942, and Plagge was drafted in 1942 and died in 1945.[66] By the end of the war Beadle's team had identified a hundred mutant genes affecting the biosynthesis of amino acids, vitamins, and nucleic acid components, and Beadle had acquired a reputation for being a highly effective research manager (Kay 1986; 1989a). In the late 1940s Cal Tech succeeded in bringing him back to Pasadena as head of their Division of Biological Sciences, and the rest of his career was spent in administrative and advisory roles. The wartime work on *Neurospora* alone, however, sufficed to bring Beadle and Tatum a Nobel Prize in 1958.

As we look back on the research strategies which Beadle pursued between 1934 and 1945, his preference for simple systems is apparent. Although development of the eye as a whole was complex, the formation of eye pigment was a nice one-dimensional phenotypic process whose temporal dimension could be conveniently ignored. Moreover, in opting subsequently for *Neurospora*, Beadle was no longer studying

66. Hans Piepho to me, 20.10.87 and 7.11.87. Kühn also lost first-rate students. On completing his Ph.D. on pattern formation in *Ephestia* in 1936, Werner Braun (1914–1972) fled to the United States, where he later became a well-known bacterial geneticist (Braun, "Some Facts about Myself," undated typescript, Beadle papers).

even *part* of a developmental process in a higher organism. It was simply the chemistry of individual gene action. The search for simplicity was entirely justified, Beadle insisted. Many important biological discoveries had been initially resisted because of "the persistent feeling that any simple concept in biology must be wrong. . . . *Moral:* Do not discard an hypothesis just because it is simple—it might be right" (Beadle 1966, 31–32). It would be wrong to think that Beadle had never been interested in the problem of development or appreciated its complexity. It is clear that he was familiar with the 1930s literature on morphogenesis and was aware that he had chosen to study "a very small part of the general reticulum of developmental reactions" because it was technically easier.[67] But as one follows Beadle's writings over this period, his interest seems to shift steadily away from the phenotype and the process by which it develops, toward the gene. In his papers with Ephrussi his aim was to illuminate the complex process whereby a sequence of specific genes' activities is triggered by particular environmental stimuli so that those genes' products appear at the right time and place for normal development. But by 1945 this context of relevance had been replaced by another: the process by which genes control chemical reactions.[68] In effect Ross Harrison had anticipated this outcome almost a decade earlier. Observing (American) geneticists' newfound interest in development, Harrison had warned that

> the prestige of success enjoyed by the gene theory might easily become a hindrance to the understanding of development by directing our attention solely to the genom [*sic*]. . . . Already we have theories that refer the processes of development to genic action and regard the whole performance as no more than the realization of the potencies of the genes. Such theories are altogether too one-sided [Harrison 1937, 372].

---

67. For Beadle's familiarity with morphogenesis, see his 1939. The quote is from Beadle 1937, 120.

68. Compare, for example, Beadle's 1946a, 1946b, 1948, or 1950a with his papers from the 1930s. An apparent exception might seem to be the section in his 1945 which is devoted to morphogenesis. Significantly, however, he makes no claim that the new biochemical genetics has shed any light on morphogenesis. His paper given at the Seventh International Congress of Genetics in 1939 (Beadle 1941) might be seen as a turning point in this process. Beadle's stated aim was still to synthesize genetics and embryology, and he discussed some of the similarities between eye-color "hormones" in *Drosophila* and the "organizer" which was then the subject of so much attention from embryologists. But the importance of his and others' work on the genetic control of specific chemical reactions, Beadle argued, was "not [that] it tells us much specifically about what genes do," but that it illustrated a *method* which he hoped would be increasingly useful to geneticists collaborating with biochemists (1941, 61).

Of course, one might well object that we lack adequate models of genes' role in development even today.[69] Perhaps Beadle simply recognized earlier than others that the time was not yet ripe for attacking the intimidating complexity of the genetics of development, and chose his research strategy accordingly?

Kühn's case suggests otherwise. Learning of the *Neurospora* work at war's end, Kühn acknowledged it as the most important development in biochemical genetics, and bacterial genetics looked similarly promising. Pigment formation, he felt, still provided a simple model of gene action, and some members of his group continued to work in this area into the 1950s (Kühn 1952). The final goal of developmental genetics, he argued, "must be to characterize biochemically both the conditions triggering gene-action and the reactions thus triggered" (Kühn 1934a, 45; see also 1938a, 111). And yet, despite his openness to the new biochemical approach, there is no sign that Kühn was ever tempted to abandon the developmental genetics of pattern formation for biochemical genetics in microorganisms. After the war he continued to work on morphogenesis in *Ephestia,* collaborating with his assistant, Albrecht Egelhaaf. In part Kühn was probably too old (aged sixty in 1945) to shift the focus of his research radically. But more than this, Kühn's first love had always been *form,* whether in animals or art. Soon after arriving in Göttingen as the new professor of zoology, he had enjoyed giving a general evening lecture series on "the problem of form," and found the ensuing discussion with colleagues from other disciplines stimulating.[70] As he often remarked, he was by nature an *Augenmensch:* a visually sensitive person.[71] When Kühn set out to bridge the "unfortunate gap," therefore, fully a decade before Beadle and Ephrussi, he chose to study morphogenesis rather than simpler problems in simpler organisms which promised quicker results.

In the 1920s and 1930s there was no way of knowing which research strategy would eventually pay off. Geneticists interested in development simply had to make a choice. Kühn chose to grapple with a classical problem, while Beadle's preference for technical ease seems to have been coupled with a predilection for conceptual simplicity. As Norman Horowitz put it, "Living matter does, in fact, operate on simple principles. The scientific problem is to discern the simplicity amid the con-

69. See, for example, Sander 1985, 380–88, or the editor's introduction in Horder, Witkowski, and Wylie 1985.
70. Kühn to Doflein, 1.1.22, Kühn papers (MPGA), and Kühn to Spemann, 24.3.22, Spemann papers.
71. Interviews with Egelhaaf and Caspari.

fusion. The psychological problem is to accept the simple solution when it is found. George Beadle has always understood these things" (Horowitz 1974, 6). Although Horowitz himself was hardly a dispassionate observer, probably everyone would agree with him that "the scientific problem is to discern the simplicity amid the confusion." But the key difference between Beadle and Kühn is certainly not that Kühn had any difficulty accepting simple *solutions,* but that he was willing to address complex *problems*. Just as Beadle's search for simplicity is endorsed by Horowitz, so Kühn's predilections seem to have been shared by Conrad Waddington. Based on an "atomistic metaphysics," Waddington argued, biochemical genetics tended to focus upon single-gene effects, and thus did not really deal with the complex phenomena which the epigeneticist addressed. On the other hand, he conceded, biochemical genetics had probably been more successful, "and all of us who want to understand living systems in their more complex and richer forms are fated to look like suckers to our colleagues who are content to make a quick (scientific) buck wherever they can build up a dead-sure payoff" (Waddington 1975, 10).

Waddington's remarks suggest that a study of the postwar reception of Beadle and Tatum's work as well as Kühn's and Henke's would be informative. Although I have not attempted anything so systematic, the evidence I have run across suggests that, at least among developmental biologists, Kühn and Henke fared rather better than did Beadle et al. While Ross Harrison admired Kühn's and Henke's work, his warning in 1937 (quoted above) suggests that he would have been disappointed with Beadle and Tatum's.[72] Viktor Hamburger also found Henke's work "extremely interesting."[73] For Guido Pontecorvo the model of the gene in which it determined the specificity of a polypeptide, though fruitful, was oversimplified. While it would suffice for analyzing the simplest systems of genetic information, "we should not expect it to work in this simple form when the information required is the vastly larger one presumably demanded by a morphogenetic process" (Pontecorvo 1959, 71; cf. 131). Most interesting of all are the views of Boris Ephrussi. When Beadle and Tatum first published on *Neurospora,* Ephrussi congratulated Beadle "on entering an unexplored field of most promising possibilities" (cited in Kay 1986, 204). Nevertheless, Ephrussi himself chose to enter a different field—cytoplasmic inheritance—

72. Harrison 1937, 374, and Harrison to L. J. Cole, 25.1.32, Harrison papers.
73. Hamburger to Gerard Pomerat, 15.2.50 (R7 RG 2 GC 1950, 717 ("German Universities A–Z"), box 504, folder 3370, Rockefeller Foundation papers).

which he hoped would shed light on the process of differentiation (Sapp 1987, chap. 5; Burian, Gayon, and Zallen 1988, 397–98). And with his book *Nucleo-Cytoplasmic Relations in Micro-organisms* (1953), as he confided to Curt Stern, he hoped "to force the geneticists to think about the problem of differentiation which they usually regard as not belonging to their province."[74] Of course, not all geneticists (especially in Germany) were guilty of such neglect. Ironically it was Ephrussi's own former collaborator, George Beadle, who—like Morgan before him—needed persuading.

2.4 Conclusion

There can be no doubting the Morgan school's importance in interwar genetics, an importance confirmed by the award of a Nobel Prize to Morgan in 1933. One might wonder, nonetheless, whether their research strategy can be regarded as typical of American geneticists as a whole. Certainly not every American geneticist shared Morgan's preoccupation with transmission genetics. Sewall Wright, L. C. Dunn, Oscar Riddle, and Edmund Sinnott, for example, worked on developmental problems between the wars,[75] but they seem to have been rather isolated. German geneticists paid tribute to Riddle as one of the very few developmental geneticists in a country "where they are still almost entirely concerned with the extension of Mendelism" (Haecker 1926, 238). And Dunn believed that his interest in physiological genetics isolated him from the mainstream of interwar American genetics. When Salome Gluecksohn-Waelsch joined him at Columbia University in 1935, it seemed to her that the role of genes in development and differentiation "was of interest to only a small, select group of people."[76] Dobzhansky, too, worked on developmental genetics during his first few years in Morgan's laboratory,[77] but he was then a newcomer to American genetics, and as we shall see in chapters 3 and 5, he felt intellectually isolated in Pasadena. Finally, R. A. Emerson's attempt to set up a network of Americans studying the physiological genetics of corn during the 1920s seems to have failed (Kimmelman, 1992). It is possible, of course, that the German-American contrast which I have drawn in this chapter will need to be revised as we learn more about

74. Ephrussi to Stern, 19.12.55, Stern papers.
75. Provine 1986, Dobzhansky 1978, Corner 1974, Whaley 1983.
76. Dunn, "Reminiscences," 352; Gluecksohn-Waelsch 1983, 244.
77. Ayala 1985; Dobzhansky, "Reminiscences," 245–47.

American genetics. By focusing so heavily upon *Drosophila* genetics, for example, historians of American genetics have perhaps overlooked the extent of developmental genetic research in the United States during the interwar period. For the time being, however, the evidence suggests that the Morgan school's approach to genetics was the dominant one in the United States.

If so, the central historiographical question is why Morgan and his followers pursued this research strategy. Garland Allen has argued (1979b) that Morgan was able to abandon embryological problems for transmission genetic ones because he saw the implications of a key analytical distinction: that between the genotype and the phenotype. While embryologists (among others) dealt with the latter, geneticists could focus upon the former. Those critics who complained that Morgan's approach to genetics neglected the developmental process, Allen believes, failed to grasp the significance of this distinction:

Neither William Bateson nor Richard Goldschmidt ever completely abandoned the older embryologically conditioned views of heredity and development as inseparable. . . . [By making this distinction] Morgan was thus able to push aside a whole set of problems (the causes of differentiation) which in the early decades of the century were proving refractory to experimental and analytical methods [Allen 1983, 841].

Once the genotype-phenotype distinction had resolved the old problem of "heredity" into two more specific ones, the key to the Morgan school's success was their decision to shelve the important but more complicated issue of development (i.e., the causes of differentiation) in order to concentrate on the narrower, simpler question of transmission genetics. In effect, Allen suggests that there was no sensible alternative:

Questions regarding the nature of genes or how they functioned physiologically were considered too difficult to get a handle on at the moment. . . . But one cannot ask for everything, and there were good reasons why Morgan and his group stayed away from questions of gene function in the period between 1910 and 1940. They restricted their work to areas where actual experiments and observations could be made. This was perhaps one of the most important and brilliant aspects of their methodology.[78]

    78. Allen 1979c, 215. See also Allen 1975, 66–67, 124, 200; Carlson 1966, 248. In several papers Jane Maienschein has offered much the same explanation: American geneticists avoided embryological problems because they sought "answerable questions, definitive results. Not grand umbrella theories which proposed to explain everything at once. . . . How best to conduct productive research? Narrow the problems and tackle

Had [Morgan] insisted on working simultaneously with the problems of the development of phenotype and transmission of the genotype, it is doubtful that the field could have developed in any clear-cut way [Allen 1985, 138].

For whereas Morgan's school "vigorously distinguished" between transmission genetics and gene action in development, "failure to distinguish between these two very different concepts of heredity led workers such as Bateson and Goldschmidt to expect too much from Mendelism" (Allen 1979a, 199).

By now it should be clear why this explanation for the Morgan school's strategy is inadequate. To begin with, I find no evidence that Goldschmidt—or any other German developmental geneticist of the 1920s—failed to understand the implications of the genotype-phenotype distinction. By 1918 at the latest, that distinction was axiomatic for every German geneticist. Although the distinction may well have allowed Morgan to *justify* his narrow research strategy, therefore, it certainly did not *necessitate* it. The genotype-phenotype distinction enabled geneticists to formulate two separate research questions: (1) What was the structure of the genotype and how was it transmitted? and (2) How was the genotype involved in specifying developmental changes in the phenotype? But by so doing, it presented geneticists with a *choice* of research strategy. Was Morgan's choice the only rational one under the circumstances? The existence of a strong developmental genetic tradition in Germany throughout the 1920s suggests otherwise. The comparative analysis in this chapter undermines any notion that Morgan's choice of strategy was dictated by some universal rationality; instead our attention is drawn to the context-dependence of research decisions, whether American or German. As we will see once again in the following chapter, most American geneticists simply found it expedient[79]—as the Germans did not—to steer clear of the more complex issues which dominated biological theory at the turn of the century.

---

accessible questions with available materials" (Maienschein 1985, 102). To account for American geneticists' skepticism toward cytoplasmic inheritance in terms of their preference for simple techniques (as Jan Sapp has; cf. section 2) is esssentially a special case of this type of explanation.

79. "Expediency" is the reason which Morgan himself gave for his decision to focus upon transmission genetics (Morgan 1917, 535).

3 Genetics and the Evolutionary Process

... writting [sic] the species book [*Genetics and the Origin of Species*] ... was real hard.... But now when it is all in past [sic] I am glad that this was done. Among other things it made me to read [sic] a lot of literature which in last years [sic] I was progressively more and more lazy to do, like a real American geneticist.[1]

The earliest Mendelians were intensely interested in evolution. Genetics, as W. E. Castle explained in the introduction to his *Genetics and Eugenics* (1916), "is only a subdivision of evolution."[2] The theme for the plenary session of the American Society of Naturalists in 1911 was "The Relevance of the Experimental Study of Genetics to Problems of Evolution."[3] Elsewhere models of heredity and mechanisms of evolution were inextricably connected in the biometrician-Mendelian controversy, which dominated the early years of genetics in Britain.[4] And Mendelians in several countries took an active role in the debate over evolutionary mechanism. The mechanism most convincing to geneticists was "mutationism," as advocated by Hugo de Vries; neo-Lamarckism was roundly criticized.[5] But from 1910 the experimental basis of de Vries's theory was increasingly called into question, and by the First World War mutationism had fallen out of favor. For the next two decades many geneticists—though not all—seem to have withdrawn from evolutionary debate, apparently disillusioned by genetics' failure to resolve the dispute over evolutionary mechanism. Although William Bateson was not one of these, his presidential address to the

1. Th. Dobzhansky to Curt Stern, 16.9.37, Stern papers.
2. Cited in Dunn 1965a, 51.
3. Program for 1911 meeting, vol. 1, American Society of Naturalists papers. In Germany at this time Oskar Hertwig was arguing that experimental biology ought to be included in the new Kaiser-Wilhelm Institute for Biology because of its relevance to evolution (Wendel 1975, 177).
4. For an introduction to the literature on the biometrician-Mendelian controversy see Olby 1989a.
5 On the popularity of mutationism see Allen 1969. On William Bateson's rejection of Neo-Lamarckism, see Koestler 1975, 50–56; on T. H. Morgan's see Allen 1978, 112–113; on Wilhelm Johannsen's see Roll-Hansen 1978a; on Erwin Baur's see Baur 1912a, and 1912b, 1913. More generally, the extent of Baur's early interest in evolution is evident from his book reviews in *ZIAV* from 1908 into the 1920s.

British Association for the Advancement of Science in 1914 captures the mood of many of his colleagues: "Somewhat reluctantly and rather from a sense of duty I have devoted most of this address to the evolutionary aspects of genetic research. We cannot keep these things out of our heads, as sometimes we wish we could. The outcome, as you will have seen, is negative, destroying much that till lately passed for gospel" (in Goldschmidt 1933b, 539). Where geneticists felt obliged nonetheless to mention evolutionary theory, as in introductory textbooks, they commonly adopted an agnostic position, criticizing various theories in the light of genetic knowledge (e.g., Baur 1911 [1919], 347; Johannsen 1909 [1926], 697). By the twenties Bateson felt that evolution had largely disappeared from the geneticist's agenda (Bateson 1922; see also Jollos 1922, 5).

Several historians have drawn attention to this process of disengagement. William Provine notes that "geneticists in the late twenties and early thirties generally paid little attention to the problem of genetics in relation to evolution" (1980, 54). More particularly, Ernst Mayr has argued that a major consequence of disengagement was that geneticists contributed relatively little to the "evolutionary synthesis" of the late 1930s and 1940s, a period of intellectual restructuring which resolved long-standing tensions between Darwinism and Mendelism and shaped the main features of modern evolutionary thought. To generalize about "geneticists," however, runs the risk of riding roughshod over local variation. As I will show in this chapter, German geneticists were much more likely than Americans to work on evolutionary problems during the interwar period.

Let us begin by looking in more detail at Mayr's thesis (1980a; 1980e). Most geneticists, he argues, were not particularly interested in evolutionary questions and were poorly informed about the evidence for macroevolutionary processes. Some of those who did address evolutionary issues were skeptical about the sufficiency of natural selection as a creative mechanism. The work on population genetics by J. B. S. Haldane, R. A. Fisher, and Sewall Wright during the 1920s and 1930s might seem exempt from these strictures. But population genetics, Mayr argues, took Mendelian genetics as a starting point and merely demonstrated the powers of selection, drift, isolation, and recombination at the populational level. The decisive step in the evolutionary synthesis, however, was the demonstration that such *micro*evolutionary selectionist models could be extrapolated to *macro*evolutionary phenomena such as speciation. Neither Fisher nor Haldane regarded speciation as a problem, and none of the three had much knowledge of natural populations or of the evidence for macroevolution.[6] On the other hand, field biologists and (broadly speaking) morphologists had relatively limited knowledge of genetics, and some of them—notably paleontologists and anatomists[7]—were especially vocal in their rejection of selection. The key to the evolutionary synthesis thus lay in bridging this gap between geneticists and field biologists/morphologists. The major contributors to the synthesis were individuals with exceptionally broad knowledge and interests: either genetically informed field biologists/morphologists such as G. G. Simpson, Bernhard Rensch, F. B. Sumner, Julian Huxley, Erwin Stresemann, and Mayr himself, or geneticists with a knowledge of systematics such as Dobzhansky.[8]

Such evidence as we have for American geneticists during the interwar period fits Mayr's thesis quite well. During the 1930s Morgan and his group liked to insist that genetics could be done without any reference to evolution; thus Dobzhansky found little enthusiasm for much of his own work among colleagues at Pasadena.[9] As he confided to a German emigré, "I have definitely left the track of the classical Drosophilaforschung [*Drosophila* research], and becoming [*sic*] more and more engaged in subjects that are somewhat abhorrent (or at least not very interesting) for the drosophila fellows here. Fortunately Sturtevant (though sometimes slightly apologetically)is going in the same direction. . . ."[10] As Sturtevant and Beadle saw it, "The rapid development

---

6. Of course Mayr acknowledges isolated exceptions to his portrayal. Having studied zoology in Berlin in the 1920s, he is aware that geneticists such as Erwin Baur and Richard Goldschmidt were very interested in evolutionary questions. The problem for Mayr's hypothesis, as we shall see, is that Baur and Goldschmidt were by no means unusual within the German genetics community.

7. Mayr to me, 20.5.83.

8. Evidence supporting Mayr's thesis can be found in the essays by Provine, Carson, Weinstein, Rensch, and Dobzhansky, all of which are in Mayr and Provine 1980. F. B. Sumner, whose work on the genetics of wild mouse populations is generally credited as an important contribution to the evolutionary synthesis, complained repeatedly during the 1920s about geneticists' neglect of systematics and ecology (Provine 1979, 228, 234–35).

9. Dobzhansky 1980a, 449; and Dobzhansky, "Reminiscences," 345.

10. Dobzhansky to Curt Stern, 26.2.36, Stern papers. Sewall Wright considered Sturtevant to be the member of Morgan's group with the greatest sophistication about evolution (Provine 1986, 166), and Curt Stern paid tribute to the breadth of Sturtevant's biological interests ("Remarks at the Memorial Meeting for A. H. Sturtevant," typescript, n.d., Stern papers). In order to keep these remarks in proportion, however, it is instructive to note what Provine has to say about the collaborative work on natural populations of *Drosophila pseudoobscura* which Dobzhansky undertook with Sturtevant in Pasadena

of [genetics] after 1900 led most geneticists to neglect the evolutionary implications—there were too many other things to be done. Evolution was not forgotten, but it seemed that more immediate and approachable problems were more profitable to attack" (Sturtevant and Beadle 1939, 363). Sidestepping the issue of evolution was justified, Morgan argued, because in the early years of the century genetics was not yet advanced enough. It would have been unfortunate, in his view, had the new and precise discipline of genetics become compromised through association with evolutionary speculation. And by the early thirties he doubted whether the time was yet ripe (Morgan 1932).

To be sure, by the late thirties some of his colleagues were more optimistic. In their textbook Sturtevant and Beadle remarked that "evolution has come back into style in genetics," and they cited Dobzhansky's *Genetics and the Origin of Species* to that effect.[11] But the discussion of evolutionary genetics in Sturtevant and Beadle's text is extraordinarily thin and fragmented. None of its twenty-three chapters is devoted to evolution, and the topic is treated only briefly in several places. There are two pages on population genetics in chapter 18, and chapter 17 discusses selection in conjunction with continuous variation and inbreeding/outbreeding. The chapter titled "Species Differences" is primarily concerned with the uses of genetics in defining species; only in the final two paragraphs do the authors address the question of speciation, suggesting that mutation giving rise to polyploidy is probably one

---

during the 1930s (Provine 1981, 21–56). Sturtevant was a keen taxonomist and especially interested in the taxonomy of *D. pseudoobscura*, but his taxonomy was of a very traditional kind, and he did not accept Dobzhansky's definition of species in terms of reproductive isolation (28, 33). And although Sturtevant was the first person to study genetic variation in laboratory stocks, he never really grasped the importance of working on natural populations (52–53). His contribution in the collaboration seems to have been confined to the genetics and cytology of the all-important chromosomal inversions in *pseudoobscura*, as well as making maps of the geographical distribution of particular inversions in various races. Furthermore, the two men approached science in very different ways: While Dobzhansky enjoyed generalizing from particular data to larger theories and was somewhat casual about experimental rigor, Sturtevant distrusted generalizations and "was an exceedingly careful and meticulous worker" (49). Finally, the two men were poles apart ideologically (a phenomenon to which we shall return in chapter 5). An extended collaboration like that required for the "genetics of natural populations" project would never have worked, and so from about 1936 Dobzhansky began to work closely with Sewall Wright.

11. Mayr (1980a, 30–32) argues similarly. Bentley Glass (interview) and James Crow (conversation, 11.7.91) recall the considerable impact which Dobzhansky's book made upon American genetics. *Drosophila* geneticists such as J. T. Patterson and Wilson Stone at the University of Texas switched from transmission genetics to evolutionary problems.

of the mechanisms responsible. Richard Goldschmidt was not impressed; the book was "pedagogically impossible and in addition so clannish that it gives to students a completely distorted picture of our science."[12]

The neglect of evolution was not peculiar to Morgan's group. Although Dobzhansky eventually left for Columbia University in 1941, he found no one teaching evolution in the Department of Zoology there. It was only in 1936 that the department had decided to resurrect the Jessup lectures on evolution, which had lapsed in 1910,[13] and they did not get around to inviting Ernst Mayr to give a seminar on evolution until the early 1950s, some twenty years after he had arrived in New York. When Mayr then moved to Harvard in 1953, he discovered that no one there had been teaching evolution for twenty-five years.[14] L. C. Dunn recalled being warned as a graduate student at Harvard during the First World War to avoid the problem of evolution because it was too speculative, and a few years later Ledyard Stebbins was told that evolution was all right for Sunday newspaper supplements, but real biology was biochemical.[15]

While most American geneticists in the 1920s and 1930s seem to have set aside the complexities of evolutionary genetics in favor of the simpler problems of transmission genetics, many of their German counterparts did just the opposite. Indeed, German geneticists often defended various forms of selectionism in the face of overwhelming skepticism from other sectors of the German biological community. In order to document this, the next section outlines some of the nonselec-

---

12. Goldschmidt to L. C. Dunn, 3.11.39, Dunn papers. Goldschmidt added that he much preferred the third edition of Sinnott and Dunn's introductory textbook, which was published the same year. Indeed, the latter's discussion of the genetics of the evolutionary process is far more informative and coherent. That both Sinnott and Dunn are quite likely exceptions to the general picture which I have been constructing of the American genetics community will come up for discussion again in chapter 5.

A comparison of Sturtevant and Beadle's text with the major German ones a decade earlier is instructive. Baur devotes one of twenty-one chapters to evolution (*Einführung*, 1911 [1930 ed.]). Haecker devotes four of thirty-six chapters to the inheritance of acquired characteristics (*Allgemeine Vererbungslehre*, 1911 [1921 ed.]), and the 1923 edition of Goldschmidt's *Einführung* (1911) contains a sixty-page chapter on Mendelism and evolution, along with additional chapters on neo-Lamarckism and the mutation theory, totaling 20 percent of the text.

13. Dunn, "Reminiscences," 865–67. Dobzhansky gave the first lectures in the new series, and they were published the following year as *Genetics and the Origin of Species*.

14. Conversation with Ernst Mayr, 30.5.84.

15. Dunn, "Reminiscences," 865–67; on Stebbins see Dobzhansky, "Reminiscences," 345.

tionist theories of evolution which were widely held by German biologists during the interwar period. In section 2 I survey briefly the work in evolutionary genetics of several major figures, turning in sections 3 and 4 once again to cytoplasmic inheritance. For in Germany, as we shall see, the plasmon theory was perceived as a solution to the problem of evolutionary mechanism, albeit in a variety of different ways. In the concluding section I assess the implications of this German work for the evolutionary synthesis, recommending several respects in which Mayr's thesis should be modified.

3.1 The Debate over Natural Selection in Interwar Germany

As several historical accounts have pointed out recently, strict selectionists were few and far between in interwar German biology.[16] It is not that natural selection was rejected altogether, merely that it was regarded as insufficient to account for all observed variations in natural populations and the fossil record. Conversely, hardly any German neo-Lamarckians endorsed monolithic explanations; by 1940, in one reviewer's opinion, almost all had conceded a partial role to selection (Ludwig 1940, 689). In both cases *multiple* mechanisms were thought to be necessary, constituting what I will call "dualist" theories of evolution.[17] Many of the reasons given for rejecting selection had already been articulated by Darwin's critics in the later nineteenth century, but they also seemed supported by more recent evidence, such as the conviction among systematists during the 1920s and 1930s that the traits which distinguished species and higher taxa appeared to be nonadaptive.[18] On the other hand, nonselectionist mechanisms of evolution did

16. On paleontologists and a few others see Reif 1983 and 1986; on idealist morphology see Trienes 1989.

17. Hamburger 1980b; Mayr 1980c; Rensch 1980 and 1983. Both Hans Kalmus (at the German University in Prague) and Hans Marquardt (at Tübingen) were taught such theories as students during the 1920s (interviews). In addition to those biologists named by Hamburger, Mayr, and Rensch and those whom I will discuss later in this chapter, Jakob von Uexküll (von Uexküll 1928) and Richard Hertwig (Jablonski 1928; see also Hertwig to Richard Goldschmidt, 19.12.27, Goldschmidt papers) also endorsed dualist theories. In "How I Became a Darwinian" (1980), Mayr recalls having adhered to such a theory, and his correspondence with Dobzhansky in 1935 seems to confirm this; see Dobzhansky to Mayr, 12.11.35, and Mayr to Dobzhansky, 25.11.35, Mayr papers (SBPK). Dualist theories were also common among French biologists at this time (Sapp 1987, 126 and 129–30; Burian et al. 1988).

18. On Darwin's critics see Hull 1973 and Bowler 1983. On interwar systematists see Provine 1983.

not seem to fit very well with Mendelism, the most plausible theory of heredity during the 1920s. It was for this reason that dualist theories of evolution usually postulated a *second* form of heredity, located outside the chromosomes, which was more amenable to some kind of evolutionary mechanism other than selection.

The German term often used to denote this non-Mendelian heredity was *Grundstock*, a word commonly used to describe the basic holdings of a library or museum. Dualist theorists found support from those embryologists who, as we saw in the previous chapter, ascribed certain fundamental traits laid down early in development to the egg, and less fundamental traits which appeared later, to the sperm. Analogously, chromosomal genes were seen as the determinants of rather trivial characters involved only in intraspecific differences (for example, eye color or bristle number in *Drosophila*).[19] Evolutionarily significant traits which distinguished higher taxa, however, were thought to be determined by a basic structure, the *Grundstock*, often located in the cytoplasm or throughout the cell as a whole. Since Mendelian genes were apparently stable in the face of environmental forces, they were believed to change via an internally generated process of mutation, thence becoming subject to selection. Microevolution would occur in this way, but macroevolution required alteration of the *Grundstock* via other mechanisms.

Although the advocates of dualist theories in the German-speaking world were primarily field biologists or morphologists,[20] the occasional geneticist was similarly inclined. Wilhelm Johannsen, disappointed at how little genetics had contributed to an understanding of evolution, felt obliged to posit the existence of a *Grundstock*:

> The great significance of the chromosomes as vehicles of recombination and linkage of genotypic elements is now established. But this does not rule out the presence of other cellular structures in the genotype.... "Mendelizing" elements usually affect abnormal or even pathological traits.... Despite Mendelism we still lack a fundamental understanding of the core [*das Zentrale*] of the organism's genotype. The more deeply embedded causation of the major differences between animal and plant classes, families and genera is actually hardly

---

19. The paleontologist Franz Weidenreich referred disparagingly to such characters as *Kinkerlitzchen* ("trifles," "itsy-bitsies") (Mayr 1980b).

20. H. J. Muller (1929) attributed such theories to paleontologists ignorant of genetics, while Friedrich Oehlkers associated them with systematists (1949). Wilhelm Ludwig (1940, 689) reckoned that most zoologists familiar with animal diversity were dualists of some kind.

addressed by modern genetics. That only chromosomal makeup is involved here seems extremely doubtful. The significance of protoplasmic structures has yet to be explored.[21]

A year later dualist theories gained further support from Hans Winkler's proposal (cf. chap. 2) that the more fundamental traits which distinguished higher taxonomic categories might be determined by a *Grundstock* located in the cytoplasm (Winkler 1924). Although saying very little about the nature of the evolutionary mechanisms which could alter cytoplasmic heredity, he suggested that macroevolution might proceed more slowly and gradually than microevolution if cytoplasmic heredity possessed a different structure from that of chromosomal genes.

Although Johannsen rejected selection from a mutationist standpoint,[22] dualist theorists more commonly coupled selection with some form of neo-Lamarckism. Like Winkler, they often assigned the *Grundstock* to the cytoplasm, where it was thought to respond to directed alteration by the environment. For example, the neo-Lamarckian anatomists (and critics of chromosome theory) Rudolf Fick (1866–1939) and Hermann Stieve (1886–1952) defended such a view by invoking the as-yet-limited evidence for cytoplasmic inheritance.[23] Similarly, the zoologist Bernhard Dürken (1881–1944)[24] doubted that heredity was confined to the chromosomes. For one thing, this was too coarse and mechanistic a conception to do justice to the intricacies of inheritance and development, and it neglected what he saw as the unity of the hereditary substance. While admitting that the evidence was still only tentative, he proposed the cytoplasm as an additional carrier of heredity

21. Johannsen 1922, 101–2. Although Danish by nationality, Johannsen was of German descent and published extensively in German.
22. Roll-Hansen 1978a, 222. That Johannsen devoted some thirty pages of the third edition of his genetic textbook to refuting various forms of neo-Lamarckism is just one of many indications that these ideas were taken seriously by German biologists—though not by geneticists—during the 1920s (e.g., Goldschmidt 1956, 164–67; Just 1921, 261). Goldschmidt reckoned in 1930 that all paleontologists accepted the inheritance of acquired characteristics (unpublished five-part lecture series on evolution, 1930, box 1, Goldschmidt papers).
23. Fick 1925, 528; Stieve 1923, 521, 533, 574–77. Although hardly a critic of the chromosome theory, Theodor Boveri also endorsed a form of neo-Lamarckism to complement natural selection (Baltzer 1967, 136–42). According to Horder and Weindling, Boveri was not convinced that Mendelism was the only form of heredity (1985, 216), but the nature and location of an alternative form are not clear.
24. Dürken was professor and head of the Department for Experimental Embryology and Genetics in the Anatomical Institute at the University of Breslau from 1921 (*Kürschners Gelehrtenlexikon*, 1928/29).

which would be much more responsive to environmental changes than were chromosomal genes. This would provide a mechanism for the inheritance of acquired characteristics (which was crucial since selection played only a subordinate role in evolution). Complaining that most neo-Lamarckian theories had little to say about the evolution of taxa above the species level, Dürken postulated a modified form of the inheritance of acquired characteristics which he believed could better do justice to "conservative" features of the fossil record, such as discontinuity and overgrowths (Dürken and Salfeld 1921, 1–45; Dürken 1928, 563–606).

The zoologist Ludwig Plate (1862–1937) developed a similar theory. Although he had been an early proponent of selection, by the 1920s Plate had come to regard selection as a sieve rather than a creative force.[25] His arguments against the sufficiency of selectionism were typical of many dualists: (a) mutations (e.g., in *Drosophila*) were usually deleterious and seemed to affect only superficial aspects of organs rather than their basic structures; (b) their undirected character made it difficult to explain the emergence of complex organs; and (c) since chromosomal genes mutated independently of one another, Mendelism could not explain the apparently coordinated and simultaneous changes in many traits which the fossil record presented. If geneticists paid more attention to phylogenetic evidence, they would see that although the experimental evidence for the inheritance of acquired characteristics was equivocal, it was impossible to account for macroevolutionary phenomena without invoking it. One could not simply reject the Mendelian chromosome theory, but it was difficult to explain how inheritance of acquired characteristics could be reconciled with Mendelism. Therefore chromosomal genes must be subordinate to another kind of heredity, the *Erbstock* (hereditary stock or main stem), which determined the fundamental and characteristic organs of each species. Because the *Erbstock* had a unitary character, the phenotypes which it determined would not segregate. Thus the *Erbstock* would respond slowly and as a whole to the shaping forces of the environment. Although he located the *Erbstock* in the nucleus (outside the chromosomes) rather than in the cytoplasm, Plate felt that the key feature of his theory was its attribution of two mechanisms of evolution to two kinds of heredity. While other dualist theorists had placed the non-Mendelian form of heredity elsewhere in the cell (among others he cited Johannsen, Winkler, and

25. Plate 1926; on Plate's early selectionist views, see Uschmann 1959. For an autobiographical sketch see Plate 1935.

Richard Woltereck), Plate regarded their theories as very similar to his own. The idea of a *Grundstock,* he concluded, was very much in the air.

Other dualist theorists were less sanguine about this combination of selection and some form of neo-Lamarckism. Although now best known for his formulation of the concept of "norm of reaction," Richard Woltereck (1877–1944) placed the emphasis in evolution upon a form of orthogenesis.[26] Although both selection and neo-Lamarckism could account for certain features of the process, how could one explain the enormous diversity of forms which inhabit more-or-less uniform and invariant environments, the unadaptive character of many traits, or the most striking feature of evolutionary change, its directionality? Resorting to teleology or a mystical "will" to explain the last of these observations was unacceptable; the answer had to lie with internal forces of some kind (Woltereck 1931a; 1932, 569–84, 597–602). This conclusion was not simply *a priori.* Citing studies of wild and laboratory populations of the tiny crustacean *Cladocera* which he had begun before the First World War, Woltereck found it difficult to reconcile the nature of heritable changes in these populations with either use-inheritance or selection acting upon chromosomal genes.[27] To solve this problem, he drew a sharp distinction between two kinds of heritable variation—"exchangeable" [*vertretbare*] and "constitutive" traits—which were altered through different evolutionary mechanisms. Exchangeable traits varied within a species, each geographical race being distinguished by a particular aggregate (or "mosaic") of such traits. Determined by chromosomal genes, they mutated independently and were subject to recombination and selection. Constitutive traits, in contrast, were common to all members of a species or higher taxon (thus specifying its *Bauplan*) and could be modified only in concert (or "collectively"). This kind of trait was determined by what Woltereck called the "matrix" (or "species plasma"), a living substance which was distributed throughout the cell although concentrated in the chromosomes. The matrix was very stable, but when it changed—in response to internal forces—all constitutive traits were modified as a whole. Otherwise the matrix served as a kind of caretaker, specifying the sequence of genes in the chromosome, replicating them, and regulating their physiological function (Woltereck 1924; 1931b; 1932, esp. 351–54, 393–95, 423–25). Since Woltereck provided no method for confirm-

26. For a sketch of Woltereck's life and work, see Zirnstein 1987 and my 'Metaphysics and Weimar biology: a study of Richard Woltereck (1877–1944),' forthcoming.

27. For a summary of this empirical work, see Woltereck 1934.

ing that the matrix actually existed, one might have expected his theory to be dismissed as speculative. Indeed, at a meeting of the German Genetics Society in 1933 several critics did complain that the matrix concept was superfluous.[28] That apart, however, the logical rigor of Woltereck's theory, along with the fact that some of his empirical work was awkward to explain in terms of selection, meant that during the 1930s his objections were taken seriously by selectionists.[29]

That some form of *Grundstock* theory was widely held by German biologists during the 1920s is evident from Richard von Wettstein's opening paper to the Fifth International Congress of Genetics at Berlin in 1927 (R. von Wettstein 1928). Of the two major approaches to evolution, he began, the limitations of nineteenth-century morphology were by then well known, but its value was in describing the characteristic evolutionary processes of differentiation, adaptation, orthogenesis, etc. Genetics' contribution had been to clarify through experiment which of the hypothesized mechanisms of evolution were tenable. Unfortunately, von Wettstein noted, that contribution had so far been largely negative; neither selection nor neo-Lamarckism had found much support. The apparent contradiction between the findings of evolutionary theorists and genetics derived from one assumption central to genetics: the stability of the gene in the face of the environment. As a result, selection had seemed to many geneticists to be the only conceivable evolutionary mechanism. But in view of the well-known objections to selection, von Wettstein respectfully asked his audience to consider whether the chromosome theory was the sole basis of heredity. Acknowledging the weaknesses of the evidence for the inheritance of acquired characteristics, he recommended nevertheless more systematic

28. See the discussion appended to Woltereck 1934.
29. According to Ernst Caspari, Woltereck was a "very important vitalist" during the 1930s (interview). The selectionist W. Ludwig paid tribute to Woltereck as a worthy opponent (Ludwig 1943, 512–13 and 517–18). Although at odds with Woltereck on various issues, Wilhelm Johannsen nonetheless found the former's empirical evidence against neo-Lamarckism convincing (Johannsen 1909 [1926], 222, 682–83). Woltereck's *Grundzüge einer allgemeinen Biologie* (1932) was an ambitious attempt to reconstruct the methodological foundations of biology, freeing it from exclusively causal and materialist assumptions. Although German reviewers were not persuaded by all of Woltereck's arguments, they did not deny the book's importance (Oehlkers 1933–34; Seidel 1936). Warren Weaver congratulated Woltereck, having heard "extremely high praise" for it (Weaver to Woltereck, 16.2.33, folder 138, box 16, ser. 242, RG 1.2, Rockefeller Foundation papers. The Foundation had awarded Woltereck a grant of $5,000 in 1932 to enable him to study the genetic basis of differences among species and races in several Philippine lakes).

exploration of the possibility of environmentally directed heritable change, especially if cytoplasmic heredity behaved differently from chromosomal.

That German geneticists still felt it necessary a decade later to criticize the *Grundstock* concept testifies to its ubiquity.[30] The extent of support for this idea elsewhere remains unclear, but scattered evidence suggests that it had its adherents among American biologists. Ledyard Stebbins recalls many botanists of the 1930s and 1940s accepting the idea that differences within a species were different in kind from those between species (Stebbins 1980, 150). And one of the reasons why Dobzhansky was at pains to refute dualist views of heredity and evolution in his classic book (1937) was the fact that they were still accepted by very well-known biologists. Accordingly, the year his book was published both Ross Harrison and Alfred Kinsey also attacked dualism.[31]

On the other hand, it would be wrong to imply that there were no strict selectionists whatsoever in Germany. One historian of German evolutionary thought has cited the botanist W. Zimmermann, the zoologist Bernhard Rensch, and a collection of articles edited by the zoologist Gerhard Heberer (Reif 1983). Rensch agrees, adding the zoologist Viktor Franz and the geneticists Fritz von Wettstein, N. W. Timoféeff-Ressovsky, and Wilhelm Ludwig, while another has mentioned the zoologist Fritz Süffert.[32] Both advocates and critics of selection seem to have agreed that support for selection was especially strong among geneticists (Woltereck 1931a, 238; 1932, 597; Ludwig 1940, 704; Plate 1932, 50). The best known of these is probably Erwin Baur (see chaps. 6 and 7). Had he lived longer, according to Stebbins, Baur would have been a major contributor to the synthetic theory of evolution in plants.[33] A staunch anti-Lamarckian before 1914, he sought to defend natural selection after the war when it seemed to him to be everywhere under threat (Baur 1911 [1919], 340–46; 1924, 145–48; 1925). His argument was that what we would now call micromutations were much more prevalent in plant populations than was commonly thought, since they are usually recessive and alter the phenotype only

30. The *Grundstock* idea "is totally false and should not be mentioned over and over" (von Wettstein 1937, 360); cf. Oehlkers 1949, 84–85; Ludwig 1940, 696; Schwanitz 1943, 466.

31. Harrison 1937, 372; on Kinsey see Gould 1983, 71.

32. Rensch 1980, 285; Hamburger 1980b, 304–5; for a sketch of Süffert's life and work see Kühn 1946.

33. Stebbins 1980, 139–40; Mayr (1980a, 39) places Baur among the handful of "enlightened pioneers."

imperceptibly. This finding was important, as antiselectionists conceded (Woltereck 1931a, 237), since it showed that there was abundant raw material for selection. Baur also had an answer to what he acknowledged as "the gravest objection that can be made to the theory of selection today," namely that micromutations seemed to have no selective value (Baur 1931, 181). Conceding that such mutations were generally selectively neutral, he argued that they would thus be able to persist in populations and combine with other similar mutations until some of these combinations had acquired selective value. In addition Baur presented important evidence that trait differences between various wild species of snapdragon (*Antirrhinum*) were almost exclusively due to Mendelian genes of the same kind as the micromutations observed in laboratory strains. He thus explained species formation in the genus *Antirrhinum* in terms of selection favoring different combinations of Mendelian alleles in geographically isolated populations.[34]

As students of Chetverikov's, N. W. and H. A. Timofeeff-Ressovsky brought the Russian tradition in population genetics to Germany when they emigrated to Berlin in 1925 and have been called "Germany's leading evolutionary theorists in the 1930s" (Adams 1980a, 267). Like Baur, the Timofeeff-Ressovskys had conducted some of the first genetic analyses of variation in wild populations during the 1920s and endorsed a theory of evolution via mutation and selection (see their 1927). Although he increasingly concentrated on mutation genetics over the next decade, N. W. continued to publish in population genetics, and from the early 1930s his assistants—S. R. Zarapkin, Klaus Zimmermann, and W. F. Reinig—worked on the genetics of variation in wild populations of *Drosophila*, beetles, and mice.[35] In the late 1930s he designed (with A. Buzzati-Traverso and C. Jucci) a large and systematic study of natural populations of several kinds of organisms in various regions of Italy, but the plan had to be abandoned upon the outbreak of war.[36]

Richard Goldschmidt is recognized as the first person to have investigated the genetics of variation among wild populations. From 1924

34. Baur 1932a. It is worth noting that about the time this paper appeared, the genetically informed ecologist F. B. Sumner was finding it difficult to persuade the editors of the American journal *Genetics* that genetic analyses of wild populations were important (Provine 1979, 234–35).

35. See the list of publications for the KWI for Brain Research in "Tätigkeitsberichte der KWG," published in *Die Naturwissenschaften* between 1928 (vol. 16) and 1938 (vol. 26).

36. Diane Paul, personal communication.

Figure 3.1   N. W. Timofeeff-Ressovsky making a point, ca. 1940 (reproduced from K. G. Zimmer [1982], p. 191; courtesy: Elsevier Science Publishers).

he published a series of papers on geographic races of the gypsy moth, based on work which he had begun in 1909.[37] Although he is commonly regarded as an arch anti-Darwinian, it is worth recalling that throughout most of the interwar period Goldschmidt explained both race and species formation in terms of selection, repudiating neo-Darwinism only in 1932.[38]

While the evolutionary work of Baur, the Timofeeff-Ressovskys, and Goldschmidt has been recognized by historians, several lesser figures are almost entirely unknown, such as the mathematical population

---

37. Goldschmidt 1924a. Goldschmidt's priority with this work is claimed by Caspari (1980, 22), Carson (1980, 24), and Provine (1980, 56). In "My Work" (typescript, n.d., carton 1, Goldschmidt papers), Goldschmidt says that he conducted studies of industrial melanism in the gypsy moth between 1909 and 1914—concluding that its spread was due to natural selection—although these were not published until 1921.

38. Goldschmidt, "My Work" (carton 1, Goldschmidt papers). His discussion in a five-part lecture series (dated 1930) of the current issues pertaining to evolutionary mechanism is entirely consistent with strict selectionism (Goldschmidt papers). According to Richmond, Goldschmidt was a neo-Darwinist from at least 1918 (Richmond 1986, 417).

geneticists Wilhelm Ludwig (1901–1959) and Klaus Pätau (1908–1975).[39] Receiving his doctorate from the University of Berlin in 1936, Pätau took advantage of a fellowship from the Rockefeller Foundation to spend a year (1938–39) at the John Innes Horticultural Institute in England before returning to Berlin as an assistant in Max Hartmann's department at the KWI for Biology. After the war he spent six months at the Institute of Animal Genetics at Edinburgh before emigrating in 1948 to the United States, where he spent the rest of his career at the University of Wisconsin, primarily as a human cytogeneticist.[40] Mathematically able, Pätau seems to have attempted to publicize mathematical population genetics among German geneticists from the late 1930s (Pätau 1939). He was a friend of N. W. Timofeeff-Ressovsky's, and the two men were planning a jointly authored series of articles on population genetics which was to appear in *Die Naturwissenschaften* toward the end of the war.[41]

After receiving his doctorate in zoology at Leipzig in 1925, Ludwig remained for several years as assistant before moving in 1929 to Halle, where he habilitated in zoology and genetics. His political views made promotion difficult during the Nazi period, but he finally obtained a chair of zoology at Mainz in 1946, moving two years later to Heidelberg, where he spent the rest of his career. Ludwig was not only mathematically inclined but also widely read in many areas of biology, and by 1945 he had published on natural selection, the cytology of crossing-over, the inheritance of right-left symmetry, and various aspects of the physiology of protista.[42] Devoting considerable attention to mathe-

39. Among the other authors in the volume edited by Heberer (1943)—Mayr (1980c, 282) and Reif (1983, 190) identify them as "selectionists"—were several other geneticists about whom relatively little is known: Franz Schwanitz (1907–?), Herbert Lüers (1910–1978), Otto Reche (1879–1966), and Wolfgang Lehmann (1905–1980). For Reche's work on the population genetics of human blood groups (and his subsequent collaboration with the SS), see Weindling 1989, 465–67, 540–41 (for brief references to Lehmann, see pp. 516 and 567). Schwanitz was an assistant at the KWI for Breeding Research from 1936 to 1945, and Lüers worked primarily on developmental and radiation genetics in *Drosophila* as an assistant in Timofeeff-Ressovsky's institute from 1935 until 1941 (*Kürschners Deutscher Gelehrten-Kalender*, 1950 and 1970 eds.). I am grateful to Dr. Marion Kazemi (Archiv zur Geschichte der Max-Planck-Gesellschaft, Berlin) for additional information on both men.
40. Abt I, Rep 1A, Nr 1539, MPG-Archiv; brief biographical details are provided in a death notice, dated 1.12.75, Curt Stern papers.
41. "Vorschlag für die Disposition einer Artikelserie in den 'Naturwissenschaften,'" enclosed with a letter from Pätau to Timofeeff-Ressovsky, dated 14.4.44, Pätau papers.
42. "Kurzer Lebenslauf," appended to a letter of 18.10.47 from Ludwig to Hans Nachtsheim, Nachtsheim papers.

matical population genetics, Ludwig argued that natural selection was a very important mechanism of evolution, and it was likely that species were frequently formed from diverging races. In any event, selection was the only mechanism so far whose existence was proven. To be sure, the reservations about selection held by so many paleontologists suggested that other mechanisms might also be involved, but Ludwig challenged selection's critics to produce hard evidence of these. Citing the work of Sewall Wright, he conceded that chance probably also played a role, but while one could calculate the effects of chance in theory, it was still very difficult to demonstrate its actual consequences in real populations.[43]

3.2 The Implications of Cytoplasmic Inheritance for Evolutionary Theory

While endorsing selection as the most plausible mechanism of speciation, however, most German geneticists acknowledged the difficulties still faced by the theory of natural selection. Erwin Baur, for example, was not prepared to claim in 1930 that mutation and selection could account for *all* evolutionary processes throughout nature, suspecting that "the problem of evolution cannot be solved for all organisms according to the same scheme."[44] He was uncertain whether selection alone could account for the evolution of complex adaptive organs such as the eye, and although his own work suggested that species differences in *Antirrhinum* Mendelized, he hesitated to rule out extranuclear, non-Mendelian bearers of heredity (Baur 1911 [1930], 395; 1924, 95). Similarly, in a review of the literature on genetics and evolutionary theory, presented at a meeting of the German Genetics Society in 1938, Timofeeff-Ressovsky argued that geneticists had shown that selection of Mendelian genes could account for microevolution (Timofeeff-

---

43. Ludwig 1940; 1943. Several strands of evidence indicate that after 1945 Ludwig was privately sympathetic to Lysenko's theories of species change and reluctant to rule out forms of neo-Lamarckism (Ludwig to Hans Nachtsheim, 18.10.47, Nachtsheim papers; Querner interview; table of contents of Ludwig papers). He joined the East German Communist Party (SED) in 1946 and had close ties to the East German Ministry of Culture (Querner interview; Ludwig's autobiographical sketch, cited above). Since no neo-Lamarckian inclinations are evident in his wartime discussions of evolutionary mechanism (e.g., Ludwig 1940), it is possible that with the onset of the cold war, Ludwig— like J. B. S. Haldane before him (Paul 1983)—found it easier to adjust his views on evolution than to modify his political commitments.

44. Baur, *Einführung* (1930 ed.), 400 (cf. 398); similar reservations were expressed in his "Untersuchungen" (1924, 148) and his "Artumgrenzung" (1932, 301–2).

Ressovsky 1939). But like Baur, he was vaguely uneasy about how to accommodate the growing evidence for cytoplasmic inheritance in a selectionist model. Perhaps, he suggested, the heritable differences apparently associated with the cytoplasm would eventually prove to be determined by chromosomal genes. Although confident that the selectionist model would ultimately prove true for macroevolution, he conceded that this was an empirical matter: "[Selectionists certainly do not want] to present natural selection dogmatically . . . as the sole explanation of evolution; rather they maintain that the mechanisms of change and differentiation known through experimental genetic analyses must first be exhaustively applied before one constructs alternative explanations based on new and as yet unprovable assumptions."[45]

Given their institutional weakness within the German biological community (see chap. 4), the caution with which German geneticists defended selection is understandable. When paleontologists maintained that natural selection could not account for major features of the fossil record, for example, it would have been imprudent for geneticists to extrapolate unconditionally from micro- to macroevolutionary mechanisms.[46] But more important is the fact that the evidence for cytoplasmic inheritance had been steadily accumulating since the twenties. By challenging the "nuclear monopoly," the German work on cytoplasmic inheritance seemed to make dualist theories of evolution more plausible than ever. That cytoplasmic inheritance was central to the debate over evolutionary mechanism during the thirties is evident from the first edition of Dobzhansky's *Genetics and the Origin of Species*. After quickly dismissing the evidence usually cited in favor of the *Grundstock* hypothesis, Dobzhansky devoted four pages to the evidence for cytoplasmic inheritance. Writing from the safety of Pasadena, he could afford to be less polite than Baur or Timofeeff-Ressovsky; most alleged instances of cytoplasmic inheritance, he insisted, were probably only predetermination. Although the evidence presented for the plasmon by von Wettstein and Michaelis was stronger, cytoplasmic inheritance was such an isolated phenomenon that it could play only a very minor role in evolution.

That Dobzhansky and other selectionists found it necessary to reject not merely the rather speculative *Grundstock* hypothesis, but also the much stronger evidence for cytoplasmic inheritance, is a reflection of

---

45. N. W. Timofeeff-Ressovsky, replying to Ludwig 1940; quoted in Ludwig 1943; 513.
46. This argument was used in Ludwig 1940, 704.

the dualist evolutionary significance which, as we have seen, was routinely assigned to the cytoplasm during the 1930s. In fact, however, the plasmon theorists themselves voiced *no* support for evolutionary dualism, perceiving the evolutionary significance of cytoplasmic inheritance in quite a different way. Throughout the thirties, in consequence, they were continually forced to dissociate themselves from evolutionary views which they did not hold. For example, in a review paper in 1934 Paula Hertwig (a selectionist) emphasized the similarity between von Wettstein's plasmon concept and the *Grundstock* hypotheses of Woltereck and Plate, ignoring the fact that von Wettstein had explicitly rejected the *Grundstock* concept on several occasions.[47] Correns, Michaelis, and Kühn were similarly critical of the concept (Correns 1921, 45; Michaelis 1931, 103–4; 1933b, 394; 1938, 456; 1941, 146; Kühn 1934a, 43). Plasmon theorists pointed out that plasmon differences were sometimes found between varieties but not between species. It was wrong, therefore, to ascribe intraspecific trait differences to chromosomal genes and trait differences between higher taxa to cytoplasmic inheritance. Instead, they argued, chromosomal genes and plasmon played an equal genetic role in the determination of all traits, regardless of the taxonomic level for which those traits were characteristic.

Advocates of cytoplasmic inheritance were also suspected of trying to make a case for the inheritance of acquired characteristics (e.g., Johannsen 1909 [1926], 616; East 1934, 409–10). In some instances this was not without foundation; in the United States both Tracy Sonneborn and the yeast geneticist, Carl Lindegren, were sympathetic (in varying degrees) to neo-Lamarckism.[48] But this was not the case for German students of cytoplasmic inheritance. To my knowledge, for example, Correns addressed evolutionary issues in a single paper in 1904, regarding selection as a sieve rather than a creative force, much as did other early Mendelians. In a review paper on Mendelism in 1921 he rejected the experimental evidence for the inheritance of acquired characteristics, adding that "the resort to long periods of time and countless repetitions of the environmental stimulus will scarcely make a differ-

---

47. Hertwig 1934, 427. For von Wettstein's refutations, see his 1928a, 189–92, and his 1934a, 34. Despite another clear statement of his position in 1937 (p. 360), von Wettstein continued to be misunderstood to have endorsed the *Grundstock* hypothesis (Melchers 1939, 252).

48. Sapp 1987, 102, 174 (on Sonneborn's struggle to dissociate his views from Lysenkoism after 1948, see 168–80). Sonneborn's student Geoffrey Beale (1982, 567–68) says that the former retained an open mind about the inheritance of acquired characteristics.

ence." More evidence is to be found in his lecture notes. In courses devoted to evolution in 1895 and 1897 he acknowledged the difficulties with natural selection but regarded it as the only major theory. In the genetics courses which he taught after 1900 he occasionally touched upon evolutionary issues, clearly rejecting the evidence for the inheritance of acquired characteristics in a lecture dated 1920.[49] Quite what Correns's view on evolutionary mechanism was—assuming he had one—is thus impossible to say; the available evidence indicates only that he was unsympathetic to neo-Lamarckism.

Michaelis never explicitly rejected the inheritance of acquired characteristics, but in 1933 he believed it unlikely that plasmon changes could be directed by environmental stimuli (1933a, 1933b). Neither Renner nor Oehlkers worked on evolutionary problems, but the latter was enthusiastic about Dobzhansky's *Genetics and the Origin of Species* (Oehlkers 1940–41). Viktor Hamburger remembers Kühn as shying away from a unitary selectionist account of evolution, and it is true that Kühn was uncertain whether a neo-Darwinist model alone would suffice to explain macroevolution.[50] But he emphasized that mutation/selection was the only known mechanism, and frequently rejected the evidence for neo-Lamarckism (e.g., Kühn 1922 [1928], 258; 1937c). Although evolution was not one of his principal research interests, Kühn was impressed by the work of Timofeeff-Ressovsky and Rensch during the 1930s, and he published an important paper in support of selection.[51]

The general features of von Wettstein's conception of evolution were laid out in a review paper (1939a) which he gave in 1938. The evidence from comparative morphology, palaeontology, and biogeography, he argued, all pointed toward Darwinism, but the ultimate proof of mech-

49. The first paper is Correns 1904; the quote is from his 1921, 48. The lecture notes are in folder 173, Correns papers.

50. Hamburger 1980b, 303–4; see also the sixth edition of Kühn's *Grundriss der allgemeinen Zoologie* (1939) or the second edition of his 1934e *Grundriss der Vererbungslehre* (1950). To a former student he wrote: "As old Weismann used to say, 'There you see it: The fittest species survives.' But to conceive how that can work in concrete cases is another thing altogether" (Kühn to H. O. Wagner, 28.12.47, Kühn papers, Heidelberg).

51. For Kühn's interest in Timofeeff-Ressovsky and Rensch, I have drawn upon interviews with H. Hartwig, W.-D. Eichler, and Viktor Schwartz. In "Über den biologischen Wert" (1934), Kühn reported certain micromutations in the flour moth which altered the pattern of wing coloration, without affecting its selective value in any obvious way. He noticed, however, that their viability was in fact altered and drew attention to the significance of such pleiotropy for selection theory. Dobzhansky regarded the paper as important enough to cite in the first edition of *Genetics and the Origin of Species*.

anism would require direct experimental evidence from genetics. The key question was whether isolation, selection, and drift, acting upon random mutation—whether in chromosomes, plastids, or plasmon—and recombination could provide a sufficient mechanism. Although more experimental evidence was necessary, it appeared likely that complex adaptive traits could emerge via selection due to pleiotropy, neutral mutations, and ploidy. In his empirical studies of evolution from the 1920s, von Wettstein had sought to extend the explanatory power of the theory of natural selection, showing how intermediate stages in the emergence of complex polygenic traits could arise through mutation and selection.[52] One strand of this work drew upon the plasmon theory. In view of the close "partner" relation between a gene and its plasmon, he suggested that the disharmony brought about by mutation in the gene might be compensated for by corresponding environmentally induced (but not directed) changes in the plasmon. Thus the plasmon would serve as a kind of mediator between random disruptive mutations and the cell's environment so that small, otherwise deleterious, mutations need not be lost immediately through selection (von Wettstein 1927b; 1928a, 202–6; 1937). Another strand in von Wettstein's research was the study of polyploidy. By placing different polyploid variants of a given species at different Alpine locations, he sought to clarify the selective value of polyploidy in certain environments. In 1943 he argued that polyploidy allowed the accumulation of recessive mutations (which would have been deleterious in a homozygous state) until combinations of selective value could occur through crossing.[53]

In the context of German evolutionary discussion, however, the plasmon theorists' support for natural selection was anomalous. No matter how frequently and forcefully the plasmon theorists sought to distance themselves from the *Grundstock* hypothesis or neo-Lamarckian mecha-

---

52. E.g., Stubbe and von Wettstein 1941. Joseph Straub, an assistant in von Wettstein's department at the KWI for Biology between 1939 and 1941, recalls his boss defending natural selection at conferences in which the majority of participants were skeptical of selection's sufficiency (Straub interview). Von Wettstein explicitly rejected neo-Lamarckian mechanisms in "Botanik, Paläeobotanik" (1939) and in "Die natürliche Formenmannigfältigkeit" (1941). Even if Melcher's recollection is correct, that von Wettstein was "not averse" to the possibility of direct alteration of the genotype by the environment ("stand diesen Ideen allenfalls gefühlsmässig nicht fern") (Melchers 1987a, 382), I find no evidence of this inclination in either von Wettstein's published or unpublished views on evolution.

53. Von Wettstein 1943. Dobzhansky had judged von Wettstein's earlier work on polyploidy important enough to cite in the first edition of *Genetics and the Origin of Species*.

nisms, the majority of German biologists during the 1930s regarded the evidence for cytoplasmic inheritance as supporting a dualist theory of evolution. In order to appreciate why cytoplasmic inheritance was perceived in this way, we must turn to the debate surrounding a related cytoplasmic phenomenon, the *Dauermodifikationen*.[54]

3.3 The Controversy over Dauermodifications

If one had attended the seventh annual meeting of the German Genetics Society in Tübingen in 1929, it might at first have seemed as though Richard von Wettstein's appeal to geneticists two years earlier had begun to take effect. The first session of the meeting was held jointly with the (German) Paleontological Society and was apparently intended to seek a reconciliation between the two fields' different perspectives on evolutionary mechanism. Although a failure in this respect,[55] the conference nicely illustrates one of Ernst Mayr's points about the evolutionary synthesis: that it required individuals who could bridge the gap between genetics and field biology/morphology. Furthermore, the conference demonstrates how differently these two sectors of the biological community perceived the evolutionary implications of cytoplasmic inheritance and other genetic evidence.

In the opening paper the neo-Lamarckian paleontologist Franz Weidenreich sought compromise (Weidenreich 1930b). Morphologists, he said, could not overlook the evidence for the gene's relative stability, but on the other hand they need not accept the evolutionary conclusions (i.e., selection) which geneticists tended to draw from such evidence. First of all, the mutations which geneticists commonly studied (e.g., in *Drosophila*) were rare and generally concerned evolutionarily trivial traits, and were thus unlikely to be as important in evolution as geneticists thought. Moreover, H. J. Muller had just shown that genes could be altered by environmental stimuli such as X-rays; it was now important to find out whether such mutation was directed. The inability so far to detect directed gene mutations after relatively short exposures (e.g., fifty generations) to environmental stimuli said nothing about what might occur after evolutionarily significant exposures. Besides, there were indications that semiheritable, directed, and adaptive changes situated in the cytoplasm (dauermodifications) arose after ex-

---

54. Anglicized as "dauermodifications," the German term translates as "lasting" or "persistent" phenotypic changes.
55. This view of the outcome was voiced in Woltereck 1931a, 231, and in Jollos 1931b.

posure to environmental stimuli and might, with longer exposures, become as stable as chromosomal gene mutations. This phenomenon would be important for geneticists to explore because the coordinate and simultaneous changes which paleontologists commonly observed throughout the whole organism during evolutionary sequences were very difficult to explain on the basis of independent random mutations in genes affecting different traits.[56]

But Weidenreich's rather cautious attempt at reconciliation brought only a brusque and condescending rejection from the Finnish geneticist Harry Federley (Federley 1930). The inheritance of acquired characteristics, he replied, had been totally demolished by genetic research. Kammerer's experiments before World War I were impressive and had offered the best evidence at that time, but had subsequently declined in quality and credibility. If genes were as malleable as neo-Lamarckians would have us believe, he argued, all of Mendelism would have to be scrapped. The attempt to sidestep this problem by speculating as to the existence of a *Grundstock* had no foundation, since chromosomal genes affected the most basic traits central to macroevolution. Furthermore, the construction of evolutionary theories solely upon observations of phenotypes was seriously flawed: "Genes or the genotype represent the essential and the constant element while the organism, the individual or the phenotype is merely something accidental, and . . . has no significance for evolution. It is for this reason . . . that geneticists define the problem of evolution as the problem of the alteration of genes" (Federley 1930, 21). Thus the key to evolutionary mechanism lay not with paleontology's speculations but with genetic experiments. Since geneticists had so far been unsuccessful in accounting for the emergence of complex adapative traits, it was understandable why paleontology had resorted to the inheritance of acquired characteristics. But paleontologists would have to face up to hard genetic facts and accept that geneticists' agnosticism on the question of evolutionary mechanism was the only defensible position.

In the ensuing discussion various participants tried to argue that the findings and methods of the two disciplines were more compatible than Federley had indicated, and Weidenreich again drew attention to the possibility of cytoplasmic inheritance as a way of resolving apparent contradictions, but Federley was unimpressed. After replying to various specific points, he concluded that discussion with neo-Lamarckians was

---

56. Weidenreich was not alone with this line of argument; Woltereck echoed most features of it in his 1931a.

pointless since they did not seem to understand genetics (Federley 1930, 43–50). Federley's impatience with field biologists/morphologists was undoubtedly shared by some geneticists,[57] but in many respects he was quite unrepresentative of the German genetics community. It is significant, for example, that in both his lecture and his reply, Federley avoided the subject of dauermodifications and did not challenge Weidenreich's claim that the randomness of mutation was unproven. In contrast, the plasmon theorists and other German geneticists were not only less estranged from their colleagues in morphology, but also, as we shall see, were rather more interested in dauermodifications and the possibility of directed mutation.

The best-known work on dauermodifications in interwar Germany was probably that of Victor Jollos (1887–1941).[58] By exposing populations of *Paramecium* over several generations to ever-increasing nonlethal levels of arsenic, Jollos could increase their resistance to arsenic to several times the normal level. After replacing the organisms in arsenic-free medium, their resistance persisted over hundreds—and sometimes thousands—of cell divisions before finally declining to the original level of sensitivity. If dauermodifications were simply phenotypic changes ("modifications") due to some gene product which was eventually diluted out by cell division, they should have disappeared within ten cell divisions or so, not one thousand. Whatever the biochemical basis of dauermodifications was, it must have replicated itself over many cell divisions. Similar effects of intermediate stability could be induced in other protozoa and in *Drosophila* by various environmental treatments.[59] Furthermore, by crossing "dauermodified" strains with wild-type strains, Jollos could show that dauermodified phenotypes were transferred not with the nucleus but with the cytoplasm. Jollos had thus produced a progressive sequence of adaptive and quasi-heritable alterations which seemed to have arisen by direct environmental action rather than by selection.

To the annoyance of selectionists, these "widely advertised claims"

57. For example, Johannsen attributed the widespread currency of neo-Lamarckian views among "natural historians" and ecologists to the fact that their "capacity to appreciate experimental analysis was remarkably weakly developed" (Johannsen 1909 [1926], 676). Curt Stern was also probably sympathetic to Federley, having himself had an unproductive exchange with Weidenreich over the nature of mutation (Stern 1929; Weidenreich 1930a; Stern 1930a).

58. Brink 1941; see also an unpublished obituary of Jollos in carton 3 of the Goldschmidt papers. For a summary of Richard Woltereck's and Emmy Stein's work on dauermodifications, see Hämmerling 1929.

59. Jollos 1924; the fullest account of these experiments is in Jollos 1939.

were "eagerly seized upon" by anti-Darwinians (Muller 1934, 459; Woltereck 1931a, 237). To paleontologists such as Weidenreich who were concerned to explain the "orthogenetic sequences" of the fossil record, the work on dauermodifications was of enormous interest.[60] For neo-Lamarckians, here at last was experimental evidence for gradual change via direct environmental action, and dauermodifications' cytoplasmic basis meant that they were still perfectly consistent with rare and random chromosomal gene mutations. Plate, for example, interpreted dauermodifications as an intermediate stage between purely phenotypic changes and stable genotypic ones; if a given environmental stimulus were allowed to act long enough, a modification of the cytoplasm would initially become a semiheritable dauermodification and eventually become a heritable mutation.[61] And Bernhard Rensch—in the early thirties a neo-Lamarckian—persuaded Timofeeff-Ressovsky to let him look for temperature-induced dauermodifications in the latter's *Drosophila* strains.[62]

Under the circumstances, it is hardly surprising that contemporaries sometimes thought that Jollos himself was a neo-Lamarckian.[63] Nothing could have been further from the truth. On numerous occasions he insisted that his work provided no support for the inheritance of acquired characteristics (see, e.g., Jollos 1931b, 171; 1933, 819; 1935a, 423–24; 1935b; 1939, 90–97). Instead he sought to play down the evolutionary implications of his work, emphasizing that dauermodifications were invariably unstable and thus of little evolutionary significance. Moreover, the fact that their phenotypes returned to normal at the same rate, regardless of how long they had been exposed to the environmental agent, should discourage neo-Lamarckians from hoping that longer exposure might eventually make dauermodifications stable. Finally, since dauermodifications were rarely adaptive—apart from re-

60. To recognize the existence of orthogenetic sequences, of course—as did Jollos, Goldschmidt, Ludwig, Sewall Wright (Provine 1986, 281, 290), and other geneticists—was not to accept orthogenesis as an evolutionary mechanism. Such sequences were based simply on observation of phenotypes; nothing was assumed about the underlying causes which had given rise to them.
61. Plate 1932. Ernst Mayr was keen for similar reasons (1980d, 414).
62. Rensch 1980, 295. Judging from the unpublished manuscript of an introductory book on genetics by the zoologist Otto Koehler (1889–1970), dauermodifications' appeal for neo-Lamarckians persisted well into the 1940s (p. 97, "Vererbung," typescript, ca. 1944, Koehler papers).
63. Jollos 1935a, 423; Alex Fabergé to Hans Grüneberg, 23.8.46 (Grüneberg papers); interview with Bentley Glass.

sistance to arsenic—it was not clear how they could be appropriated in support of evolution by use-inheritance (Jollos 1931a, 267–69; 1935a, 416). Like the plasmon theorists, therefore, Jollos had to swim against the tide of biological opinion, struggling to counter what he regarded as unjustified interpretations of his work. Unlike so many of his admirers, Jollos is best described as a selectionist. As early as 1921 he was at pains to defend what he called "the much abused doctrine of evolution by chance" (*Zufallslehre*) at a time when "the most superficial attacks upon Darwinism enjoyed the widest distribution and uncritical acceptance (Jollos 1922, preface). Neo-Lamarckism had been clearly refuted, he argued, and natural selection looked increasingly tenable.

Like von Wettstein, Ludwig, and other German geneticists, Jollos took seriously the apparent discrepancies between the theory of natural selection and the data of macroevolution and tried to resolve them. Perhaps he listened closely at the joint conference with paleontologists at Tübingen in 1929. For although he rejected Weidenreich's neo-Lamarckian explanation for orthogenetic sequences, he accepted the latter's claim that there was as yet no genetic evidence that mutation was a random process. If mutation were nonrandom, perhaps selection could account for these sequences? At the same time, H. J. Muller was conceding that since X-rays were artificially produced, his own pathbreaking work on X-ray mutagenesis might have little relevance for mutation in nature and thus for evolution. Furthermore, even levels of natural radiation seemed too low to account for observed mutation rates in nature (Carlson 1981, 158; Plough and Ives 1934, 269). It is for this reason that Richard Goldschmidt's claim in 1929—to have produced mutations in *Drosophila* through brief exposure to elevated temperatures characteristic of the fly's natural habitats—was received with much excitement.[64] Although no one was initially able to replicate Goldschmidt's results, *Drosophila* was the best genetically analyzed organism available. In the autumn of 1929, therefore, Jollos decided to use Goldschmidt's method to look for nonrandom mutations in particular genes (Jollos 1930).

Over the next few years Jollos presented evidence not only for a sixfold increase in the mutation rate using heat, but also for several ortho-

64. The paper in question was Goldschmidt, "Experimentelle Mutation," which appeared in the July 1929 issue of *Biologisches Zentralblatt*. Everyone at Woods Hole was "höchlichst erregt" over this work (Karl Belar to Curt Stern, 21.8.29, Stern papers), and a year later Muller was "considerably excited" about reports of Jollos's confirmation of Goldschmidt's work (Muller to Curt Stern, 14.11.30, Muller papers).

genetic sequences (Jollos 1931a; 1931b). For example, a mutation in the wild-type red-eyed fly produced a "dark eosin" eye color; further heat exposure induced a further mutation in the same gene, altering the eye color to "light eosin," and subsequent mutational steps led in succession to yellow, ivory, and white. No reverse mutations were ever observed, nor did the wild type ever mutate to any eye color but "eosin." Jollos explained these mutational sequences in terms of "directed mutation," by which he did not mean that a repeated environmental stimulus had induced progressive and *adaptive* shifts in a gene. There was, for example, no adaptive relationship between exposure to heat and gradual shifts of eye color from red to white; thus Jollos emphasized that his data provided no support for use-inheritance. Instead mutation was directed only in the sense that a given locus had a tendency to mutate in a particular direction rather than randomly. Through this kind of mutational inertia, initial mutational steps which might be of no selective value would mutate further in the same direction until the resulting phenotype was extreme enough to possess a selective value, whether positive or negative. At this point the mutations giving rise to such a phenotype would be consolidated (or eliminated) by natural selection. The existence of directed mutation, he argued, would eliminate a very important difficulty with the theory of natural selection (Jollos 1930, 553).

Although various geneticists tried in vain to demonstrate an increase in the mutation rate due to heat treatment, this aspect of Jollos's results was eventually confirmed by the American *Drosophila* geneticists Plough and Ives, who acknowledged the evolutionary importance of his work (Plough and Ives 1934; 1935). They were unable, however, to replicate Jollos's more controversial claims of directed mutations, commenting that these had met with a "good deal of skepticism" (1934, 269). Indeed, Morgan's group had been critical even before Jollos's *Drosophila* results were published.[65] Although Jollos replied in English to a short preliminary communication of Plough and Ives's results, he seems never to have responded to the full critique which they published the following year.[66]

65. Jollos had first learned of these criticisms through a letter from Jack Schultz to Curt Stern (Jollos to E. E. Just, 20.8.30, Just papers).
66. His reply to Plough and Ives (1934) was published in the Dutch journal *Genetica* in 1934. The only other papers which he published that year—in *ZIAV* and in *Biologisches Zentralblatt*—sought to downplay the evolutionary significance of his and others' work on dauermodifications. As far as I can tell, the only occasion thereafter on which Jollos

Quite why Jollos failed to answer his critics is difficult to say. Some felt that his work on *Drosophila* was simply sloppy.[67] On the other hand, even the most meticulous experimentalist would have found it difficult to work in the professional and personal circumstances which gradually destroyed Jollos after 1933. When Nazi "Aryanization" laws cost him his job at the KWI for Biology in the autumn of 1933, his prospects at first looked fairly promising. Enjoying a high reputation, not only among German geneticists but in Britain as well,[68] he was able to move almost immediately to F. A. E. Crew's department at the University of Edinburgh until early 1934, when an apparently better job lured him to the University of Wisconsin.[69] For his first six months in the United States, however, he lacked the incubators crucial for heat induction of mutants, and from the autumn of 1937 Wisconsin was no longer able to provide him with laboratory facilities.[70] Once his fellowship from the Emergency Committee in Aid of Displaced German Scholars ran out in the summer of 1935, Jollos was unable to find an academic job. It was not that he lacked American supporters; several geneticists familiar with his work—H. S. Jennings, Tracy Sonneborn, and L. C. Dunn—

---

returned to the question of directed mutation was a lengthy review paper on modification, dauermodifications, and mutation (1939, 40–42). Conceding that his *Drosophila* results (*Genetica* 1934) needed to be confirmed, he felt nonetheless that at least his evidence for the unequal frequency with which particular genes mutated in different directions was convincing. He made no reference to Plough and Ives's critique.

67. Bentley Glass (interview) and H. J. Muller took this view (see log of R. A. Lambert's conversation with John Whyte, 20.7.36, folder 2005, box 163, ser. 200D, RG 1.1, IEB papers), and R. A. Brink—Jollos's colleague at Wisconsin from 1934—recalls Jollos being "hopelessly inept" in the laboratory (Millard Susman to me, 26.5.83).

68. On his German reputation see Hämmerling, "Mein Wissenschaftlicher Lebenslauf," p. 11 (Curt Stern papers); interview with Charlotte Auerbach; Oehlkers 1930–31, 381; Richard Hertwig to Richard Goldschmidt, 11.12.33, Goldschmidt papers; and the unpublished obituary of Jollos by Goldschmidt in the latter's papers, carton 3.

69. See letters to the Academic Assistance Council in 1933 in support of Jollos from J. B. S. Haldane, Julian Huxley, R. N. Salaman, and J. Gray (Jollos file, papers of the Society for the Protection of Science and Learning). Crew had hoped that Jollos would enhance Edinburgh's reputation as a center of genetics and was annoyed that American organizations with their greater financial resources were able "to pick and choose and take the best and leave us with only the unsuccessful and the untried" (Crew to Academic Assistance Council, 20.1.34, SPSL papers).

70. In the spring of 1934 the National Research Council declined Jollos's application for $400 to purchase laboratory equipment (F. B. Hanson diary, 21–29.8.34, folder 2004, box 163, ser. 200D, RG 1.1, IEB papers; "Brief Summary of the Jollos Case," typescript, probably ca. November 1939, folder 2005, box 163, as above; L. J. Cole to Ross Harrison, 29.5.34, Harrison papers).

believed that his experiments, with dauermodifications on protozoa at least, were both correct and of great theoretical importance.[71] But the evidence for dauermodifications had never been taken as seriously in the United States as it was in Germany,[72] and if Jollos was to improve his American reputation, he would need access to laboratory facilities in order to answer the criticisms of his *Drosophila* work. Without an academic job, Jollos struggled to support his family with occasional income from lecture tours, short-term employment, and temporary assistance from the National Coordinating Committee for Aid to Refugees and Emigrants Coming from Germany, but he became increasingly desperate. More than once he was spared eviction from his apartment only through charitable donations from his Wisconsin colleagues. Following two heart attacks in 1937, Jollos's health rapidly deteriorated, and he died in 1941 at the age of fifty-four.[73]

3.4 The Relationship between the Plasmon and Dauermodifications

The work on cytoplasmic inheritance and on dauermodifications had developed independently with largely different aims and methods. While plasmon theorists had analyzed preexisting stable heritable differences in higher animals and plants, Jollos and others had induced semistable phenotypic changes in protozoa and other simple organisms.

71. Dunn, "Reminiscences," 406; H. S. Jennings to Henry Allen Moe, 13.3.40, and Jennings to Alfons Dampf, 11.4.40, both in Jennings papers; Sonneborn to F. B. Hanson, 11.2.40, folder 2005, box 163, ser. 200D, RG 1.1, IEB papers. The geneticists T. H. Goodspeed and L. J. Cole, as well as the embryologist Ross Harrison, also thought highly of Jollos (see Goodspeed to Cole, 12.12.33, and Cole to the National Research Council, 9.4.34, both in folder 2004, box 163, etc., and Harrison to Cole, 27.11.39, Harrison papers).

72. Jollos felt that his work had been better received in Germany than in the United States, and Sonneborn agreed (Jollos to Sonneborn, 14.2.40, Sonneborn papers; and Sonneborn to F. B. Hanson, 11.2.40, folder 2005, box 163, ser. 200D, RG 1.1, IEB papers).

73. The best sources on Jollos's "career" after 1933 are folders 2004 and 2005, box 163, ser. 200D, RG 1.1, IEB papers, as well as the papers of L. C. Dunn, but there is also some correspondence in the papers of H. S. Jennings and Tracy Sonneborn. In addition to a rather abrasive personality which won him few friends, Jollos faced formidable obstacles in the United States. Like other emigrés after 1933, he had to contend not only with anti-Semitism, but also with many universities' reluctance to hire foreigners at a time when young American academics were also out of work. As if that were not enough, some of the Jewish charitable organizations whom he approached for financial assistance resented the fact that he had been brought up as a Protestant and was unfamiliar with Jewish traditions.

Nevertheless, both of these groups were concerned with quasi-genetic structures located in the cytoplasm, and in view of the centrality of the cytoplasm to evolutionary debate, it is hardly surprising that the precise relationship between cytoplasmic inheritance and dauermodifications was widely discussed from the late twenties (Oehlkers 1931, 231).

Neo-Lamarckian *Grundstock* theorists regarded them as related facets of the same phenomenon. Prolonged exposure to an environmental stimulus would convert phenotypic modifications to semistable dauermodifications, which in turn would eventually become stable mutant forms of the plasmon. Predictably, Jollos rejected both the *Grundstock* hypothesis and dualist theories of evolution; more surprising perhaps is his rejection of the plasmon theory. Phenotypes attributed to a plasmon, he argued, were no different from dauermodifications; eventually plasmon phenotypes would begin to decline, just like dauermodifications. The cytoplasm, in Jollos's view, had no independent genetic significance; it was simply a substrate for chromosomal gene action (see chap. 9) (Jollos 1924, 372–73; 1931a; 1935a; 1939, 83–84).

Despite Jollos's dismissal of their work, the plasmon theorists were not in a position to reciprocate, since dauermodifications offered a potentially useful model for explaining how plasmon differences had evolved.[74] But plasmon theorists had to be very careful in discussing the relationship between cytoplasmic inheritance and dauermodifications. To regard the two phenomena as fundamentally similar was to invite Jollos's charge that plasmon effects would prove unstable and were thus of no more evolutionary significance than dauermodifications. The plasmon theorists' discussion of this issue, therefore, displays a variety of strategies designed to contain this potential threat while sometimes exploiting the connection with dauermodifications in other ways. Von Wettstein took the hard line: Sharply distinguishing dauermodifications from the plasmon, he argued that dauermodifications were merely instances of predetermination. For him the genetic independence of the plasmon was of paramount importance, and he was critical of fellow plasmon theorists whenever they suggested that certain plasmon effects wore off—even if only partially—under the influence of foreign nuclei.[75]

---

74. Well into the post-1945 period, for example, Alfred Kühn cautiously voiced the possibility—while emphasizing that it was not yet proven—that dauermodifications, if exposed to environmental stimuli long enough, might eventually become stable. Such a mechanism would aid the survival of intermediate stages of complex adaptive traits (e.g., Kühn 1922 [1961], 281).

75. Von Wettstein 1937, 361–62. Though no friend of the plasmon theory, Joachim Hämmerling acknowledged that the cytoplasmically determined phenomena which von

In contrast, Peter Michaelis sought to account for the emergence of both dauermodifications and stable plasmon mutations in terms of a unitary model relying only upon selection. He began with the observation that certain changes in plasmon-determined phenotypes in *Epilobium* can either disappear in the following generation (mere transient phenotypic changes), or persist over several generations before declining (like dauermodifications), or persist indefinitely (like mutations). In order to account for all three phenomena with a single model, he borrowed the idea of intracellular selection with which Baur (and subsequently Renner) had explained certain forms of variegation (see chap. 2). Thus in 1935 Michaelis conceptualized the plasmon as a population of mutable genetic units whose replication rates were influenced by environmental factors (Michaelis 1935a). Where this population consisted of *several* kinds of units, their replication rates might be differentially influenced by a given environmental change, such that the daughter cells' plasmon would be of a slightly different composition than the parents'.[76]

His model could account for a number of observations. For one thing, it could explain the fact, well known among students of cytoplasmic inheritance, that cytoplasmically determined traits were highly responsive to environmental conditions (Michaelis 1929, 315). Another advantage was that the rapid and progressive shift of plasmon phenotypes in response to particular environmental factors, which made cytoplasmic inheritance and dauermodifications so attractive to neo-Lamarckians, could instead be explained by random rare mutations of plasmon units which then became subject to intra- and intercellular selection. Over a period of time this process could produce a succession of progressively varying plasmon phenotypes, resembling the orthogenetic sequences from the fossil record (Michaelis 1949). But this apparently "directed" character of the plasmon's response to the environment would simply be due to a shifting distribution of the plasmon's constituents. Furthermore, the model could also explain why a given phenotypic change could behave like a phenotypic modification, a dauermodification, or a mutation. The differing degrees of stability displayed by these

---

Wettstein had demonstrated were stable and thus probably different in kind from dauermodifications (Hämmerling 1929).

76. To what extent Michaelis drew upon population genetic theories of the 1920s and 1930s in developing his model is not clear. He was certainly aware of the similarity between natural selection shifting gene frequencies within a population's gene pool and environmental factors affecting the replication rates of genetic variants within a plasmon population (Michaelis 1938).

three categories would be due to the speed with which the original plasmon composition was reestablished following return of the plant to the original environment. In the case of modifications the shift-back in replication rates of "normal" and mutant plasmon units would regenerate the original mix within a generation. Dauermodifications would arise where this shift-back took many generations. Stable plasmon mutants would occur where one type of plasmon constituent had replicated so slowly in the altered environment as to be lost altogether from the cell (Michaelis 1949; 1954). In addition, like von Wettstein and Jollos, Michaelis wanted to show how complex polygenic structures could evolve via natural selection. Analogous to arguments for the evolutionary significance of polyploidy, he suggested that if the plasmon was a population of mutable units, it could "mask" harmful cytoplasmic mutants until a combination had emerged which offered a selective advantage. Similarly, the plasmon's capacity to shift its composition rapidly meant that disharmonies arising through disruptive chromosomal mutations could be "neutralized" through compensatory alteration of the plasmon (Michaelis 1954, 366–70, 391).

3.5 Conclusion

The foregoing account of German evolutionary debate between the world wars sustains Ernst Mayr's account of the evolutionary synthesis in several respects. Between pure selectionists at one pole and neo-Lamarckians at the other, there was a broad intermediate category occupied by dualist theorists who generally explained microevolution by selection, and macroevolution by the inheritance of acquired characteristics or other mechanisms. As Mayr has pointed out (1980a, 4–6), claims for the existence of "soft" inheritance—frequently situated in the cytoplasm—played a major role in defending this middle ground. On the other hand, Mayr's tendency to dismiss geneticists' contribution to the evolutionary synthesis, on the grounds that most of them either were uninterested in evolution altogether or advocated rather simplistic selectionist models which took no account of the complexities of macroevolution, applies rather better to American geneticists than to the Germans. *Pace* Mayr (Mayr 1980c, 280–81), many German geneticists worked on evolutionary problems during the interwar period, and most of them endorsed the theory of natural selection. The defense of selection by Baur, Timofeeff-Ressovsky, and (up to 1932) Goldschmidt was not just armchair theorizing by those who could think of no other mechanism consistent with genetic theory, but rather the result of empirical studies of natural populations. Similarly, Jollos and the plasmon theorists were not content with complacent pronouncements that mutation and selection were the only mechanisms of evolution which enjoyed experimental support. Instead they devoted some of their empirical work to strengthening the selection theory's ability to explain certain features of macroevolution.

The selectionist sympathies of most plasmon theorists are unfortunately obscured in Mayr's account of interwar evolutionary debate, since he does not distinguish clearly the (selectionist) evolutionary views of the plasmon theorists from those of the dualists who *appropriated* cytoplasmic inheritance for *anti*selectionist purposes. That he tends to conflate "soft inheritance" (a conception of heredity thought to be amenable to neo-Lamarckian mechanisms) with cytoplasmic inheritance, for example, is evident in his remark that

the weight of the accumulating evidence had become so convincing by the 1930s and 40s that even the last advocates among geneticists of the occurrence of some form of *non-Mendelian* inheritance were either converted or became silent. For another thirty years a belief in *soft inheritance* could occasionally be encountered among nongeneticists . . . but as a viable scientific theory it was dead [Mayr 1982, 793 (emphasis supplied); cf. 786–90].

"Soft inheritance" may have been dead by the 1940s, but as we have seen, cytoplasmic inheritance was alive and kicking. Jan Sapp similarly misconstrues the evolutionary views of the plasmon theorists. Knowing that several of the French and American geneticists who worked on cytoplasmic inheritance were sympathetic to neo-Lamarckism, he assumes that the same was true for Germans such as Correns and von Wettstein.[77] And knowing how keen paleontologists of orthogenetic or neo-Lamarckian persuasion were to find evidence of environmentally induced directional and heritable variation, he assumes that Jollos's search for "directed mutation" was prompted by similarly dualist evo-

77. As evidence for Corren's evolutionary views, Sapp cites only a letter from Diter von Wettstein (Fritz's son), who wrote: "I don't think that they [Correns and the well-known neo-Lamarckian Richard von Wettstein] really differed in their views on evolution. . . . " (Sapp 1987, 74). The evidence which I presented above in section 2 suggests the opposite, just as it contradicts Sapp's statements that Correns endorsed the *Grundstock* concept and that Fritz von Wettstein held neo-Lamarckian views (ibid., 72–74). On the contrary, in the diary of his visit to the United States in 1938 von Wettstein notes with some surprise that following his lecture on evolution at the University of Iowa, the emigré embryologist Emil Witschi sought to defend neo-Lamarckism: "Diskussion über Lamarckismus. Witschi steht merkwürdig nahe solchen Gedankengängen. Trotz Genetik ist er durch die Hormonforschung stark dorthin gekommen" (von Wettstein 1938 diary, 32).

lutionary assumptions (Sapp 1987, 61–62). As we have seen, however, Jollos had no sympathy with neo-Lamarckism, and one major proponent of orthogenesis regarded Jollos as enemy rather than ally (Woltereck 1934, 189). Although the form of neo-Darwinism which prevailed from the 1940s regarded mutation as a randomly directed process, there is nothing in Jollos's notion of "directed mutation" which is necessarily inconsistent with natural selection as the sole sufficient mechanism of adaptation. In those of Jollos's papers which I have read I find no support for any other evolutionary mechanism.

Measured against today's neo-Darwinism, of course, some of the German geneticists' advocacy for natural selection looks a bit lukewarm. But if Gould's "hardening thesis" is correct, even the celebrated architects of the evolutionary synthesis were rather less single-mindedly selectionist during the thirties and early forties than by the 1950s (Gould 1983; cf. Provine 1983; Beatty 1987). If Jollos's interest in nonrandom mutation disqualifies him from being classed as a selectionist, then the same should apply to Wright and Dobzhansky by virtue of the importance they assigned at that time to drift. To categorize the evolutionary views of Jollos, von Wettstein, or Correns as "dualist" is to take them out of historical context.

To be sure, a few eminent German geneticists were critical of natural selection. It is well known that Goldschmidt began to repudiate his earlier view of speciation via selection during the 1930s, culminating in his *Material Basis of Evolution* (1940). Much less well known is Valentin Haecker, whose theory of *Pluripotenz* postulated a relatively plastic conception of the hereditary material with which he tried to account for certain forms of the inheritance of acquired characteristics.[78] Nevertheless, more important than (most) German geneticists' support for selection is the fact that they chose to address evolutionary problems *at all*. Although he disagreed with the thesis of Goldschmidt's *Material Basis of Evolution*, Dobzhansky was in no doubt about its importance. Praising the "impressive array of evidence" drawn from many fields, he recommended that everyone interested in evolution should read the book for its analysis of the weaknesses of the theory of selection.[79] Gold-

---

78. Haecker 1918, chap. 25; 1911 [1921], chap. 14–17; 1925b. For a brief account of Haecker's views on evolution, see Rensch 1983, 31–32.

79. Dobzhansky 1940b; more recently H. L. Carson has expressed the same opinion (Carson 1980). The seriousness of the challenge which Goldschmidt posed to neo-Darwinists in the 1940s has recently been argued by Michael Dietrich in "Goldschmidt's Heresies," paper presented at the meeting of the International Society for History, Philosophy, and Social Studies of Biology, Evanston, Illinois, July 1991.

schmidt was not the only German geneticist familiar with the evidence relevant to macroevolution. Correns, Michaelis, and von Wettstein possessed extensive knowledge of systematics, ecology, and biogeography.[80] Jollos's colleagues at Wisconsin were impressed with the "extraordinary extent" of his biological knowledge and interests.[81] And despite his own legendary mastery of the biological literature, Alfred Kühn admired the breadth of Timofeeff-Ressovsky's knowledge.[82] Selectionists or not, German geneticists had a familiarity with many areas of biology which equipped them—like the celebrated architects of the evolutionary synthesis—to attempt a reconciliation of genetic theory with the findings of field biology/morphology.[83]

To emphasize this is not to deny that some American geneticists were also interested in taxonomy and evolution. One thinks immediately of Wright and Dobzhansky, but also of E. R. Babcock, who began working in the 1930s on the genetics of evolution in the plant genus *Crepis,* then read widely in paleontology and other areas before publishing a major monograph in 1947.[84] Nor is it to deny the extensive interest in evolution displayed by American agricultural geneticists *before 1915* (Kimmelman 1987) or by a number of American *Drosophila* geneticists *after* the appearance of Dobzhansky's *Genetics and the Origin of Species*. It is simply to claim that during the 1920s and 1930s evolutionary genetics was more weakly developed in the United States than in Germany. As one émigré to the United States put it, "Some of our [American] colleagues consider [the genetics of populations] old-fashioned, not enough advanced, and akin to such terrible things as taxonomy and morphology. But I happen to be interested in just the

80. On Correns see Renner 1961b and von Wettstein 1938. On Michaelis, including his early work in ecology, see W. Stubbe 1987 (also interview with Stubbe). On von Wettstein see chapter 7.

81. Brink 1941. See also L. J. Cole to Ross Harrison, 25.4.35 and 21.11.39, Harrison papers; and Lowell E. Noland to Jollos, 23.2.40, H. S. Jennings papers.

82. Interview with W.-D. Eichler. H. J. Muller, too, was surprised at the extent of Timofeeff-Ressovsky's knowledge of natural populations (Adams 1980a, 269). On Kühn as "universal genius" (the term is Hamburger's in his 1988, 101), see chapter 7.

83. That this was also Haecker's aim—even if he adopted rather unconventional theories of heredity to that end—is clear from both the preface to his *Pluripotenzerscheinungen* and its subtitle: "Synthetic Contributions to Genetics and Evolution."

84. Stebbins 1980; Hagen 1984, 258–63. It is worth recalling, however, that Wright had little knowledge of natural populations (Provine 1981, 42; 1986); that Dobzhansky was Russian-educated; and that the leader of a team of ecological geneticists who began work at the University of California at Berkeley in 1933–34—Jens Clausen—was recruited by Babcock from Denmark (Stebbins 1980, 142–43).

old-fashioned stuff, and perhaps after all it will give some bits of information interesting even for the modern biologists."[85]

The relationships between geneticists and members of other biological specialties are clearly important if we are to understand the emergence of the evolutionary synthesis, as well as the subsequent impact of the synthesis (or lack of it) upon the biological sciences as a whole. Nevertheless, very few historians have yet considered this problem.[86] One way in which this issue might be illuminated is to look at the way in which problems of mutual interest were treated. Although I have hardly attempted to do this systematically, some American geneticists' writings on the species problem display a variety of standoffish attitudes, ranging from a reluctance to grapple with the issues, at one extreme, to a condescending dismissal of systematics, at the other. At a symposium on "The Utility of the Species Concept" in 1921, for example, the geneticist George Shull argued that the extent of morphological differences between species correlated poorly with the degree of genetic difference between them. Since there were no natural breaks in the range of visible phenotypes displayed by different groups of plants or animals, the morphological species concept was merely a convenient way to label organisms. The only natural groups were the geneticist's biotypes, not the taxonomist's species (Shull 1923). T. H. Morgan took a less aggressive tone and was reluctant to tell systematists how to do their jobs, suggesting that the systematist, the evolutionist, and the geneticist might well need to develop distinct species definitions to meet their respective purposes (Morgan 1923b, 246). But rather like Shull, he closed his paper with a reminder that since mutations in different

85. Dobzhansky to Curt Stern, 16.9.37, Stern papers. Robert Kohler's recent argument that 'the imperative to produce . . . overrides commitments to resolve particular problems or to establish theoretical positions' accords well with both his own and my account (chapter 2) of George Beadle's adoption of *Neurospora* (Kohler, 'Systems of production: Drosophila, Neurospora, and biochemical genetics', *Historical Studies of the Physical and Biological Sciences*, vol 22 (1991), 87–130 at 128). By contrast, his attempt to extend the same argument to Dobzhansky's development of an evolutionary genetics using *Drosophila* is unconvincing, in part because it ignores the fact that Dobzhansky was the product of a (Russian or European) research tradition whose training and aims were quite distinct from those of American *Drosophila* geneticists (Kohler, 'Drosophila and evolutionary genetics: the moral economy of scientific practice', *History of Science*, vol 29 (1991), 335–75). If the comparative analysis in Part I of this book is correct, the essentially opportunistic, 'productivity-driven' research strategy which Kohler evidently believes to be a general feature of scientists' behavior, was instead a strategy peculiar to American genetics—and perhaps other American disciplines—at this time.

86. See Hagen 1984.

genes could yield essentially identical phenotypes, "the whole argument from comparative anatomy built upon the descent theory seems to tumble in ruins."[87]

By 1939 some of the geneticists taking part in two symposia on speciation, held at meetings of the American Association for the Advancement of Science, were more prepared to take systematics seriously.[88] Sewall Wright was under no illusion that genetics alone held the key to the problem; even a complete account of genetic differences between related species would tell us little about the process by which they arose (Wright 1940). Similarly, Dobzhansky warned against simple models in which accumulated genetic differences between groups necessarily entailed reproductive isolation (Dobzhansky 1940a). And Warren Spencer was as critical of mathematical population geneticists' "academic discussions of natural selection on populations indefinitely large and in breeding equilibrium" as he was of morphologists' "purely historical discussions of evolution when the process is taking place all about us" (Spencer 1940, 310). Far better, he insisted, to study speciation in natural *Drosophila* populations. These attitudes must have been gratifying to Ernst Mayr, whose paper on speciation in birds opened with an appeal for cooperation on the grounds that no one specialty could solve the problem on its own:

Evolution is a very complicated and many-sided process. Every single branch of biology contributes its share of new ideas and new evidence, but no single discipline can hope to find all the answers or is justified to make sweeping generalisations that are based only on the evidence of its particular restricted field. This is true for cytology and genetics, for ecology and biogeography, for paleontology and taxonomy. All these branches must cooperate. . . . It is obvious that the taxonomist will not find out very much about the origin of new genetic characters nor about their transmission from one generation to the next. On the other hand, the taxonomist will be able to give answers to certain questions which are not attainable by the geneticist since speciation is not a purely genetic process [Mayr 1940, 249].

87. Compare the conciliatory stance taken by the German geneticist Otto Renner about the same time. In a paper on inheritance in species hybrids he warned against widening the existing rift between experimental genetics and speculative systematics. It would be no bad thing, he concluded, if geneticists would provide assistance to those geographically oriented systematists who were interested in hereditary relationships between taxa. And "since systematists can hardly dispense with the species concept, we geneticists ought to leave it alone" (*sollten auch wir Genetiker ihm sein altes Recht noch lassen*) (Renner 1924a, 345).

88. The proceedings were published in *American Naturalist* 74 (1940): 193–321.

But other geneticists' contributions illustrate precisely the parochialism which Mayr was trying to discourage. Two geneticists discussed whether the genetic differences between morphological species were quantitative or qualitative in character (Irwin and Cumley 1940), but (like the authors of two other short papers in the session) they simply reported their genetic results, saying nothing about their relevance for speciation as a process. More revealing was L. J. Cole's brusque and confrontational introduction to the symposium:

Discussions of evolution had reached the point of being relatively sterile and largely academic until the newer knowledge of the cytological and genetic mechanisms of heredity exercised its revivifying influence. The older methods of observation and speculation have since given way to definitely directed experimentation and logical exposition [Cole 1940, 197].

Admittedly, Cole's arrogance was little different than that displayed by Harry Federley at the Tübingen conference of 1929. But in general the kind of cooperation which Mayr envisaged was far more common in Germany. When Baur founded what was to become the major genetics journal, he recruited not only Correns and Haecker as coeditors, but also the systematist-evolutionist Richard von Wettstein (1863–1931) and the paleontologist-evolutionist Gustav Steinmann (1856–1929).[89] On Steinmann's death, rather than appointing a geneticist, Baur appointed another paleontologist.[90] And unlike those of its American counterparts—the *Journal of Heredity* and *Genetics*—the German journal's name signaled the importance assigned to evolution: *Zeitschrift für induktive Abstammungs- und Vererbungslehre* (*Journal for the Inductive Study of Evolution and Heredity*). Among the founding members of the German Genetics Society were evolutionists such as Woltereck and Plate, and Richard von Wettstein was elected second president of the society.[91] When Germany played host to the Fifth International Con-

---

89. Having both von Wettstein and Steinmann on the editorial board cannot have been an easy option; both were neo-Lamarckians (on the latter see Reif 1986, 98–100). Neither Baur nor Correns had been keen on having a paleontologist on the board, but von Wettstein had urged it (Baur to Correns, 31.12.09, and Correns to Baur, 22.1.10, both in the Baur letters, Sammlung Darmstädter).

90. The title page of vol. 52 (1929) of *ZIAV* indicates that the new paleontologist was Jaworski from the University of Bonn.

91. See "Deutsche Gesellschaft für Vererbungswissenschaft," *ZIAV* 27 (1921): 229–33. The third president of the society was Richard Hertwig, whose dualist predilections were mentioned above. From about 1925 those elected to the presidency were individuals whose principal research interests were genetic.

gress of Genetics in 1927, it was once again von Wettstein who was asked to give the opening address (R. von Wettstein 1928). In view of the rapidly expanding and increasingly specialized knowledge in genetics, von Wettstein argued, it was necessary to stand back occasionally and assess how such knowledge related to the broader problems of biology. Conceding the right of geneticists to define their conceptual territory as they saw fit, he urged them nonetheless to address the problems of evolution. Given his prominence within their discipline, German geneticists could not ignore von Wettstein's views, however much they might have liked to. Nor was von Wettstein the only nongeneticist then calling for cooperation between geneticists and field biologists/morphologists.[92] In Germany geneticists themselves repeatedly emphasized the importance of communication and cooperation across specialty lines.[93] At a joint meeting of the German Zoological and Botanical Societies in 1938, for example, Fritz von Wettstein (Richard's son) summarized the evidence on evolutionary mechanisms from comparative morphology, paleontology, biogeography, and genetics. Just as Ernst Mayr was to do the following year, von Wettstein called for cooperation between experimentalists and those who used the comparative method. An appreciation of each others' problems and methods would create the basis for an eventual solution to the problem of evolution.[94]

Why should German geneticists have been both better informed about, and more interested in, macroevolutionary issues than were their American counterparts, as well as more concerned to maintain a dialogue with field biologists/morphologists? And more generally, what is the significance of the German case for our understanding of the evolutionary synthesis? According to Mayr (1980a, 40–41), the architects of the synthesis were all biologists who were "willing" and who "took the trouble" to learn about specialties other than their own. His language suggests that the synthesizers were exceptional individuals, and

92. E.g., Weidenreich's paper to the 1929 conference (section 3); preface to Dürken and Salfeld 1921; Woltereck 1931a, 251.

93. E.g., Correns 1921, 49; Kühn 1934c; von Wettstein 1939b, 9–10; Ludwig 1940, 700; Ludwig 1943, 518; Haecker 1925b, v–vi.

94. Von Wettstein 1939a. This was not just empty rhetoric, designed to appease general botanists and zoologists. Within a few years of arriving in Berlin, von Wettstein succeeded in organizing within the staid Prussian Academy of Sciences a "Working Party on Evolution" whose perspective was altogether more modern than those of the botanical and zoological projects previously sponsored by the academy. He also managed to get Timofeeff-Ressovsky onto the Working Party, even though the latter was not a member of the academy (Grau, Schlicker, and Zeil 1979, 312–13).

there is little doubt that they were indeed persons of exceptional ability, energy, and vision. But the analysis in this chapter suggests that the range of knowledge and interest so characteristic of the synthesizers did not simply vary randomly among individuals. Instead it was associated with particular contexts: It was more common among German geneticists, for example, than among American ones. The same idea is implicit in Mayr's remark that Julian Huxley and E. B. Ford were the products of a "school" at Oxford (Mayr 1980a, 11, 37, 39). Mark Adams's work on Soviet evolutionary genetics during the period suggests that the key to such breadth lay in the structure of the *institutions* in which geneticists were educated and employed. Most Soviet "geneticists" had extensive knowledge of natural populations and resented being told to stay away from evolutionary problems (Adams 1980a, 269; Dobzhansky 1980b, 240). The distinction between experimentalists and naturalists, so familiar to us from twentieth-century Anglo-American biology, was far weaker in the Soviet Union (Adams 1968). In his Institute for Experimental Biology, Kol'tsov sought to unite (rather than to replace) the older morphological tradition with the newer experimental one and to avoid narrow and analytic approaches to biological problems in favor of synthetic ones. A premium was placed upon broad-based training, then specialization, followed by collaboration among specialists within research teams. Filipchenko and Vavilov were also broadly educated and interested in evolution (Adams 1980b). Clearly, Soviet breadth, so consequential for the evolutionary synthesis, was the product of institutions which *integrated* genetics with older descriptive traditions and did not allow geneticists to specialize in the Morganian manner, pushing aside the big complicated problems in order to concentrate on easier ones. In the next chapter we shall see that structural differences of this kind between German and American institutions of higher education and research go a long way toward explaining why German geneticists were more interested in both developmental and evolutionary problems than were their American counterparts.

4 Demarcating the Discipline: Germany versus the United States

By now it should be clear that German and American geneticists tended to work on different kinds of problems. The contrast is summarized in table 4.1, which compares the principal research interests of about twenty-five major figures in each genetics community during the interwar period. The table shows that a much higher proportion of the American community worked in transmission genetics rather than in other areas, while the majority of Germans worked in developmental and/or evolutionary genetics.[1] As we have seen, a substantial number of

---

1. See chapter 5 for relevant references, including obituaries. My categorization of the Americans should be regarded as provisional; it is largely based on the as-yet-limited secondary literature and may, therefore, need to be revised as new work on the history of American genetics appears. For example, in *A Progressive Era Discipline,* Kimmelman (1987) emphasizes that many of the Mendelians working in American land-grant institutions were interested in evolutionary issues. Her analysis, however, covers only the period up to 1915, during which time geneticists everywhere were still interested in the question of evolutionary mechanism, since deVries's mutation theory had not yet been discredited. It remains to be seen, therefore, whether American interest persisted after 1918. Marcus Rhoades recalls that up to about 1920 most maize geneticists were trying to solve transmission genetic problems (Rhoades 1984, 11).

It might be objected, however, that the contrast evident in table 4.1—as throughout chapters 2 and 3—is an artifact which arises because I have compared fish with fowl. That is, I have been comparing American geneticists with a group of German biologists, most of whom would be better described as zoologists or botanists whose research interests included genetic problems. This raises what is an important historiographical issue for any prosopographical study: namely, on what criteria does one decide to include particular individuals within the sample population? In short, who counts as a "geneticist" in the German context? I must confess that apart from excluding those working on human genetics (cf. preface), I have not used *explicit* criteria in making these decisions. (If Vogt's [1982] account of standard practice within intellectual history is accurate, I seem to be in good company.) I would, however, defend my sample as a representative group of German geneticists on several grounds. First, it includes selected senior figures in the major institutions in which genetics was researched during the interwar period (and these institutions are unlikely to be in dispute): the KWIs for Biology, Breeding Research, and Brain Research; Berlin Agricultural College; and the institutes of botany and/or zoology at Göttingen, Freiburg, Tübingen, and Jena. By "senior," I mean those individuals holding tenured positions, those with the title of *Extraordinarius,* and those assistants for whom records exist. Individuals were selected whose primary research interest involved the study of heredity, even if they occupied positions in zoology, botany, or some other discipline. Second, relatively isolated individuals at other institutions were

American geneticists defined the central problems of the new discipline so as to *exclude* precisely those more complex and traditional issues which the Germans regarded as fundamental: development and evolution. While the Germans demarcated the new discipline broadly, Morgan et al. drew the boundaries very narrowly, relegating the problems of development and evolution to neighboring fields such as embryology and systematics. Why?[2]

A similar German-American contrast has been observed in other disciplines during the interwar period. In psychology the German Gestalt school, for example, was characteristically concerned with problems inherited from philosophy: Perceptual experiments were used to illuminate epistemological issues. In the United States, by contrast, behaviorists were hostile to theoretical discussion of the mind. When

---

included, again on grounds of their research (e.g., Schwemmle, Winkler, Haecker). Having scanned the principal journals in which genetic research was published during the interwar period (mainly ZIAV, *Biologisches Zentralblatt, Berichte der Deutschen Botanischen Gesellschaft, Die Naturwissenschaften, Handbuch der Vererbungswissenschaft*), and having interviewed twenty-five persons with first-hand knowledge of various geneticists from this period, I am confident that no figure of any significance has slipped through my net. Given their substantial interest in matters genetic, Ludwig Plate and Richard Woltereck might be regarded by some historians as deserving of inclusion. I chose not to include them in my sample because their main research interest was the *general* question of evolutionary mechanism, rather than the *genetics* of the evolutionary process. By including Hans Winkler in my sample, however, I have admittedly violated my own criteria, for only a minority of his published work was in genetics. On the other hand, his 1924 review paper on cytoplasmic inheritance attracted considerable attention among geneticists (cf. chaps. 2 and 3), and a few years later he published an important book-length critique of the theory of crossing-over (Winkler 1930). Apart from Winkler, all of the individuals in my German sample would have accepted the designation "geneticist" (thought not as an exclusive label). That such usage may strike modern readers as rather loose, derives from the fact, as we shall see in this chapter, that what came to be called "genetics" was organized quite differently in Germany and the United States.

2. One possibility which readily springs to mind is that the intellectual relationships between geneticists and embryologists in Germany might have been especially close. The evidence suggests not. Johannes Holtfreter, an assistant in the Department of Experimental Embryology at the KWI for Biology from 1928 to 1933, shared an apartment during the late 1920s with the (cyto)geneticists Karl Belar, Curt Stern, and Joachim Hämmerling in the attic of the institute, but they did not discuss their work. During the day geneticists and embryologists at the KWI rarely saw one another, apart from lunchtimes and at the café, and shared seminars were few (Holtfreter, 1981 interview). Several sources confirm Holtfreter's view that Hans Spemann, the doyen of German embryology during the interwar period, was largely ignorant of, and uninterested in, genetics (e.g., Joseph Straub interview; Hamburger 1980b; Fankhäuser 1972). And according to Holtfreter, Otto Mangold (head of experimental embryology at the KWI, 1923–1933) was little different.

Table 4.1 Research Interests among German and American Geneticists, ca. 1920–1935

	German	American
Transmission and cytogenetics	Erwin Baur + students Curt Stern Hans Nachtsheim Paula Hertwig Hans Bauer Edgar Knapp J. Schwemmle Hans Kappert Otto Renner Elisabeth Schiemann Emmy Stein Hans Stubbe	T. H. Morgan E. M. East Raymond Pearl Lewis Stadler H. J. Muller C. B. Davenport Calvin Bridges A. H. Sturtevant George Beadle R. A. Emerson E. G. Anderson Ralph Cleland Milislav Demerec T. S. Painter W. E. Castle George Shull A. F. Blakeslee J. T. Patterson Wilson Stone Karl Sax D. F. Jones Marcus Rhoades
Physiological/ developmental genetics	Joachim Hämmerling Carl Correns Alfred Kühn + students F. Oehlkers + students Ernst Lehmann Karl Henke Hans Winkler	T. M. Sonneborn Edmund Sinnott L. C. Dunn H. B. Goodrich Oscar Riddle

members of the Gestalt school arrived in the United States—virtually all of them forced to leave Germany after 1933—they were seen as "intruders, alien to the prevailing psychological atmosphere" (Mandler and Mandler 1969, 375; cf. Arnheim 1972). From the Germans' point of view, the Americans seemed to ask limited questions that could produce measurable answers in controlled experiments, thereby avoiding the complexity of the mind. Although Gestalt psychology was not rejected out of hand in America, it changed as it was assimilated; in the course of adapting to the American environment, Wolfgang Köhler felt, Gestalt psychologists narrowed their concerns.[3] In biochemistry, too, Germans often dealt with general biological processes such as growth

---

3. Köhler 1953, 121. On Gestalt psychology in America see also Ash 1984. On the selective American reception of German psychology a generation earlier, see Dolby 1977.

Table 4.1 *continued*

	German	American
Evolutionary genetics	Victor Jollos Erwin Baur	Ernest Babcock H. S. Jennings Warren Spencer
Developmental and evolutionary genetics	F. von Wettstein Richard Goldschmidt Valentin Haecker N. W. Timofeeff-Ressovsky Peter Michaelis Georg Melchers Friedrich Brieger	Sewall Wright Th. Dobzhansky

Note: Notice that the meaningful comparison is of the numbers working on various research problems *within* each country. To compare the number of Germans working on a given problem with the number of Americans is misleading, since I have not attempted to control for the size of the total genetics community in each country.

Except for Stubbe and Schiemann, most of Baur's students are not listed separately, since most of them left genetic research for plant breeding after receiving their degrees. Similarly, Henke is the only one of Kühn's students to be listed separately, since he was the only one to have established an independent research career by 1935.

The table covers only the period to 1935. Had it extended to 1940, the American concentration in transmission genetics might have been weaker, since by the late 1930s various American *Drosophila* workers had moved into other areas (e.g., Beadle, Patterson, and Stone).

Some geneticists in both countries worked on the history and evolution of various domesticated plant species (Elisabeth Schiemann, Roy Clausen). I have not classified them as evolutionary geneticists since, as far as I know, this work was not intended to shed light upon the general processes by which evolution occurs in nature.

---

and respiration. In America, however, biochemists focused upon more narrowly medical problems: vitamins, hormones, human nutrition, and the development of rapid assay techniques for clinicians (Kohler 1982).

In both cases historians have attributed these contrasts to differing institutional contexts in Germany versus the United States. In the United States, Kurt Danziger has argued, psychology freed itself from philosophy relatively early and sought legitimation as a provider of expertise to primary and secondary education (Danziger 1979). In Germany, as Mitchell Ash has shown (1980), psychology remained institutionally dependent upon philosophy until the Second World War. Psychologists commonly held chairs in philosophy and taught courses in the history of philosophy and in epistemology. Accordingly, German psychologists were far less free than Americans to distance themselves from the complex issues of mind so central to philosophy in order to

142 The Peculiarities of German Genetics

focus on simpler, experimentally accessible puzzles. In biochemistry, as Robert Kohler has demonstrated, Germans chose to pursue more broadly biological or chemical problems because the only available academic posts were in institutes of physiology or organic chemistry. In the United States, however, biochemical research had a medical orientation because the field was institutionalized as a separate discipline in medical schools by the 1920s. For both of these disciplines, therefore, theoretical breadth arose through institutional subordination to older established disciplines.

As we shall see in this chapter, the same was true for genetics, due to the different structure and dynamics of the German and American university and research systems. Between 1870 and 1914 the rapid expansion of American research institutions, along with a university structure that encouraged specialization, created institutional conditions in which practitioners in new fields were free to demarcate that terrain as narrowly as they wished. In Germany, however, a different pattern of expansion prior to the First World War, aggravated by financial crisis and stagnation after 1918, hindered the establishment of new universities or departments. Moreover, since the structure of the German university discouraged specialization, it was in geneticists' career interests to define their discipline very broadly.

4.1 Patterns of Growth in Higher Education and Research

As the American Walter Landauer observed in 1924, there was a great deal of genetic *research* going on in Germany, but "vanishingly few" laboratory *courses* (Landauer 1924). The situation in genetics in Germany was indeed paradoxical. By the outbreak of the First World War, as we have seen, foreign observers regarded genetics as relatively well established there. Research was well catered for in the Kaiser-Wilhelm Institute for Biology, which was comparable in size to the Carnegie Institute of Washington's Station for the Experimental Study of Evolution.[4] And after the war there were some important institutional developments. In 1921 the German Genetics Society was established, fully ten years earlier than the Genetics Society of America, and by 1930 it was substantially larger than its American counterpart.[5] In 1927

---

4. On the KWI for Biology, see chapter 6; on the Station for the Experimental Study of Evolution see Allen 1986.

5. In 1929 the German Genetics Society listed 413 members ("Verzeichnis der Mitglieder," *ZIAV* 54 [1930]: 108–18). In 1933 the Genetics Society of America listed 338 members ("List of Members," *Records of the Genetics Society of America* 2 [1933]: 15–29).

the establishment of the Kaiser-Wilhelm Institute for Breeding Research made further provision for genetic research (cf. chap.6). Nevertheless, until 1945 there was but a single chair—at the Berlin Agricultural College—devoted exclusively to genetics in the twenty-two German universities and four agricultural colleges. (Many far smaller countries had done as much or more by this time.)[6] Beyond this, there were two joint chairs in which genetics was combined with another subject (embryology, animal breeding). Of six junior posts allocated to "pure genetics," five of them were untenured.[7] The situation in the United States was strikingly different. By 1938 there were at least twenty-eight full professorships of genetics or plant breeding, the great majority of them in the agricultural faculties of state universities.[8] By 1933 there were at least ten departments of genetics or plant breeding at American universities.[9] By 1916 genetics was being taught at

---

The difference in size is significant in view of the fact that the population of the United States at that time was nearly twice as large as that of Germany, and there were far more universities and colleges in the former.

6. On the department of genetics established at Oslo in 1916, see Roll-Hansen 1989. By 1926 there were also chairs of genetics in both Denmark and Sweden (Erwin Baur to Minister für Landwirtschaft, 27.9.26, ZStA-Merseburg, Rep 87B, Sign 20282). In 1919 the Netherlands' first chair of genetics was created at Gröningen (Schiemann 1949).

7. Lehmann and Beatus 1931–32. The nine remaining posts which these authors list were either earmarked for eugenics or anthropology or located in research institutions. I have not counted a chair of genetics which was established at the newly annexed University of Strassburg in 1941.

8. There were two chairs at California/Berkeley, two at Wisconsin (plus two associate and three assistant professorships), five at Cornell (in plant breeding, plus one assistant professorship), two at Harvard, two at Iowa State, two at California Institute of Technology (plus two assistant professorships), one at Pennsylvania State (plant breeding), one at Johns Hopkins (biometry and vital statistics), one at Maryland (joint with agronomy), one at Kentucky, two at Minnesota (joint with agronomy), one at the state college of Washington (joint with agronomy), one at Bucknell (joint with botany), one at West Virginia (joint with agronomy), one at Texas A&M, one at Kansas State, one at Illinois (plus one associate and one assistant professorship), and one at Princeton (joint with botany). In addition, there were two assistant professorships at Connecticut State and one associate professorship (joint with farm crops) at Florida (*Minerva: Jahrbuch der gelehrten Welt* for 1930, 1934, 1936, and 1938; membership list of the Genetics Society of America for 1933, cited in n. 5 above). This is possibly an underestimate of the actual number of chairs, since *Minerva* did not list all American universities. I thank Diana Gilbert for sharing these data with me.

9. Since it is difficult to establish the number of genetics departments from standard reference works such as *Minerva*, the figure of ten departments is almost certainly an underestimate. On the departments of plant breeding/genetics at Wisconsin, Cornell, and California/Berkeley, see Kimmelman 1987. Judging from the membership list of the Genetics Society of America for 1933, there were also departments at Iowa State, Min-

fifty-one American colleges and universities, fifteen of which had 140 research students among them (Caullery 1922, 91). By comparison, twenty years later, perhaps one-half of Germany's twenty-six universities and agricultural colleges offered courses in genetics. During the 1920s the only significant centers of postgraduate training in genetics were at Göttingen, Tübingen, and the Berlin Agricultural College.[10]

Part of the explanation for the comparative institutional strength of American genetics lies in the sheer scale of institutional expansion in the United States from 1880 to World War I. New fields (such as genetics) which happened to be emerging in the midst of such growth, Cravens has argued (1977; 1978, 18–33), were at a great advantage. With a larger and expanding range of academic niches to be colonized, American geneticists could dictate more favorable terms of employment, including institutional autonomy for their discipline. Between 1870 and the First World War the growth of existing colleges and universities, plus the establishment of new ones, created a seven fold increase in the number of academic posts.[11] Particularly important for research were the graduate schools, above all in private universities blessed with generous endowments. With the $35 million which John D. Rockefeller donated to the University of Chicago during its first decade, for example, the University was able to create separate, experimentally oriented departments of anatomy, botany, neurology, physiology, and zoology, each with its own Ph.D. program and research budget. The favorable terms of employment offered by President William Rainey Harper included generous facilities and funds for research as well as light teaching loads at the graduate level, all of which attracted special-

---

nesota, and West Virginia. While working in the United States in 1926–27, Hans Nachtsheim visited additional departments at the Universities of Connecticut, Maine, Kansas State (in plant breeding), and Illinois. In addition, there were "laboratories of genetics" in the Departments of Zoology and Field Crops at the University of Missouri (Nachtsheim, "Die Genetik in den USA," report to Ministerium für Landwirtschaft, ZStA-Merseburg, Rep 87B, Sign 20283).

10. My estimate of the number of institutions teaching genetics is based upon the data presented in Lehmann 1938. Prior to the First World War several important figures in the genetics community (e.g., Goldschmidt, Jollos, and Nachtsheim) had been trained at Richard Hertwig's Zoological Institute at Munich. By 1918, however, these geneticists had moved on; Hertwig was approaching retirement age, and his successor in 1925 (Karl von Frisch) was interested in other problems. To my knowledge, no student received a doctorate in genetics at the Institute during the interwar period.

11. Liebersohn 1985, 172. Growth rates varied from one university to another. Between 1899 and 1919, for example, the increase in the numbers of academic staff at sixteen elite research universities ranged from two- to fourfold (Geiger 1986, 272).

ists.¹² Growth continued after the war as well. The number of graduate students increased nearly eight fold between 1900 and 1930, and the numbers of staff at several elite research universities nearly doubled between 1919 and 1929.¹³

In Germany, too, the university system began to expand after 1870, albeit more slowly than in the United States. Between 1870 and 1914 student numbers increased about fourfold (versus sevenfold in the United States).¹⁴ In order to cope, several of the German states increased their spending on higher education by more than tenfold during this period (McClelland 1980, 307). Nevertheless, in 1910–11 the Prussian Ministry of Education estimated that the average annual university budget in Germany was less than one-third that of comparable American universities. Similarly, the total annual expenditures of the top twelve American universities were about twice the total budgets of *all twenty-one* German universities.¹⁵ More significant than the (modest) size of the increase in German spending was the *way* in which it was spent. Rather than increasing the number of universities, departments, or chairs, the German education ministries spent more on each existing university, leaving its institutional structure largely intact. For example, while the number of American colleges and universities increased from 563 in 1870 to 951 in 1910 (about 70 percent), the number of German universities increased from only 19 in 1870 to 22 by 1914 (a 16 percent increase).¹⁶ Within any given university, moreover, a greater share of

12. Cravens 1978, 23–24; Miller 1970, 152. On the creation of the graduate schools, see Veysey 1965.

13. On student numbers nationally see Coben 1979, 229. Among the sixteen research universities whose development Geiger traces, only Cornell had fewer staff (by 8 percent) in 1929 than in 1919 (Geiger 1986, 272). The others showed increases varying from 16 percent (Illinois) to over 80 percent (Michigan, Chicago, Johns Hopkins).

14. For Prussian student numbers see Paulsen 1921 (vol. 2), 696; on Germany as a whole, see McClelland 1980, 239–40.

15. Johnson 1990, 18; Feldman 1990. In physics, for example, between 1900 and 1910, laboratory expenditure, capital expenditure on new physical plants, and the number of posts grew more than twice as fast in the United States as in Germany (Forman, Heilbron, and Weart 1975, 8).

16. The newcomers were Strassburg (1871) and Frankfurt (1914). Münster was upgraded from a "theological-philosophical academy" to a university in 1902, through the addition of faculties of medicine and law/social sciences (Riese 1977, 321). I have not counted the eleven polytechnic institutions which were upgraded to *Technische Hochschulen* during this period, since the biological sciences were so weakly established there. In 1920, for example, there was a single chair of zoology and four chairs of botany in this sector of higher education (von Ferber 1956, 209). Of the approximately fifty geneticists whose careers I have followed, only Friedrich Oehlkers, Valentin Haecker, and Peter

the increase went toward the cost of equipping and running institutes than toward academic salaries.[17] As a result, the number of professors grew far more slowly than the number of students. The number of chairs in zoology, for example, increased only slightly: from twenty in 1890 to twenty-two in 1910 (and in botany, from twenty-seven to twenty-eight, respectively). Student/staff ratios increased accordingly,[18] and professors were forced to spend a greater proportion of their time on teaching and administration, leading to alarm in some quarters about the universities' declining research output.

One reason for this pattern of expenditure is that it was a relatively cheap way of providing for both the growth in student numbers and the emergence of new fields. The advocates of new specialties undoubtedly bombarded their minister of education with requests for new specialized chairs, but ministers were not always keen, particularly when the specialty was a laboratory science. It was one thing to establish a second chair in a given humanities discipline or in a nonlaboratory science, since that entailed providing the new professor with his own seminar room (and perhaps a library). A second chair in a laboratory science, however, was something else.[19] Either both professors would have to share the existing institute's facilities—an arrangement which was rarely attempted before 1945—or a separate institute would have to be provided for the newcomer at great expense.[20] Thus a variety of

---

Michaelis worked in such institutions. In each case they took up these posts at a relatively early stage of their careers, moving on to a university or Kaiser-Wilhelm Institute after a few years.

17. At Berlin, for example, institute costs increased about fourfold between 1880 and 1910, while salaries less than doubled (Paul Weindling 1991, 156). The picture at Heidelberg was much the same (Riese 1977, table 37). On Prussian expenditures in physics and chemistry on personnel versus institutes, see Lundgreen 1983, table 11.

18. At Berlin the number of students per total academic staff increased from 11.6 in 1894 to 17.2 in 1907 (vom Bruch 1980, 107). At Heidelberg the ratio increased from 7.4 in 1880 to 15.7 in 1914 (Riese 1977, table 19).

19. In 1900, for example, there were eighty-eight chairs for civil law, fifty-five for classical philology, and forty-two for philosophy in the twenty German universities. By contrast, the number of chairs for most laboratory sciences (plus clinical disciplines like surgery, gynecology, or ophthalmology) varied between twenty and twenty-five. Mathematics, with forty-four chairs, was the exception which proves the rule (von Ferber 1956, tables I and II; cf. Johnson 1985, table 2). Burchardt confirms that multiple chairs were common in law, theology, and history before the First World War (1988, 191).

20. During his visit to the United States in 1909–10, the academic plant breeder Kurt von Rümker was much impressed by the extent of specialization in American agricultural sciences. On his return, he recommended that the *single* institutes at selected German universities, which were then responsible for teaching all areas of agricultural

cheaper solutions were preferred. If an expert in the new field was available locally, for example, one could award him or her the title *Honorarprofessor* in exchange for some part-time teaching. Or one could hire a young (and unsalaried) *Privatdozent* in a related institute to teach the new subject on a contractual basis (*Lehrauftrag*). Since the contract guaranteed only a modest salary but neither pension rights nor tenure, the ministry could decline to renew the contract whenever finances became tight.[21] If more permanent provision for the new field was deemed essential, the ministry could create a new (tenured) associate professorship (*Extraordinariat*) attached to an *existing* institute. Even this was far cheaper than endowing a new chair and institute, not least because by 1918 the salary in question was less than that of a train conductor or a secondary school teacher.[22] Sometimes the new *Extraordinarius* became head of a division (*Abteil*) within the institute which had been created to cater to the new specialty. This was often the case in chemistry. By 1910, for example, the chemical institute at Heidelberg had divisions for inorganic, organic, physical, technical, and pharmaceutical chemistry.[23]

One consequence of this policy was that although the number of professors grew by only about 40 percent between 1870 and 1914, the number of junior posts (*Privatdozenten* and *Extraordinarien*) increased by threefold to fourfold, thus keeping pace with the growth of the student body.[24] For young biologists, however, the situation was much worse, since there was no net increase in the number of tenured posts

---

science, be replaced by several specialized institutes. The Prussian ministry is said to have been appalled at the financial implications of von Rümker's proposal, and new institutes were not established until the 1920s (Böhm 1988, 13–14). On the ministry's attempts to economize in Baden, see Riese 1977, 111–12, 118–19, 132–34, 142.

During the 1870s several chemical institutes were led by two professors as codirectors. Their successors rejected this arrangement, however, so that by 1890 all but one of the universities' chemical institutes had a single director (Johnson 1985, 508). In genetics the American model of several professors sharing the chairmanship of a department/ institute appears to have been introduced for the first time in the early 1960s at Cologne by Max Delbrück (Fischer 1985, 218–27).

21. In Baden around 1910 a "modest salary" meant up to 800 marks per semester at a time when *Extraordinarien* were earning 2,000 to 2,500 marks per year (Riese 1977, 112). On the *Honorarprofessoren* see Spranger 1930, vol. 3, 21.

22. In his *Gedanken zur Hochschulreform,* C. H. Becker (Prussian minister of education during the 1920s) drew this salary comparison (1919, 32).

23. Riese 1977, 142. On similar arrangements at Munich and Berlin, see Johnson 1985.

24. On German universities as a whole, see Busch 1959, 76–77. At Heidelberg the number of professors in the science faculty increased from eight in 1880 to nine in 1914, while junior posts nearly trebled (Riese 1977, table 21).

between 1900 and 1910.[25] In any event, the growth of junior posts over the longer term offered little consolation to young academics since most of these posts were untenured and many were unsalaried.[26] Thus the growth of German higher education between 1870 and the First World War differed from its American counterpart both quantitatively and qualitatively. Not only did the German system grow more slowly, but the manner in which the state funded that expansion tended to increase the number of insecure and ill-paid jobs within the existing institutional structure. In Germany, therefore, the outlook was bleak for that younger generation of experimental biologists who were just beginning to enter the job market at this time.

The gap between German and American higher education widened yet further after the war. State spending on the universities declined from 1918 and did not regain its 1910 level until 1930. At first there were fears that one or two universities would have to close.[27] It was obvious to Abraham Flexner in 1930 that the German university's most serious problem was that a near-total lack of capital spending had halted the creation of new chairs (Flexner 1930, 332). It is not that there was no growth whatsoever in the immediate postwar period; by the mid-1920s new chairs had in fact been created for a range of medical and agricultural specialties, perhaps prompted in part by wartime disease and hunger.[28] But most other new specialties were not so lucky; no German university, Flexner noted, had a chair in genetics. The financial situation meant that plans to endow a tenured *Extraordinariat* in genetics at Berlin in 1918 had to be cut back, yielding instead only a teaching contract.[29] There was, of course, already a chair of genetics occupied by

25. Between 1900 and 1910 one chair was established in zoology and three in botany, but the number of associate professorships (*Extraordinariate*) remained constant in zoology and declined by four in botany (von Ferber 1956, 209).

26. In 1907 26 percent of the *Extraordinarien* in the sciences outside Prussia were unsalaried, and among the salaried the majority had teaching contracts rather than tenure (Eulenburg 1908, 57, 134–35). The *Privatdozent* did not become a salaried position until 1939 (Burchardt 1988, 169).

27. Jastrow 1930, 282. In 1923 it was rumored that the largely privately financed University of Frankfurt might have to close, and Hamburg was also facing difficulties (Gerald Feldman, personal communication).

28. Pfetsch and Zloczower 1973, 33, 62; Weindling 1989, 332; Böhm 1988. Sociology was another special case; Germany's first chairs were established in the 1920s— rather later than in France, Italy, or the United States—thanks in part to the Prussian minister of education (C. H. Becker), who envisaged a role for the new discipline in the democratizing process (M. Neumann 1987, 299).

29. O. Hertwig to Geh. Regs. Rat Krüss, 1.3.19, ZStA-Merseburg, Rep 76, Vc, Sekt 2, Tit 23, Litt A, Nr 112, Bd II.

Erwin Baur at the Berlin Agricultural College, but by 1920 the *institute* which the Prussian Ministry of Agriculture had promised Baur upon his appointment to that chair in 1914 had still not been built. Frustrated, Baur began to consider offers not only from the Universities of Utrecht and Uppsala, but also a lucrative one from a large seed company.[30] Elsewhere the situation was much the same. Although institutions in Hamburg and Cologne were elevated in 1919 to university status, they offered little scope for innovation. Only Hamburg made even modest provision for genetics: a teaching contract in human genetics located in the anatomical institute. When the chairs for both systematic botany and plant physiology at Göttingen became vacant in 1921, the ministry of education reduced one of the posts to an *Extraordinariat*. As the Prussian government's financial position became steadily worse over the next few years, the ministry eliminated one of the botanical chairs.[31]

Conditions at the Kaiser-Wilhelm Institute for Biology were equally poor. As Goldschmidt wrote to a colleague in 1920, "The danger exists, that German science will become extinct for economic reasons. In any event the next ten years will be a critical period."[32] The Kaiser-Wilhelm Society's finances were in bad shape. The Society had lost the capital which it had invested in war bonds, and wealthy individuals were no longer so enthusiastic about contributing to the support of pure science. By 1920 worsening inflation forced the Society to appeal to the Reich and Prussian governments to cover its deficit, and by 1922 the state's contribution to the Society's income was up from 20 percent in 1918 to 50 percent (Wendel 1975, 227–28). In return the Prussian government urged the Society to solicit funding from the private sector more intensively and to cut back wherever possible what it called "less urgent" activities, citing the Institute for Biology as a case in point (Feldman 1990). Over the next two years the Society did in fact manage to attract enough industrial funding to establish a series of new institutes, most of them for applied sci-

30. Baur to Ministerium für Landwirtschaft, 17.2.22, ZStA-Merseburg, Rep 87B, Sign 20281. Irritated by constant delays in the construction schedule for the new institute, plus the ministry's reluctance to meet his annual budget requests, Baur used a variety of threats (e.g., that he would sell off stocks of breeding animals, fire all nonscientific staff, leave the college, etc.) to get his way. The new institute was finally completed in 1925 (ibid., Sign 20281 and 20282).
31. See the correspondence in the Göttingen University Archive, AZ: Math-NW Fak, 18 and 23.
32. Richard Goldschmidt to Julian Huxley, 29.11.20, Huxley papers. On the financial plight of the KWI for Biology at this time, see Saha 1984, 231.

ences.[33] But the KWI for Biology's basic research had little appeal for such sponsors (cf. chap. 6). When Hans Spemann left the Institute for a chair at Freiburg in 1919, therefore, his Department of Experimental Embryology was closed, and Spemann's post was left unfilled for four years.[34] Meanwhile Goldschmidt was inquiring with American colleagues about his job prospects there, and Correns explored his chances of getting the chair in botany at Leipzig.[35]

If the financial crisis at war's end inhibited institutional growth, the monstrous inflation of 1923 forced an actual reduction in junior posts. In 1924 the governments of Prussia and other German states economized by reducing the size of the civil service (which included academics) by 25 percent. Many institutions attempted to achieve this reduction by firing postdoctoral teaching or research assistants. At the Berlin Agricultural College four out of the thirty assistantships were eliminated; in the science faculty at Freiburg four of ten assistantships were cut.[36] In the Botanical Institute at Jena the cuts forced the geneticist Peter Michaelis to share his meager assistant's salary with another colleague. Since Friedrich Brieger's assistantship in the same institute had to be scrapped, the director was relieved to learn that Brieger had been awarded a fellowship from the International Education Board which would allow him to work in genetics at American institutions for two years.[37] At Göttingen in 1923 the science faculty introduced 40 percent salary cuts for assistants, and the Botanical Institute lost one assistantship. Fritz von Wettstein's first few years as director of the institute were difficult. On his arrival in 1925, he, an assistant, and their wives painted the institute's library themselves. The laboratory facilities

33. These included institutes for aerodynamics (1919), fiber chemistry (1920), metal research (1920), leather research (1922), and coal research (1922) (Henning and Kazemi 1988).

34. Minutes of the meeting of the Kuratorium of the KWI for Biology, 13.3.19 (Fischer Papers), 17.12.20 (ZStA-Merseburg, Rep 92, Sign C82), 21.11.22 (ZStA-Merseburg, Rep 76, Vc, Nr 112). On the appointment of Spemann's successor at the KWI for Biology in 1923, see Spemann to Carl Correns, 15.10.23 (Spemann papers).

35. Jacques Loeb to Richard Goldschmidt, 16.2.20 and 11.5.20, Goldschmidt papers. In 1920 Correns notified the Kaiser-Wilhelm Society that he had been offered the chair of botany at Leipzig (ZStA-Merseburg, Rep 76, Vc, Sekt 2, tit 23, Litt A, Nr 112, Bd II, pp. 171–72). Although he declined this call, three years later colleagues were still advising him to seek a university post (Correns to Hans Spemann, 11.1.23, Spemann papers).

36. Minister für Landwirtschaft to Rektor of Berlin Agricultural College, 12.5.24, ZStA-Merseburg, Rep 87B, Sign 20105; Nauck 1956, 38.

37. W. Stubbe 1987; Otto Renner to "Herr Geheimrat" (probably Fritz Haber), 21.3.24, Goldschmidt papers.

were unable to cope with the rapid influx of doctoral candidates—from two in 1925 to nineteen in 1926–27—and the lecture hall was so small that von Wettstein had to give the lectures in his introductory course twice.[38]

By 1924 the inflationary crisis was over, and scientific institutions began to recover from the cuts.[39] Fellowships from the International Education Board not only allowed a number of young German geneticists to work in the United States from the mid-1920s, but also brought European and American geneticists to German laboratories.[40] Nevertheless, the longer-term job prospects for young geneticists remained poor.[41] In 1925 a representative from the new Egyptian university offered Victor Jollos and Karl Belar chairs, requesting a reply within twenty-four hours. "We will probably decline," Belar wrote to a colleague, but he was wrong; Jollos accepted and was professor of zoology in Cairo from 1926 to 1929.[42] For his part, Belar took leave from the

38. Melchers 1975, 480; University of Göttingen Archive, AZ: Math-Naturw Fak, 41; also AZ: XVI V C wa - Pflanzenphysiologisches Institut. On the lost assistantship see von Wettstein to Kurator, 22.6.25; this is located in a third file in the Archive, covering the period 1886–1937, whose cover is missing but whose AZ is probably V.C.w.a.1.

39. Although the number of *Privatdozenten* in various faculties at Freiburg had dropped by anywhere from 20 to 40 percent between 1922 and 1923, most of that loss had been recovered by 1924 (Nauck 1956, 38). While on leave in Japan from 1924 to 1926, Goldschmidt was impressed by the recovery of the German economy and thus the financial health of the Kaiser-Wilhelm Society (1960, 229–30).

40. The Germans who held IEB fellowships were Curt Stern (1924–26 and 1932–33), Hans Nachtsheim (1926–27), Karl Henke (1930), and Friedrich Brieger (1924–26). On German recipients in general see IEB papers, folder 150 ("Fellowships 1929–1939"), box 16, ser. 717D, RG 1.1. On Stern see folder 997, box 60, subser. 3, ser. 1. On Nachtsheim see folder 1254, box 75, subser. 4, ser. 1. On Brieger see folder 659, box 45, subser. 3, ser. 1. On Henke see Rockefeller Foundation papers, folder 321, box 48, ser. 717, R7-RG-2GC-1954.

Erwin Baur had hosted two Europeans with IEB fellowships by 1929 and was keen to accommodate more if the Rockefeller Foundation would oblige; see IEB papers, folder 1266 ("Ernst Oehler, 1927–30"), box 76, subser. 4, ser. 1; and folder 867 ("Felix Mainx, 1927–28"), box 54, subser. 3, ser. 1. In 1925 both Erwin Baur and Victor Jollos received IEB fellows from the United States (Baur to Ministerium für Landwirtschaft, 18.9.25, ZStA-Merseburg, Rep 87B, Sign 20282; Karl Belar to Richard Goldschmidt, 10.6.25, Goldschmidt papers).

41. "Hopeless" was the way Richard Hertwig described the situation in a letter to Goldschmidt (13.2.24, Goldschmidt papers).

42. Belar to Richard Goldschmidt, 10.6.25, Goldschmidt papers; Brink 1941. Jollos was not the only one to look to the Third World for help; in 1929 Günther Just was considering a temporary post in China (Just to Goldschmidt, 12.1.29, Goldschmidt papers).

KWI for Biology for a research fellowship with T. H. Morgan in 1928 and was soon offered a teaching post at the California Institute of Technology. In addition Columbia University offered him a professorship in zoology with a light teaching load and a salary twice what he had been paid in Berlin. Although he hoped he could use the American offers to bargain for a higher salary at the KWI for Biology, Belar was advised by the Kaiser-Wilhelm Society that, given its financial circumstances, he should accept Columbia's offer. Perhaps, they suggested, he could return in a few years if the situation improved. For his pains, Belar was awarded merely a professorial title (*nicht beamteter ausserordentlicher Professor*). The Society's situation did not improve, and Belar remained in Pasadena until his unexpected death in 1931.[43]

The partial recovery from 1924 was short-lived; by the end of the decade the depression had begun to inflict much worse cuts on the German scientific community. The budget of the Emergency Council for German Science (*Notgemeinschaft der Deutschen Wissenschaft*), the government research-funding body established in 1920, was immediately cut: by 12 percent in 1929, 11 percent in 1930, and a further 25 percent in 1931. As a result, between 1928 and 1933 the Emergency Council reduced its expenditure on fellowships by 50 percent and on experimental research by 80 percent.[44] Government cuts also had consequences for the Kaiser-Wilhelm Society. In 1931 all staff at the KWI for Biology took a 6 percent cut in salary, and various of the library's journal subscriptions were canceled. The following year the Institute's budget was cut by another 9 percent, and Correns had to dismiss one of his assistants. No longer able to afford heat or light for his greenhouses, Correns (and Goldschmidt) applied to the Rockefeller Foundation for support.[45] They were not the only German biologists

43. Curt Stern to Kaiser-Wilhelm-Gesellschaft, 2.8.28; Stern to Alfred Kühn, 11.11.29; Max Hartmann to "Herr College," 13.11.29; and Karl Belar to Stern, 6.10.29 and 22.12.29, all in Stern papers; Caspari, personal communication, and "Karl Belar," forthcoming in *Dictionary of Scientific Biography*.

44. Feldman 1987, 270; S. M. Gunn to Max Mason, 19.1.31, folder 36 ("Program and Policy, 131, 1933–38"), box 7, ser. 717; and A. Gregg to Richard M. Pearce, 15.11.29, folder 150 ("Fellowships 1929–39"), box 16, ser. 717D (both in RG 1.1, Rockefeller Foundation papers); cf. Zierold 1968, 70, 38–39. On the fellowship scheme see Rektor and Senat of the University of Breslau to Rektor/Senat of all German universities, 5.11.31, copy in University of Göttingen Archive, AZ: Math-Naturw Fak 41. On the history of the Emergency Council during the 1920s, see Schroeder-Gudehus 1972.

45. In October 1931 an editorial in *Der Biologe* referred to a more than 20 percent cut in the Kaiser-Wilhelm Society's budget, with worse things to come by the winter

Demarcating the Discipline 153

to be bailed out by Rockefeller at this time. Elsewhere in Berlin large sums from the Foundation also made possible the establishment of a new KWI for Cell Physiology (1930) and a new building for the KWI for Brain Research (1931).[46] In the universities, salaries and institute budgets were cut, and rumors again circulated of impending university closures.[47] Hans Spemann's Zoological Institute in Freiburg had a single laboratory for research staff and a budget that failed to cover the cost of research materials. Although the Rockefeller Foundation's field officers regarded Spemann as possibly Europe's leading zoologist, his institute was "one of the smallest and least imposing" they had seen. Since far more American students wanted to visit than he had space for, Spemann applied to the foundation in 1931 to finance the expansion of his institute.[48] In Munich, Karl von Frisch had approached the In-

("Die Notlage der deutschen Biologie," vol. 1, 17–18). On the cuts see Correns to Curt Stern, 12.11.30, Stern papers; "Jahresrechnungen für die Rechnungsjahre 1930, 1931, 1932" and "Haushaltsplan des KWIB in Berlin-Dahlem für 1933," both at ZStA-Merseburg, Rep 76, Vc, Sekt 2, Tit 23, Litt A, Nr 112, Bd II. On Correns's appeal to the Rockefeller Foundation, see folder 109 ("KWI for Biology, 1931–37"), box 13, ser. 717D, Rockefeller Foundation papers.

46. On the KWI for Cell Physiology see folder 46 ("KWI for Anthropology 1933: KWI for Biology 1929–35"), box 4, ser. 1.1, RG 6.1, Rockefeller Foundation papers. On the Institute for Brain Research see Spatz 1962. In 1931 the Foundation resolved to make up some of the shortfall in state funding by granting the Kaiser-Wilhelm Society $80,000 per year for 1932 through 1934, on the condition that the money go to particular designated KWIs, including that for biology (folder 519, box 64, ser. 717, RG 2, Rockefeller Foundation papers). On the Foundation's crucial role in building the KWI for Physics from 1934, see Macrakis 1986.

47. At the Berlin Agricultural College salaries were cut by over 20 percent in 1932 ("Nachweisung über die den Lehrstuhlinhabern im Rechnungsjahre 1932 gezahlten Grundgehälter," dated 27.4.33, Rep 87B, Sign 20096; and Rektor to Ministerium für Landwirtschaft, 7.10.31, Rep 87B, Sign 20105, both in ZStA-Merseburg). At the *Technische Hochschule* Darmstadt the Botanical Institute's budget had been steadily reduced since 1929, and by 1931 the director (the geneticist Friedrich Oehlkers) anticipated substantial salary cuts (Oehlkers to Ralph Cleland, 2.1.31, Cleland papers). In 1931 all institutes at the University of Munich suffered a 10 percent budget cut, with salary cuts on the way (folder 134 ["University of Munich, 1930–1933"], box 14, ser. 717D, RG 1.1., Rockefeller Foundation papers).

On university closures see Kelly 1973, 46; Correns had heard that the University of Rostock was under threat (Correns to Steinbrinck, 26.12.31, Steinbrinck papers).

48. Folder 120 ("University of Freiburg, Spemann, 1931"), box 13, ser. 717D, RG 1.1, Rockefeller Foundation papers. When the biochemist Hans Krebs was at Freiburg in 1931, his research costs were met by the Foundation (Krebs 1981, 46). In Marburg the Zoological Institute's budget was cut by nearly 25 percent in 1932, forcing one junior staff member to approach the Foundation for $150 (roughly equal to 10 percent

ternational Education Board in 1927 to finance a new building for his zoological institute, since the Bavarian Ministry of Education could only afford to pay for the land plus running costs. On a visit to Munich two years later, Rockefeller officials observed that the courtyard of von Frisch's institute was shared with a monastery, and that small laboratories suitable for two workers were often occupied by four or five. Reckoning that von Frisch was one of the leading zoologists in Europe, the Foundation met his request, and by 1933 Munich had acquired what was then regarded as the finest zoological institute in Germany.[49]

At Göttingen the science faculty was forced to cut four assistantships (about one-tenth of those in the faculty) and reduce all institute budgets by 7 percent. In order to continue employing these four assistants, however, Max Born got his fellow professors to agree to contribute a fixed percentage of their salaries to a rescue fund. Finances in Alfred Kühn's Zoological Institute were tight. Apart from 15 percent reductions in the salaries of those *Privatdozenten* on teaching contracts, the institute's budget was cut by 10 percent in 1932, again by 10 percent in 1933, and by 3 percent in 1935. Kühn complained to the Prussian ministry's representative that he was having to pay for instruments out of his own pocket, and that the library was increasingly dependent upon his personal journal subscriptions. He was thus grateful to the Rockefeller Foundation for grants between 1934 and 1937. Foundation officials had great respect for Kühn's work on the physiology of gene action and seem to have felt that they were getting a bargain. As one field officer commented, the grants to Kühn were "not very large" but seemed to be "the difference between scientific life and death for him."[50]

Of course, the depression did not spare American science. Federal scientific agencies' budgets declined, and the number of National Research Council fellowships was reduced by 70 percent between 1933

---

of his salary) to buy apparatus (C. Schlieper to W. E. Tisdale, 9.10.32, folder 132 ["University of Marburg, Zoological Institute"], box 14, ser. 717D, RG 1.1, Rockefeller Foundation papers).

49. Folders 133 ("University of Munich 1929") and 134 ("University of Munich, 1930–33"), box 14, ser. 717D, RG 1.1, Rockefeller Foundation papers.

50. W. E. Tisdale to Warren Weaver, 11.2.37, folder 123 ("University of Göttingen, A. Kühn, 1934–37"), box 13, ser. 717D, RG 1.1, Rockefeller Foundation papers. Although these grants of some $3,000 per year were only about half the Natural Sciences Division's average annual project grant (Kohler 1979, 273), they enabled Kühn to employ two assistants and three technicians. On the financial situation in Kühn's institute, see University of Göttingen Archive, AZ: XVI.V.C.v.4.2. On the science faculty's finances, see AZ: Math-Naturw Fak. 41; Becker 1987, 414.

and 1936. According to Geiger, the research universities were relatively unscathed for the first few years, continuing to open new laboratories. But from 1932 appropriations for the state universities were slashed, and private universities' endowments collapsed.[51] As a result, academic salaries in American higher education were reduced by an average of 15 percent between 1931 and 1933 and twice that much at some universities. On the other hand, the pay cuts were largely offset by deflation so that purchasing power remained the same. And following a 1.5 percent decrease in the total number of academic jobs between 1931 and 1933, the number of teaching staff in the higher education system as a whole grew by 36 percent from 1930 to 1939, with the elite research universities growing even faster.[52] Although American academics quite rightly felt that the depression was a difficult period, therefore, the available evidence suggests that the effects upon German science were yet more devastating.

To summarize, we have seen that during the period 1870 to 1914 the higher education system (in particular, the number of tenured posts) was growing much more slowly in Germany than in the United States. After the First World War German growth ground to a halt, punctuated by severe cutbacks in 1923 and the early 1930s. In the United States, apart from a relatively brief period of contraction after 1929, the universities continued to grow rapidly. To what extent can these contrasting rates of growth account for the different ways in which German and American geneticists defined the key problems of their discipline? If limited finance and slow institutional growth had been the only obstacle confronting genetics in Germany, one might at first have expected that German genetics would differ from its American

51. On cuts in federal science spending, see Kevles 1979, 236–37, 250. Between 1929 and 1933 state appropriations at the Universities of Wisconsin and California dropped by 30 percent and 15 percent, respectively, while endowment income at Harvard and Yale declined by 18 percent and 30 percent (Geiger 1986, 248; on expansion between 1929 and 1932 see 246–47).

52. As a National Research Council fellow, the geneticist George Beadle took a pay cut of 33 percent over the period 1931 to 1934 (Kay 1986, 85), and salaries at the University of Texas were chopped by 30 percent (Carlson 1981, 191–92). On salary cuts, numbers of posts, and the effects of deflation, see Weart 1979, 307–12; Geiger 1986, 248–50; and Kevles 1979, 274. In 1933 the embryologist Emil Witschi felt that there was "almost no hope for new positions" at American universities (Witschi to Curt Stern, 6.4.33, Stern papers), but things quickly changed. Between 1929 and 1939 three of the sixteen elite research universities that Geiger studied more than doubled their teaching staffs, while another three increased their staffs by over 50 percent (Geiger 1986, 272).

counterpart merely in *scale:* There would be fewer German geneticists or they might be less productive, but the kinds of genetic research need not have differed. Clearly this model is too simple; "genetics" meant different things in the two countries, and there were, if anything, *more* geneticists in Germany than in the United States. But quantitative differences sometimes have qualitative consequences. Rapid growth of resources within an educational system, for example, makes institutional innovation easier. Instead of competing with better-established fields for scarce resources, new fields can set up independently of older ones. Geneticists who inhabit chairs or departments *designated* for genetics, for example, will be freer to define their own research agenda. This process may well have contributed to the narrower conception of genetics in the United States, but it is at best a partial explanation. For as we saw in section 1, most American geneticists—including the Morgan school—did not work in departments of genetics or plant breeding, but rather in departments of zoology, botany, agronomy, etc. The most important difference between these two university systems, as we shall see in the next section, was a qualitative one: The ways in which German and American universities were *organized* made the former far less adaptable than the latter. Metaphorically speaking, adolescent specialties in Germany were kept at home within the parent discipline well into young adulthood, while their American counterparts were setting up their own households. Even if German higher education had grown as rapidly as in the United States, therefore, the German university would still have been much slower to grant independence to new fields. And it is precisely this extended "adolescence" which provides the key to understanding why German geneticists demarcated their discipline so broadly.

4.2 The Effect of University Structure upon Specialization

As we saw in chapter 1, progressive specialization within the academic community encountered less resistance in the United States than in Germany. One reason for this is that the structure of the German university better enabled its members to contain the pressures for institutional innovation.[53] In this section I will discuss the structural differences between German and American universities at three levels: (1) the relations between the university and its wider constituencies;

---

53. As we will see in chapter 8, German resistance to specialization also had an ideological dimension.

(2) the power of the faculty to influence university development; and (3) the power of the professor within the institute or department.

On the most general structural level, historians of the American university have often noted its "service" orientation—that is, its responsiveness to external interest groups. Private universities had to be flexible and pragmatic about new courses, departments, and faculties, in order to attract both students and philanthropy.[54] Even in state universities, the power of legislatures to oversee university policy constituted a channel by which local interest groups could affect the directions in which such universities developed. Another such channel was the elective system, which left students free to choose many of their courses, even in the major subject. First introduced at Harvard in 1869, electives shifted control over curriculum from staff to students. By honoring the principle of freedom of choice within the university, the elective system also established a mechanism for accountability to a wider public. In research-oriented universities, staff members presumably sought to gear their teaching to their research whenever possible; courses that proved popular with undergraduates enlarged the pool of potential graduate students and were assumed to enhance the teachers' prospects of promotion.[55] In this way student perceptions could affect the fate of new specialties.

Arguably the most important constituencies in the emergence of genetics were agricultural. In Denmark, Sweden, the Netherlands, and Britain, for example, agricultural institutions provided crucial support for the new Mendelism.[56] As we saw in the previous section, the only German institute for genetics before 1945 was at the Berlin Agricultural College, and in the United States (as Charles Rosenberg pointed out many years ago) almost all of the departments of genetics/plant breeding by 1939 were in the agricultural faculties of state universities.[57] Before World War I this agricultural connection had provided

54. Veysey 1965, chaps. 6 and 7; Hawkins 1979; Ben-David and Zloczower 1962, 73–75; Hofstadter and Metzger 1955, 380–83.

55. Veysey 1965, 320 and 233. At the University of Chicago, William Rainey Harper felt that small class enrollments condemned a teacher (Maienschein 1988, 163).

56. On Scandinavia see Roll-Hansen 1978a; on the Netherlands see Robert Visser, paper given to the Anglo-Dutch meeting for History of Science, Amersfoort, April 1987; on Britain see Olby 1989b; Palladino 1990; and Margaret Deacon, "The Institute of Animal Genetics at Edinburgh University: the first twenty years," typescript, n.d., Science Studies Unit, Edinburgh University.

57. Rosenberg 1976, chap. 12. By 1921 there were geneticists working in at least fifteen agricultural colleges (Jones 1921). When Hans Nachtsheim came to the United States in 1926–27 to work in Morgan's laboratory, he proposed to spend two months

American geneticists with institutional settings in which they could meet quasi-professionally and publish (Kimmelman 1983). As early as 1902, at the second International Congress for Plant Breeding and Hybridization in New York, various British Mendelians were astounded by the scale of plant-breeding research then being conducted in the United States' "vast system of experiment stations."[58] The origins of that system went back to the Morrill Act in 1862 and its successor in 1890, which provided federal land to the states for colleges of engineering and agriculture. The resulting "land-grant universities" began to develop faculties of agriculture in which various biological sciences were taught. In the 1890s, for example, the University of Wisconsin established a doctoral program in biology and by the early 1900s had hired its first geneticist.[59] With the passage of the Hatch Act (1887) and the Adams Act (1906), the federal government provided each state annually with funds earmarked for agricultural research, thus granting scientists in experiment stations a measure of autonomy. With federal funds on offer, almost all states quickly established a station.[60] In addition to the stations and land-grant universities, rapidly growing federal expenditure on agriculture made the United States Department of Agriculture another major employer of geneticists. By 1916, one historian has concluded, "no comparable agricultural research organization existed anywhere else in the world" (Dupree 1987, 183).

To hire geneticists is one thing; to allow them to work on problems of basic genetics is quite another. Nevertheless, the available evidence suggests that the first generation of American geneticists entered a sell-

---

visiting "the most significant places" in genetics. Of the twenty-five institutions he listed, all but five were experiment stations, agricultural faculties, or departments of the U.S. Department of Agriculture (Nachtsheim to Wallace Lund, 21.8.26, folder 1254 ["H. Nachtsheim, 1925–27"], box 75, subser. 4, ser. 1, International Education Board papers).

58. C. C. Hurst, quoted in Kimmelman 1987, 54; on this theme more generally see her chap. 3. I estimate that approximately 90 percent of the American participants at the 1902 International Congress were employed at an experiment station, a land-grant university, or the U.S. Department of Agriculture. By the Fourth International Congress at Paris in 1911, this proportion was lower (50 percent), though still significant ("Proceedings of the International Conference" [1904]; de Vilmorin 1913).

59. Cravens 1978, 26–27. Even some private universities founded faculties of agriculture. Harvard's Bussey Institution, for example, eventually became a graduate school of applied biology and a major center of genetic research before 1914 (ibid., 21).

60. According to Kimmelman, there were already more than forty agricultural experiment stations by 1887, and within three years at least half of them were engaged in plant-hybridization work (1987, 20, 29–30). On the early history of the experiment stations, see Rosenberg 1976, chaps. 8–10.

er's market and were able to demand a considerable degree of autonomy from their agricultural employers. Rapid expansion of the agricultural research network between 1890 and 1920 meant that the supply of candidates in relevant applied sciences was insufficient, forcing experiment station administrators to hire those trained in basic sciences, such as genetics (Rossiter 1979, 216–18). After receiving his doctorate in botanical genetics from Harvard in 1904, for example, Alfred Blakeslee looked for a job in which he would not have to teach undergraduate botany. At the experiment station attached to Connecticut Agricultural College he found a post in 1907 that allowed him to develop a center for genetics research. In the same year Raymond Pearl was delighted to find a similar position at the Maine Experiment Station, where he enjoyed the freedom to pursue pure science for eleven years. Although Sewall Wright would have preferred a job at a major university, he accepted a post at the United States Department of Agriculture in 1915. Working conditions there cannot have been too bad; he remained for ten years, turning down offers from the Universities of Maine and Texas. Although L. C. Dunn had never handled a chicken before, he took up a post as "poultry geneticist" at Connecticut's Agricultural Experiment Station in 1920. The station's director gave him the freedom to work in genetics, as long as it had some applications to poultry. For the next nine years Dunn remained there, working on the effects of inbreeding, the inheritance of many traits, and on developmental mutants, turning down many job offers which were unable to improve his working conditions. Although many of the geneticists at land-grant universities worked on maize, they were not engaged solely in varietal improvement. Many made important contributions to basic genetics, not least at Cornell, where Barbara McClintock, Marcus Rhoades, and George Beadle were working in cytogenetics.[61]

The utility which farmers and other agricultural clients perceived in genetics may have helped in recruiting students. For one thing, the fact that genetics was normally taught in *agricultural* faculties probably gave the new discipline access to more students. Of the students at land-grant universities in 1914, 13 percent were studying science while 37 percent were studying agriculture (Caullery 1922, 127). Many of the first generation of graduate students in Wisconsin's Department of

61. The French geneticist Maurice Caullery noted the favorable working conditions in agricultural institutions on his visit to the United States in 1916 (Caullery 1922, 164). On Blakeslee see Cravens 1978, 25–26. On Pearl see Kevles 1980, 452. On Wright see Provine 1986, 98 and 160. On Dunn see Dobzhansky 1978 and Bennett 1977. On maize geneticists see Rhoades 1984 and Keller 1983, 48.

Experimental Breeding came from agricultural backgrounds. A. H. Sturtevant recalled that about 1910 he was excited by an early book on Mendelism, "for I could see that the principles could be applied to the inheritance of colors in the horses whose pedigrees I knew so well." George Beadle's father, a Nebraska farmer, was not convinced of the value of higher education, but finally, in 1922, allowed his son to enroll at the state college of agriculture, where Beadle studied entomology, ecology, and genetics. From a casual survey of forty members of the interwar American genetics community, I found that roughly 28 percent were the children of farmers, a further 15 percent had spent a part of their childhood on farms, another 28 percent were born in small towns in the South or Midwest, 5 percent were born in small towns in the East, and 25 percent were born in major urban areas. Fully three-quarters of this sample, therefore, came from rural backgrounds.[62]

In Germany, however, with the notable exception of Baur's institute at the Berlin Agricultural College, genetics' agricultural connections were quite weak. Although German experiment stations had provided the model for their American counterparts, the German stations' research reputations declined from the end of the nineteenth century. By the 1920s Hans Nachtsheim reckoned that genetic research there was "almost totally lacking."[63] To be sure, by 1930 there were chairs of animal breeding at each of the four agricultural colleges and at all but one of the nine university institutes of agriculture.[64] And agricultural scientists made up a sizable proportion (28 percent) of the membership of the German Genetics Society (versus 39 percent in the Genetics Society of America).[65] But apart from Baur's group, exceedingly few of

---

62. The information was extracted from the *Dictionary of Scientific Biography*, as well as obituaries in *Genetics* and *Biographical Memoirs of the National Academy of Sciences*. On Wisconsin's students see Kimmelman 1987, 275. On Sturtevant see "A. H. Sturtevant—Biographical Notes," typescript, folder 7.2, box 7, Sturtevant papers. On Beadle see his "Recollections" (1974). Before 1914 scientists in experiment stations "often had rural backgrounds and sympathized with agriculture . . ." (Danbom 1986, 253).

63. Nachtsheim to T. H. Morgan, 29.5.25, folder 1254, box 75, subser. 4, ser. 1, International Education Board papers. See also Rossiter 1975; 1979, 242.

64. These chairs were frequently joint with dairy science, and chairs of plant breeding existed at only five of these institutions (*Deutsches Universitätskalender* for winter semester 1929–30; *Minerva: Jahrbuch der gelehrten Welt* for 1930).

65. Nineteen percent of the German Genetics Society worked at agricultural colleges or university institutes of agriculture (versus 26 percent of the Genetics Society of America), and 9 percent of the Germans worked at state agricultural experiment stations (versus 13 percent of the Americans). These figures are based on the membership lists for the German Genetics Society for 1929 (*ZIAV* 54 [1930]: 108–18) and the GSA for

these professors of breeding contributed anything to basic genetics. Despite their numbers, they published very little in either the principal genetics journal (*ZIAV*) or in the main outlet for review papers and monographs in genetics (*Handbuch der Vererbungswissenschaft*).[66] The institutional segregation of genetics from agriculture can also be seen in the committee structure of the Emergency Council for German Science: In 1921 grants in genetics were administered by the Biology Panel, while research in animal breeding was funded by the Panel for Agriculture, Forestry, and Veterinary Medicine.[67]

Just as agricultural scientists played little role in the genetics community, so also were German geneticists far more ignorant of agricultural problems than their American counterparts. On his return from a year in the United States, Nachtsheim concluded that "we are in a rather embryonic stage yet in my country, especially as far as the application of genetics in the field of plant breeding and animal breeding are concerned."[68] To some extent this failure to link theory and practice reflected the family backgrounds and work experience of German geneticists. Whereas 28 percent of American geneticists were the children of farmers (see above), only 2 percent of German geneticists were.[69] Similarly, although as much as one-half of the American genetics community spent some part of their careers working in agricultural experiment stations or the United States Department of Agriculture, less than 20 percent of the German geneticists active during the interwar period

---

1933 (*Records of the GSA* 1 [1933]: 15–29). Members giving private addresses or foreign institutional addresses (including Austria, German-speaking Switzerland, or the German University in Prague) were eliminated from the denominator.

66. Scanning single volumes of *ZIAV* for 1910, 1914, 1919, 1924, 1928, 1932, and 1936, I found a total of two articles from authors working in agricultural colleges or institutes (other than Berlin's). Of twenty-four volumes of the *Handbuch der Vererbungswissenschaft* which had been published by 1938, only one was written by a professor from an agricultural institution.

67. Bestand R73, Akte 119, papers of the *Notgemeinschaft der Deutschen Wissenschaft*. It is not clear how long this funding arrangement continued. If geneticists were still being funded in 1932 by the biology panel rather than that for agriculture, forestry, and veterinary medicine, they were at a clear disadvantage: The former's budget for experimental research was less than half that of the latter (Zierold 1968, 69).

68. Nachtsheim to Wallace Lund, 20.4.27, folder 1254, box 75, subser. 4, ser. 1, International Education Board papers. Unlike the Americans, Nachtsheim felt, German geneticists had hardly any interest in applied genetics (Nachtsheim, "Die Genetik in den USA," ZStA-Merseburg, Rep 87B, Sign 20283).

69. Georg Melchers was the only farmer's son in a sample of forty-three German geneticists whose fathers' occupations were known (Melchers 1987a).

were employed at comparable institutions.[70] For these reasons (and others to be discussed in chapter 8), German geneticists failed to exploit the agricultural connection for the development of their discipline.

The second important structural difference between the German and American university concerns the faculty's role in academic policy-making. In a "business-minded" society whose universities had to sell themselves to prospective students and donors, it is not surprising that American university presidents were entrepreneurial figures with considerable power (Jastrow 1906; Beach 1968; Bledstein 1976, 302–9). Selected by the board of trustees or board of regents (in private and state universities, respectively), presidents were formally accountable to the boards rather than to the faculty for their decisions. Before World War I faculty members were not generally consulted on matters of academic policy, nor were they represented on the board.[71] Since board members were commonly members of the business community and other local notables whose support for academic values was unreliable, the American professoriate felt disenfranchised and particularly resentful during these years. Possessing the power to establish new departments or faculties—in order, perhaps, to accommodate a donor's wishes—and to control staff appointments and promotions, the president was a key ally for any new discipline.

In 1889 the president of Clark University, G. Stanley Hall, convinced of the necessity for a graduate program in biology, simply recruited eight staff members whom he found interesting. Similarly, the

70. For the American sample I consulted the 1938 and 1944 editions of *American Men of Science* for E. M. East, M. M. Rhoades, Barbara McClintock, L. J. Stadler, H. S. Jennings, D. F. Jones, L. C. Dunn, T. H. Morgan, George Beadle, A. H. Sturtevant, H. J. Muller, Calvin Bridges, George Shull, Bentley Glass, R. A. Emerson, W. E. Castle, Sewall Wright, C. B. Davenport, Raymond Pearl, and A. F. Blakeslee. Among the (roughly forty) German geneticists of the interwar period, with whose careers I am familiar, only seven were ever employed at agricultural colleges: Valentin Haecker (cf. chap. 2), Erwin Baur, and five of Baur's assistants (cf. chap. 6). Hans Kappert and Gerta von Ubisch were the only ones to have had industrial experience (in seed companies) prior to taking up academic posts (Kappert 1978, von Ubisch 1956). It is no accident that Kappert was Baur's successor as professor of genetics at the Berlin Agricultural College. Prior commercial experience or service in a government agricultural agency was far more common among those appointed to chairs of animal or plant breeding.

71. It is not clear to what extent this remains the case. According to Kenney, since regulations at the Massachusetts Institute of Technology do not require faculty approval for the establishment of collaborative organizations, the affiliation agreement between the Whitehead Institute (for molecular biology) and MIT was never circulated to the faculty (1986, 51, 53).

presidents of Harvard, Columbia, and the University of Chicago gave particular support to those biologists on their staffs who were working in medically related areas (Pauly 1984). At Cornell University between 1903 and 1913 the dean of the College of Agriculture created departments of plant pathology, plant physiology, agronomy, poultry husbandry, agricultural economics, farm management, pomology, floriculture, agricultural engineering, and experimental plant biology. Creation of the first two enabled the last of these departments to focus more narrowly upon plant breeding, and it was renamed accordingly. At Wisconsin the dean of the Agricultural College mobilized support from an influential dairyman on the Board of Regents (who happened to be keen on scientific breeding) in order to set up the Department of Experimental Breeding (along with five other new departments) in 1909. At Berkeley the dean of the Agricultural College created seven new divisions in 1906, more than doubling the staff within six years. His successor, enthusiastic about plant breeding, established the Division of Genetics in 1913.[72] At Harvard during the 1920s the geneticist E. M. East had to cope with a department head (W. M. Wheeler) who "hates all experimental work." This of itself was not a serious problem, East felt, because "Wheeler can't hurt me except to keep my salary down temporarily." The real danger lay in Wheeler's influence—or anyone else's—with Harvard's president, James Lowell: "Four or five years ago the French government sent that damned fool Blaringhem over to Harvard as exchange professor. Blaringhem told Lowell that my work wasn't good."[73] By 1935 East was more optimistic because Harvard's new president (James Bryant Conant) was "highly sympathetic" (quoted in Provine 1986, 188).

That universities were organized differently in Europe is evident from Maurice Caullery's study of American universities. On the basis of a visit to America in 1916, Caullery concluded that boards had excessive power and that the president tended to be "a tyrant, good or bad."[74] German universities had nothing corresponding to a board of trustees, and their rectors (who were academics elected by their col-

72. Kimmelman 1987 (on Cornell, 202, 208; on Wisconsin, 248–53; on Berkeley, 300–303, 308–10, 317–20).
73. East to Richard Goldschmidt, 17.6.24, 23.8.24, and 28.7.24, Goldschmidt papers.
74. Caullery 1922, 41–52 (on p. 46). For Bertrand Russell's critique of the power of American university presidents, see Feinberg and Kasrils 1973 (vol. 1), 132–33, 310–11, 314.

164   The Peculiarities of German Genetics

leagues) possessed few powers. Decisions on appointments were made by the faculty concerned, and requests for new posts or institutes were negotiated directly between the relevant faculty and the Ministry of Education. The German faculty was thus far more powerful than its American counterpart, and it made the key decisions on the development of new disciplines.[75] And as we saw in the introductory chapter, the early-twentieth-century German professoriate was distinctly uneasy about specialization. Some argued against the establishment of a second chair in a given discipline on the grounds that it would upset the balance of power in small faculties; others voiced ideological reservations about specialization (see chap. 8). But there is little doubt that an important underlying cause of the professoriate's resistance was straightforwardly mercenary. Although (full) professors (*Ordinarien*) were salaried, a substantial proportion of their income came from fees of various kinds: capitation fees (levied upon each student who enrolled in their courses), laboratory fees for each student occupying bench space, examination fees (e.g., for each medical student who had to pass an oral examination in zoology or botany), and doctoral examination fees. Fee income could vary widely from one discipline, university, or individual professor to the next, depending on their access to students, but it was not unusual for fees to double a science professor's income.[76] When it came time to decide who would teach which courses in a discipline, it comes as no surprise that the professor chose to teach the well-attended introductory courses, leaving the "less important" spe-

75. On the administrative structure of the German university, see Spranger 1930. My portrayal simplifies the German situation to some extent, since officials in the ministries of education possessed the formal power to overrule faculty recommendations as well as to press for particular academic innovations. Friedrich Althoff, the relevant official in the Prussian ministry from 1882 to 1907, was extremely effective in imposing such changes (vom Brocke 1980). But during the period 1870–1933, Althoff seems to have been the exception rather than the rule. In Baden during the 1860s the minister of education made many such decisions against the faculty's vote, but his successors between 1868 and 1915 were much more cautious about doing this (Riese 1977, 286). In any event, Max Weber's comment in 1911 is apt: "The United States have an Althoff at every university. The American university president is such a man" (Weber 1973, 599).

76. Forman, Heilbron, and Weart (1975, 2) and Kelly (1973, 4) estimate that over the period 1900–1914 the typical professor's income was about equally derived from salary and fees. At Wurzburg in 1912, for example, Theodor Boveri's fee income added 8,000 marks to his salary of 10,000 (Paul Weindling 1991, 179). Occasionally professors (usually in large law or medical faculties) received as much as 80 percent of their income from fees (Flexner 1930, 355). In Berlin's philosophical faculty in the 1880s the doctoral examination fee was 340 marks, nearly 10 percent of a full professor's salary (Burchardt 1988, 184–85).

Demarcating the Discipline 165

cialist courses to junior staff.[77] As critics of the university's institutional rigidity pointed out, the fees system tended to block the establishment of new institutes. Faculty membership before 1918 was restricted to full professors, and no professor was likely to promote the creation of a specialist chair/institute in his discipline since it would compete with his own institute for both students and financial resources. If a new specialty emerged, it was intellectually preferable as well as cheaper, professors argued, to provide for it within an existing institute through the creation of a junior post and/or a new subdepartment (*Abteil*).[78] Since ministries of education were trying to economize, even before 1914, they usually concurred.

However much American department heads might have liked to block the formation of new departments in similar fashion, they simply lacked the power to do so. Even in the face of faculty opposition, therefore, new departments of genetics could be established by the university president's decision, as long as there was a likely demand for their teaching/research and influential backers could be recruited to finance it. And within the confines of their own departments American geneticists were free to define their discipline as narrowly as they wished. Nonetheless, although such departments were far more common in the United States than in Germany, it should not be forgotten that less than 10 percent of American geneticists worked in genetics departments. Nearly 60 per-

77. "Less important" was how Eduard Spranger described the subjects taught by *Extraordinarien* (1930, 21). At Jena between 1921 and 1932, for example, the professor of botany (Otto Renner, a geneticist) taught courses in general botany and in plant systematics five times, but a course on reproduction and inheritance only once (*Deutsches Hochschulkalender*). At Göttingen in 1937–38 the professor of zoology (Karl Henke, a geneticist) taught the introductory zoology course (to ninety-five students), while his junior staff taught comparative vertebrate anatomy (fourteen students) and evolutionary theory (six students). In the spring, to be sure, Henke did teach genetics, leaving insect biology—about which Henke knew a great deal—to a junior colleague, perhaps because the former attracted forty-eight students while the latter attracted none (Henke to Kurator, 11.10.38, University of Göttingen Archive, AZ: XVI.V.C.v.1.II.).

78. Where very large institutes were organized into several divisions (as in chemistry), each headed by an assistant who supervised students' research and taught in that area, the professor was still entitled to all of the laboratory fees (Bock 1972, 216; Riese 1977, 222–23). On the very high fees enjoyed by chemistry professors and their resistance to the creation of new chairs which could jeopardize that income, see Burchardt 1975, 108–9; Riese 1977, 113, 116; Kohler 1982, chap. 2; and Johnson 1985, 517–18.

Prussian university reform in 1923 opened up faculty decision making to wider participation, but since the junior staff in each faculty were only allowed to elect three representatives to vote at faculty meetings, the professors still constituted a large majority (Dingler 1930, 216).

cent worked in departments of zoology, botany, or biology, where many more jobs were available.[79] Even in these latter departments, however, younger American geneticists enjoyed much more autonomy than their German counterparts. To appreciate this we must turn to the third important structural difference between German and American universities: the power of the professor within the department or institute.

A number of American graduate schools, notably Johns Hopkins, had drawn upon the model of the German university. They borrowed, however, selectively. One departure from the German model was the department system. The department tended to disperse power among its members, partly because of the presence of several full professors in larger departments, but also through the election of chairpersons, a practice initiated by the University of Chicago as early as 1911. Where capitation fees were levied, they went into the university's general fund rather than the professor's pocket. Large introductory courses thus possessed few charms for full professors and were left to younger staff to teach, sometimes on a team basis. Each staff member, however, offered one or more advanced courses based upon his or her specialty, and as long as a course attracted students, it was retained in the curriculum, whether it addressed "big" problems or not. Although new departments were created less frequently after 1900, the structure of the department—as a relatively nonhierarchical aggregate of specialists—allowed it to grow incrementally, incorporating new fields as necessary.[80]

At Johns Hopkins, for example, the head of the biology department (H. Newell Martin, a physiologist) envisaged a department which would contain representatives not only from physiology, but also from

79. At the fifty-two American universities listed in the 1930 edition of *Minerva: Jahrbuch der gelehrten Welt*, there were 13 chairs of genetics or plant breeding and approximately 150 chairs in botany plus zoology. Accordingly, 8 percent of the members of the Genetics Society of America worked in departments of genetics versus 8 percent in botany, 13 percent in biology, 12 percent in applied biology, 6 percent in medical sciences, and 20 percent in zoology (*Records of the GSA* 2 [1933]: 15–29). At land-grant universities, genetics was also taught in departments of animal husbandry, horticulture, and agronomy, among others (Kimmelman 1987, 215, 257–58, 314, 364).

80. Veysey 1965, 59 and 321–24; Bledstein 1976, 300–301. On capitation fees see Forman, Heilbron, and Weart 1975, 45. Americans studying at German universities were often surprised to find that the professor gave the introductory courses (Jarausch 1988, 189). On his arrival at Chicago in 1925, Sewall Wright found that all of his colleagues in the Department of Zoology preferred graduate to undergraduate teaching. At first Wright taught various introductory courses, but by 1932 all of his courses were specialist ones (Provine 1986, 168–74).

comparative anatomy and zoology, human anatomy, botany, and biology. In 1876 he hired W. K. Brooks, a younger morphologist, to help teach the introductory course. He allowed Brooks freedom to develop a research school of morphology, in accord with student demand. Within two years Brooks had persuaded the university's president and trustees to found a marine zoological laboratory and was offered his own specialized courses in morphology and embryology. Soon Brooks and Martin had together organized a departmental journal club and were jointly editing a house journal. By 1891 Brooks was promoted to full professor and had produced as many research students as his department head (Benson 1985; 1979, 72–102; Maienschein 1987, 178). If Hopkins proves to have been typical, the department system certainly did not hinder specialization.

German visitors to the United States during the 1930s agreed. While Karl von Frisch and Fritz von Wettstein admired the opportunities which talented young scientists enjoyed within the departmental structure, they expressed reservations about its tendency to produce "narrow over-specialization."[81] The structure of the German institute had quite the opposite effect. Since the professor was normally appointed head of the institute by the ministry of education, rather than elected as chair by his peers, he was not accountable to his colleagues in the institute,[82] and they remained dependent upon him in many respects. On completing the doctorate, for example, young academics had to find an institute where they could conduct postdoctoral research for several years. Only then, assuming they had the support of the institute's director, could they qualify (via the process of *Habilitation*) as unsalaried university lecturers (*Privatdozenten*). This promotion did not, however, greatly increase their independence, since the right to teach was valid solely within the granting university. And although a *Privatdozent* was guaranteed access to teaching rooms within the university, the same did not apply to laboratory space or equipment. Thus unless a *Privatdozent* was wealthy enough to set up his own private laboratory—which occurred fairly often in the mid-nineteenth-century but was usually prohibitively

81. Von Frisch 1967, 116 (cf. 115–18); von Wettstein, "Über deutsches und amerikanisches Forschen," in *Das Jahrzehnt 1930–1940 im Spiegel der Arbeit des Stifterverbandes der Deutschen Forschungsgemeinschaft (Notgemeinschaft der Deutschen Wissenschaft)* (1940), 171–79, typescript, Archiv zur Geschichte der Max-Planck-Gesellschaft, Sign: B1621.

82. Nor was a professor responsible to the science faculty for the way in which he ran his institute. This is because the status of "professor" (as a member of the science faculty) was formally separate from that of "institute director" (which was negotiated between the professor and the ministry) (Riese 1977, 222).

168   The Peculiarities of German Genetics

expensive by the early twentieth—he was obliged to cultivate favor with the professor.[83]

The only income to which the *Privatdozenten* were automatically entitled were the capitation fees from their lectures, but since the specialized courses which they taught attracted relatively few students, such fees were hardly adequate. In order to survive financially, those *Privatdozenten* lacking independent wealth had to find additional income. Few of them were fortunate enough to obtain teaching contracts (which would have allowed them a modicum of independence).[84] Many were forced to take posts in the institute as assistants (*Assistenten*), helping the professor with teaching, research, or the administration of a museum or botanical garden.[85] The assistant often had a heavy workload on a salary far lower than the professor's, and their contracts could be terminated on six months' notice.[86] Although they were entitled to

---

83. On the *Privatdozent*'s lack of entitlement to laboratory space, see Burchardt 1988, 163, 168. The high cost of antagonizing the professor explains why *Privatdozenten* avoided offering general or introductory courses, even though they were formally entitled to teach whatever they liked within their discipline (Eulenburg 1908, 114).

In the zoological institute at Würzburg, Theodor Boveri did not allow "his" *Privatdozenten* to supervise doctoral students' research (Baltzer 1918, 59). Even where *Privatdozenten* could have their own research students, they were not entitled to take part in the students' doctoral examinations at Prussian universities until 1924. Even then, the professor was also allowed to examine the candidate (C. H. Becker to Dr. Wellmann, 20.6.24, ZStA-Merseburg, Rep 76, Va, Sekt 1, tit IV, Nr 61). For a full account of the professional disadvantages and indignities suffered by nonprofessorial staff see Nauck's 1937 biography of the embryologist Franz Keibel.

84. In 1924 less than 20 percent of *Privatdozenten* nationally held teaching contracts, although the proportion was substantially higher at the new universities (Remme 1926, 28). In 1931, 15 percent of the *Privatdozenten* in Berlin's philosophical faculty (which included the sciences) held teaching contracts (*Kalender der deutschen Universitäten und Hochschulen*).

85. In 1907, 43 percent of all *Privatdozenten* in the sciences (and 50 percent of those in medical fields) were assistants (Eulenburg 1908, 65). According to Nauck (1937, 86), *Privatdozenten* in anatomy in the mid-1930s normally held assistantships. Ferber's data indicate that 37 percent of *Privatdozenten* in the biological sciences about this time were assistants (von Ferber 1956, 87–88). The proportion employed as assistants was probably substantially higher by the 1920s. After the war there were articles in the press about the financial crisis among *Privatdozenten,* partly because those who had lived off independent wealth before 1914 were ruined by the war and subsequent inflation. Calls for at least more senior *Privatdozenten* to receive a salary had had little impact by 1930 (Dingler 1930; Nauck 1956, 55; Spranger 1930, 24). The *Privatdozent* did not become a salaried civil servant until 1939 (Seier 1988, 260).

86. In 1908 the salary of an assistant in Baden was increased to a maximum of 1,500 marks, versus approximately 4,000 for a professor (Riese 1977, 170–72; on contracts see 162). By the early 1930s the assistant's salary in Bavaria started at 3,000 marks (Karl

spend part of each day on their own research, that research had to be "in the interests of the institute," and some assistants were forbidden by the professor to work on certain topics. Others were required to obtain prior approval from their professors before submitting manuscripts for publication, and in some cases the assistant's work was published under the professor's name.[87]

Of course, promising *Privatdozenten* could be promoted—assuming they enjoyed the support of their professors—to the post of associate professor (*Extraordinarius* or *ausserordentlicher Professor*). But promotion solved few of their problems. For one thing, since promotion usually conferred nothing more than a title, the *Extraordinarius* was frequently obliged to remain in the professor's employ as an assistant.[88] Even that minority of *Extraordinarien* who enjoyed a salary and tenure, however, were not fully independent within the institute. Although the *Extraordinarius* was entitled to supervise doctoral candidates, he or she was not allowed to take part in the student's oral examination until 1924. Similarly, while the *Extraordinarius* in principle had a right to research facilities, in practice these were allocated at the discretion of the professor.[89]

---

von Frisch to Curt Stern, 12.1.33, Stern papers), but by then a professor's salary was in the vicinity of 10,000 marks.

87. On the general subordination of assistants and *Privatdozenten* to the professor, see Bock 1972, 142–47, 185–91; Busch 1959, 126. On particular cases see Forman, Heilbron, and Weart 1975, 53, 125; Lang 1980, 6. On assistants in chemistry who were not allowed to publish under their own names, see Fruton 1985, 318; Burchardt 1980, 335.

88. On the professor's influence upon promotions see McClelland 1980, 268. In 1924 three-quarters of the *Extraordinarien* in German universities had only the title, and at several universities the proportion was over 90 percent (Remme 1926, 28). In 1931 approximately 90 percent of the *Extraordinarien* in botany and zoology at German universities had only the title (von Ferber 1956, 209).

In 1907 about one-third of the *Extraordinarien* in the sciences held assistantships (Eulenburg 1908, 65). Two of the three *Extraordinarien* in zoology at Heidelberg around 1900 had to take up assistantships (Goldschmidt 1956, 9, 11). In 1910 over half of the *Extraordinarien* in botany and zoology held assistantships (von Ferber 1956, 73).

89. On the restrictions faced by *Extraordinarien* for theoretical physics, see Riese 1977, 147–48. As a tenured *Extraordinarius* at Leipzig until 1909, Correns had difficulty in getting the professor of botany to grant him access to a plot of land in the botanical garden so that he could conduct breeding experiments (Goldschmidt 1950, 314). George Shull was surprised at how little space Correns had, compared to his own facilities at the Station for Experimental Evolution at Cold Spring Harbor (Shull, "Report on a European Tour" [1908], Shull papers).

During the years prior to the First World War, German nonprofessorial staff organized to press for improved working conditions (Riese 1977, 153–92; vom Bruch 1980,

Erwin Baur's career illustrates some of the frustrations of junior staff. When he came to Berlin in 1903 as senior assistant in the Institute of Botany, Baur had to take on a great deal of administrative work, since his seventy-five-year-old professor did nothing beyond lecturing. Though the professor gave him free rein to run the institute, Baur lost enthusiasm for doing a professor's job on one-quarter of a professor's salary and was happy to turn over some of the burden to a new assistant in 1908. Since the institute's garden was inadequate even for teaching purposes, Baur decided in 1905 to lease a tract of land and some greenhouse space near his house in order to conduct his genetic research with fewer constraints. When a new professor was appointed in 1910, Baur (aged thirty-five) was uneasy about his future in the institute. Although he was to be given the title of *Extraordinarius,* he knew that he was still basically an assistant, "and I am fed up with being an assistant, especially because the construction of the new institute is probably going to entail a lot of work."[90]

Baur was not the only young geneticist who was desperate for independence.[91] But achieving independence in the German university required a chair, and chairs of genetics were extremely rare. This undoubtedly discouraged many young biologists from working in genetics at all. Carl Correns, Baur felt, was a good example: Despite the quality of his work, he had been passed over for chairs in botany several times in favor of lesser lights. At forty-three, Correns was so discouraged with his prospects that he considered switching from genetics back to plant physiology, where his chances for advancement would have been far greater. "Experimental genetics," Baur concluded, "is not in very high esteem, especially in Germany, so that it is rather risky for a young botanist to concentrate on this specialty."[92] But for those

---

122–28). Despite certain reforms achieved in Prussia in 1910 and in Germany as a whole in 1919, the overall extent of structural change was "nearly insignificant" (Ringer 1969, 76–77).

90. Baur to Carl Correns, 31.12.09 and 25.7.10, Sammlung Darmstädter; von Rauch 1944, 77–93.

91. Richard Goldschmidt felt the same. After several years as an assistant (aged thirty-two), running the zoological institute in Munich and looking after all of the professor's doctoral students, "I longed to be my own master" (Goldschmidt 1960, 68). Aged forty-seven and still an assistant in the Botanical Institute at Heidelberg, the geneticist Gerda von Ubisch was anxious to secure an independent post (L. von Ubisch to Richard Goldschmidt, 22.7.29, Goldschmidt papers).

92. Baur to C. Schröter, 22.11.08, quoted in von Rauch 1944, 234. On Correns's discouragement, see Correns to C. B. Davenport, 7.2.07, Davenport papers. Between the wars Otto Renner often complained about the small numbers of young people entering

Demarcating the Discipline 171

willing to take the risk, the best chances were in botany or zoology, where a substantial number of chairs became vacant during the 1920s and early 1930s.[93] Since German professors were expected to teach the introductory courses single-handedly and supervise a wide range of dissertation topics, however, a great deal of emphasis in appointments was placed upon the breadth of a candidate's knowledge of botany or zoology.[94] It was thus a career imperative for young geneticists to remain as broad as possible. At the Botanical Institute in Freiburg, Oehlkers urged his *Privatdozenten* to offer courses in a variety of fields in order to enhance their appointability. In the botanical institutes at Berlin and Heidelberg other *Privatdozenten* working in genetics did the same. By 1927 Goldschmidt's student Jakob Seiler reckoned that unless he could find a job in genetics, he would either have to set about broadening his knowledge of zoology or return to schoolteaching. Hans Grüneberg's situation was similar. By the time he received his doctorate in 1929, he

---

genetics (Renner to Richard Goldschmidt, 1.5.24, Goldschmidt papers; Renner to Ralph Cleland, 2.1.34, Cleland papers). His own students much preferred dissertation topics in plant physiology to those in genetics, presumably because jobs in botany would be easier to obtain (Renner to Cleland, 5.1.30, Cleland papers). Perhaps this is why Hans Kappert was allegedly the only one of Correns's doctoral students whose dissertation was in genetics (Kappert to ?, 17.1.66, box 2, Kappert papers).

93. In the mid-to-late 1920s there were chairs of zoology available at Tübingen, Breslau, Halle, Marburg, Münster, Munich, Berlin, and Cologne, and two in Vienna (Karl Belar to Richard Goldschmidt, 10.6.25, and Max Hartmann to Goldschmidt, 26.4.25, both in Goldschmidt papers, plus various sources). In the late 1920s and early 1930s chairs in botany became vacant at Halle, Stuttgart, Darmstadt, Frankfurt, Berlin, Erlangen, Freiburg, Göttingen, and Munich (Carl Correns to Steinbrinck, 26.12.31, Steinbrinck papers, plus various sources). Throughout the interwar period only one chair of genetics became vacant: at the Berlin Agricultural College, when Baur moved to the KWI for Breeding Research (cf. chap. 6).

94. See, for example, Hans Spemann's recommendation for H. Braus (dated 11.4.19, Spemann papers); Carl Correns's recommendation for Fritz von Wettstein (Correns to Spemann, 27.6.29, Spemann papers); von Wettstein's recommendation for Theodor Schmucker (von Wettstein to Reichserziehungsminister, 25.5.35, Geheimes Preussisches Staatsarchiv, Rep 76, Sign 88); Spemann's discussion of the candidates for Göttingen's chair of zoology (Spemann to Ernst Ehlers, 18.11.19, Ehlers papers); Alfred Kühn's judgment of Karl von Frisch (Kühn to Franz Doflein, 2.4.21, Kühn papers/MPG-Archiv); and countless obituaries (cf. chap. 5).

At Göttingen during the early 1920s the faculty of science sought to fill two botanical chairs, one of which had been occupied by a systematist and the other by a "general" botanist. Candidates whose research and teaching were "broad" (*vielseitig*) were regarded as potentially appointable to a chair; "specialists" were said to be more appropriate for an *Extraordinariat*. Eventually the geneticist Fritz von Wettstein was appointed to the chair because of his breadth (University of Göttingen Archive, AZ: Math-Naturw Fak. 18).

had already published ten papers, and his work on *Drosophila* had attracted the attention of Calvin Bridges and N. W. Timofeeff-Ressovsky. But the best position he could find was as an assistant in anatomy at Freiburg. In offering him the job the professor of anatomy acknowledged Grüneberg's reputation as a promising young geneticist but made it clear that the appointment could only be made on the condition that Grüneberg commit himself to anatomy and redirect his research appropriately.[95]

The same pattern of institutional dependence is evident in some of the German organizations which funded genetic research. At the Prussian Academy of Sciences, for example, one grant proposal in genetics was refereed in 1921 by the Academy members Karl Heider (1856–1935) and Willy Kükenthal (1861–1922).[96] Although both recommended funding the proposal, neither worked in genetics or in any other area of experimental biology.[97] Admittedly, the Academy was a much less important source of funding for genetic research between the wars than the Rockefeller philanthropies or the Emergency Council for German Science. But in the latter, at least during the 1920s, genetics remained institutionally dependent upon botany and zoology. Grant proposals in basic genetics were dealt with by the council's Biology Panel. Although the panel seems to have looked favorably upon genetics, its members were nominated by the German Zoological and Bo-

---

95. The geneticist Cornelia Harte, one of Oehlkers's *Privatdozenten* in Freiburg during the 1940s, taught plant morphology, cytology, and statistics (interview with Harte). In Berlin, Friedrich Brieger taught not only genetics, but also ecology, developmental physiology, and bacteriology (Brieger file, Society for the Protection of Science and Learning). In Heidelberg, Gerda von Ubisch taught not only genetics, but also evolution, plant physiology, and plant development (von Ubisch file, Society for the Protection of Science and Learning). On Seiler's career situation, see Seiler to Goldschmidt, 11.12.27, Goldschmidt papers. On Grüneberg's career see W. von Mollendorf to Hans Grüneberg, 17.10.32, and Grüneberg's curriculum vitae, both in Grüneberg papers.

The principle was a very general one; at Würzburg, Theodor Boveri urged his student Hans Spemann to choose dissertation and *Habilitation* topics in descriptive (rather than functional) invertebrate morphology because there were many more jobs in zoology than in anatomy (Lynn Nyhart, paper presented at a meeting of the History of Science Society, Raleigh, North Carolina, 31.10.87).

96. The proposal was submitted by Paula Hertwig; see sheet dated 13 and 14 Feb. 1921, Sign II-VIIb2, Bd 90, papers of the Preussische Akademie der Wissenschaften.

97. Although sympathetic to the Mendelian chromosome theory, Heider—then professor of zoology at Berlin—worked primarily on the descriptive embryology and phylogeny of invertebrates (Kühn 1935; Ulrich 1960). Kükenthal, at that time director of the Natural History Museum at the University of Berlin, was also supportive of experimental biology and genetics, but his own research had been in descriptive embryology, systematics, and comparative anatomy (Jahn et al. 1982, 694–95).

tanical Societies throughout the 1920s, and a number of critics objected that panel members were too old.[98]

Under these circumstances, it becomes easier to understand why Baur and Correns chose a systematic botanist and a paleontologist—both of them neo-Lamarckians—as coeditors of the first German genetics journal; or why two eminent nongeneticists were among the first three presidents of the German Genetics Society; or why an aging systematist was invited to deliver the keynote address at the International Congress of Genetics in Berlin in 1927.[99] Given the new discipline's institutional weakness, it was plainly expedient to choose such figureheads. But more important, it should by now be clear why geneticists focused their research upon development and evolution, the classic problems of established disciplines. Lacking the institutional resources and autonomy enjoyed by their American peers, German geneticists were much less free to define the scope of their discipline as narrowly as they might have wished.

It might be objected, however, that this argument lays too much importance upon the structure of power within the universities. After all, two of the most important centers of genetic research at this time lay outside the universities: in the KWI for Biology and the KWI for Breeding Research. Surely the Kaiser-Wilhelm Society was established precisely in order to get around the universities' institutional conservatism, providing for those specialties which the universities had ne-

98. Records of the Biology Panel's funding decisions in 1927–29 and 1933–34 indicate that about 20 percent of the grant proposals approved were in genetics; no proposals in genetics were rejected (Notgemeinschaft papers, R73, folders 107 and 117). But the German Genetics Society was not consulted at the first election of panel members in 1922. It is not clear from the records whether the Society was consulted at the next election (which did not take place until 1929 [Zierold 1968, 60]), but it was consulted in 1931 about the 1933 election (Notgemeinschaft papers, R73, folders 123, 125, and 130). By the 1930s grant proposals in genetics were being routinely refereed by geneticists (see the grant applications submitted by Hans Nachtsheim, Peter Michaelis, and Georg Melchers: ibid., folders 13328, 13149, and 13053, respectively).

The 1929 election was held in response to widespread criticism from younger academics that the Council was dominated by "a clique of old men" (Zierold 1968, 61, 110, 114, 150). For complaints about the age of Biology Panel members, see the draft letter from Schmidt-Ott to Rektoren, 19.11.27 (Notgemeinschaft papers, R73, folder 125) and E. Schaffnit to Notgemeinschaft, 14.10.32 (ibid., folder 131).

99. The presidents of the German Genetics Society in 1922 and 1923 were the botanist Richard von Wettstein (aged fifty-nine) and the zoologist Richard Hertwig (aged seventy-three) (see the Society's proceedings in *ZIAV,* vols. 30 and 33). On *ZIAV*'s founding editors, see chap. 3. It was not until 1935 that *ZIAV*'s editorial board consisted exclusively of geneticists (*ZIAV* 68).

glected?[100] And surely the Society's institutes were less hierarchically organized and thus more conducive to specialization? At first sight the Institute for Breeding Research seems to sustain this objection: It was a stronghold of transmission and agricultural genetics, resembling many American institutions (cf. chap. 6). On reflection, however, it is clear that the Kaiser-Wilhelm institutes failed to provide an alternative career structure that would have allowed geneticists to escape the constraints of the university system. For the KWIs were conceived as temporary institutions, tailor-made for talented senior scientists. To ensure flexibility, the Society offered no guarantees that a given institute would continue after its head retired, nor that nontenured staff would be redeployed in related institutes.[101] The assistantships were conceived as short-term posts where young scientists would concentrate upon research for two to three years before taking up academic posts.[102] Thus virtually all of the young geneticists working in the KWI for Biology became *Privatdozenten* and offered courses at the University of Berlin.[103]

Moreover, if one looks at the way in which the Institute for Biology was established, it is evident that the Kaiser-Wilhelm Society's failure to counteract the rigidity of the universities was more fundamental. From the start both the Prussian Academy and professors at the University of Berlin had expressed strong opposition to the creation of the Society, seeing it is as a competitor for research funding. In response,

100. Wendel 1975, 170–85. Unfortunately I have not been able to draw upon an extensive collection of essays which came to my attention only after completion of this book: Rudolf Vierhaus and Bernhard vom Brocke (eds.), *Forschung im Spannungsfeld von Politik und Gesellschaft: Geschichte und Struktur der Kaiser-Wilhelm-/Max-Planck-Gesellschaft* (Stuttgart: Deutsche Verlags-Anstalt, 1990).

101. At the Institute for Breeding Research in 1928 only the director, Erwin Baur, held a tenured position (i.e., was a *Wissenschaftliches Mitglied* of the Kaiser-Wilhelm Society). At the Institute for Biology there were only two tenured posts for geneticists (or three if one counts Max Hartmann as a geneticist), since only department heads enjoyed tenure (Harnack 1928).

102. Glum 1928; Glum to Dr. Leist, 28.1.33, and Max Planck to Preussischer Kultusminister, 14.3.33, both in ZStA-Merseburg, Rep 76, Vc, Sekt 2, Tit 23, Litt A, Nr 112, Bd II.

103. Fritz von Wettstein, Victor Jollos, Karl Belar, Joachim Hämmerling, and Curt Stern were all *Privatdozenten* in botany or zoology at the University of Berlin (*Minerva: Jahrbuch der gelehrten Welt* for 1930; *Kalender der deutschen Universitäten und Hochschulen* for 1924–25 and 1931–32). Richard Goldschmidt had always told Stern that an academic post was desirable, and in 1928 Stern was prepared to hang on at the KWI for Biology for a decade, if need be, until he had obtained such a post (draft letter from Stern to Goldschmidt, attached to a draft letter from Stern to Karl Frisch, 30.1.33; and Franz Schrader to Stern, 30.6.28, both in Stern papers).

Demarcating the Discipline 175

both the Prussian Ministry of Education and the first president of the Society, Adolf Harnack, tried to buy off the opposition, stressing that the Society's activities would be closely coordinated with the Academy and University and that the Academy would have some say over the running of Kaiser-Wilhelm institutes.[104] The Institute for Biology, for example, ran into considerable opposition from Oscar Hertwig, professor of anatomy in Berlin. Hertwig was critical of experimental embryology (for which a department had been planned), and he resented the fact that the man likely to be appointed director of the institute (Theodor Boveri) would be working on problems similar to his own but enjoying much better research facilities. Apparently in order to sweeten Hertwig, the Society gave him a position on the Institute for Biology's supervisory committee (*Kuratorium*).[105] Since the Academy had been granted the statutory right to appoint a representative to the institute's supervisory committee, Hertwig was succeeded in 1920 by another aging Academy member, Karl Heider, who served on the committee throughout the 1920s. The institute's director was to be appointed by the supervisory committee, on the advice of an academic advisory committee (*Wissenschaftlicher Beirat*), and the latter was also entitled to advise on and approve the appointments of department heads. At the institute's founding in 1914 the advisory committee was dominated by aging zoologists and botanists—many of whom, in Correns's view, had no competence in experimental biology—representing the University of Berlin, the Prussian Academy, and three other academies of science. Correns and his codirector, Hans Spemann, were worried about the supervisory committee's power and argued that an advisory committee was unnecessary, but they succeeded only in getting some of the latter's powers curtailed.[106] Thus an older, traditionally oriented generation of

---

104. Wendel 1975, 116, 125–34; Burchardt 1975, 113–14. The best general account of the structural constraints upon the Society's institutes is Johnson 1990, 159–65.

105. Wendel 1975, 175, 331; Goldschmidt 1956, 79–80; Paul Weindling 1991, 178, 181–82. On the eve of the institute's founding, there were hopes that Richard Hertwig might be able to persuade his brother, Oscar, to calm down (Boveri to Goldschmidt, 3.12.12, Goldschmidt papers).

106. On the composition of the management and academic advisory committees see Nr 1551, 1552, and 1553, Rep 1A, Abt I, MPG-Archiv; Harnack 1928, 194–95. On the powers of these committees see "Satzung des KWIs für Biologie," Nr 1533, Rep 1A, Abt I, MPG-Archiv. On the advisory committee's right to approve the director's appointment of department heads, see Emil Fischer to Kaiser-Wilhelm Gesellschaft, 12.4.19, Fischer papers; minutes of the management committee meeting, 17.12.20, Nr 1553, Rep 1A, Abt I, MPG-Archiv. On Correns's fears of these committees' powers and his

botanists and zoologists was represented on both of the committees which oversaw the activities of the KWI for Biology.

Finally, there is good evidence that the Kaiser-Wilhelm Society's failure to negotiate a more favorable relationship to the University of Berlin was a major reason why Theodor Boveri eventually declined to become first director of the institute.[107] In his early proposals for the organization of the Institute for Biology, Boveri emphasized the importance of close ties to the University of Berlin. In order that students could be attracted to experimental biology, the institute's director should be a professor in the philosophical faculty and should have the right both to examine those students and to propose candidates for *Habilitation* to the faculty. A few months later Boveri discussed these proposals with Berlin's professor of zoology, Franz Schulze (1840–1921), but was concerned by the latter's reaction. It looked as though the candidates whom Boveri put forward would have to habilitate in a *specialized* area of zoology, which would jeopardize their chances of subsequently getting a chair of zoology. On the other hand, if Boveri's protégés were to habilitate in *general* zoology, then Schulze (as professor of zoology) would insist on the right to recommend them to the faculty. Since Schulze's research had been in comparative anatomy and descriptive embryology of a very traditional kind, Boveri was worried that experimental biologists would have difficulties habilitating in general zoology. When he asked Emil Fischer (who was handling the negotiations on behalf of the Kaiser-Wilhelm Society) to try to change Schulze's mind, Fischer tried to assure him that there was no problem, adding, "We do not want to antagonise either the Faculty or the Academy at present. Neither will tolerate any pressure via the Ministry of Education or any other route, I can assure you. So please do not take

---

limited success in reducing them, see minutes of the meeting of the management committee on 3.12.14 (Nr 1552, Rep 1A, Abt I, MPG-Archiv) and the draft constitution of the institute which was discussed at that meeting ("Satzung des KWIs für Biologie" [Entwurf], folder: "KWIB 1913–1918," box 1, Emil Fischer papers).

107. To my knowledge, this was first pointed out in 1918 by Boveri's colleague at Würzburg, Wilhelm Wien (Wien 1918, 154–55). Most recent historical accounts of the founding of the KWI for Biology, however, have made surprisingly little of it. Burchardt cites Boveri's failing health as the main reason why he resigned (1975, 115). Others are aware of Boveri's worries about the institute's relation to the University of Berlin, but place little or no emphasis upon them (Wendel 1975, 181, 337; Weindling 1991, 178–81). In a paper devoted to another topic, Jacobs singles out the relationship to the University as the most important reason for Boveri's resignation, but discusses the issue only briefly (Jacobs 1989, 240–41). The most detailed account so far is Johnson 1990, 164–65.

the issue to the Ministry."¹⁰⁸ In his letter of resignation a few months later, Boveri again made reference to "an issue which is very important to me, namely the relations between the Institute and the University."¹⁰⁹ Although the expressed intention of the Kaiser-Wilhelm Society was to aid those newer specialties which the universities had neglected, therefore, the Society's institutes were in fact conceived as little more than *feeders* of the university system rather than genuine alternatives to it. As a result, very few geneticists were able to make a career in the KWI for Biology.¹¹⁰

4.3 Conclusion

Joseph Ben-David's pioneering work has demonstrated the impact of institutional structure upon specialty formation and upon rates of growth of science in various countries since the mid-nineteenth century. But institutional context, he believed, can only help or hinder scientific advance; it cannot shape it. Thus the hierarchical structure of European institutes was "alien to the pursuit of science." The intellectual substance of science is "determined by the conceptual state of science and by individual creativity—and these follow their own laws, accepting neither command nor bribe" (Ben-David 1968, 36; 1971, 12). Com-

---

108. Emil Fischer to Theodor Boveri, 3.2.13, Fischer papers, box 4, microfilm 1. In this letter Fischer also suggested that Boveri ought not to insist upon being present at his own students' doctoral examinations; if Schulze gave them difficulties, they could always transfer to another university!

109. Boveri's plans for the institute (dated 25.9.12) are appended to a letter from Adolf Harnack to Emil Fischer, 4.11.12, folder: Harnack no. 1, box 19, Fischer papers. On Boveri's worries about Schulze, see Boveri to Fischer, 24.1.13, Fischer papers. On Schulze see Tembrock 1958–59, 193. Schulze was one of those Berlin professors who wanted the Kaiser-Wilhelm institutes to be set up under the wing of the Prussian Academy (Laitko et al. 1987, 353). Boveri's resignation letter is Boveri to Herr Ministerialdirektor (Schmidt), 3.5.13, folder: Harnack no. 1, box 19, Fischer papers.

110. I have found only two cases of assistants at these institutes who were too specialized to qualify for a university chair and who succeeded in making a career within the Kaiser-Wilhelm Society. Joachim Hämmerling (KWI for Biology) may have been a case in point (Sund 1984, 203). There is, at any rate, no mention of Hämmerling's declining an offer of a chair in either his autobiography ("Mein wissenschaftlicher Lebenslauf," Stern papers) or in the obituaries of him by Schweiger and Harris. Peter Michaelis (KWI for Breeding Research, see chap. 9) was probably another. Correns argued that Ludwig Armbruster, a biologist working on bees who had his own department during the first few years of the KWI for Biology, ought to be granted tenure by the Society, since he was too specialized to have much hope of getting a chair. The request was denied (minutes of the Kuratorium meeting on 2.6.18, file titled "Trendelenburg 1914–1919," Emil Fischer papers).

parative analysis of German and American genetics between the wars, however, demonstrates that the demarcation of a discipline from neighboring fields can vary widely from one country to another. Which problems are defined as central to a discipline is determined not by an internal logic but by the structure of power in and around that discipline. To be institutionally dependent on another discipline is to be intellectually dependent as well. The key to understanding why genetics was defined more broadly in Germany than in the United States, I have argued, lies in the differing structures of their respective universities and other research institutions. And the qualitative effect of university structure upon the scope of genetics was intensified by quantitative differences in the rate of institutional expansion before 1914, and especially by German economic crises after 1918.[111]

To be sure, alternative explanations are conceivable. For example, Garland Allen has suggested (1985, 141) that the emphasis upon transmission genetics in the United States was due to the strong influence of agricultural concerns there. This hypothesis is plausible, not least because the same connection can be found—albeit less often—in Germany (cf. the discussion of the Baur school in chap. 6). On the other hand, it is noticeable that many of those "atypical" American geneticists who worked in developmental or evolutionary genetics spent a considerable part of their careers working in agricultural institutions (e.g.,

---

111. Readers familiar with the recent historical literature on cytoplasmic inheritance will have encountered this model before, albeit in highly truncated form (Sapp 1987, 58–59). Although Sapp cites a number of secondary sources in support of the idea that German institutional structures have influenced the development of various disciplines, most of these authors do not in fact develop such an argument. To my knowledge, the first systematic attempt to develop such a model and apply it to the German and American genetics communities was presented in my paper at the International Congress for History of Science, Berkeley, California, 1985 (subsequently published as Harwood 1987).

In any event, in terms of its historiography Sapp's *Beyond the Gene* is quite remote from the approach taken here. Despite frequent references to "institutional power" and the "struggle for authority," *Beyond the Gene*—like the discourse analysis of Pierre Bourdieu which inspired it—is basically intellectual history given a sociological gloss. Claims about the institutional structure of the American genetics community, for example, are for the most part either unsubstantiated or derivative (e.g., 45–47, 52–53). Since the existing secondary literature on the institutions of genetics is still so sparse, the only way to make the book's (altogether plausible) claims convincing would be by using records of departments, faculties, funding bodies, journals, professional societies, etc. Apart from papers of the Rockefeller philanthropies, however, the only unpublished sources cited in the book appear to have been the personal papers of Sonneborn, Ephrussi, H. J. Muller, and F. R. Lillie. If *Beyond the Gene* is at one level supposed to be a study in the sociology of science, therefore, it is a sociology with the institutions left out.

Ernest Babcock, L. C. Dunn, Edmund Sinnott, Sewall Wright). To test this hypothesis we need a comparative analysis of genetics at various kinds of American institutions: in the agricultural faculties of the landgrant universities as well as in the biological departments of the private universities.

If the theory which I have developed here is correct, however, it ought to apply to other disciplines. The literature on the forced emigration of large numbers of German academics to the United States after 1933 is illuminating on this score. The ways in which German and American academics from a wide range of disciplines perceived one another's strengths and weaknesses were remarkably consistent. The emigrés were struck by their American hosts' preoccupation with method and measurement; the Americans were amazed by their guests' predilection for theorizing on a grand scale. When the emigré physicists (including many eminent theoreticians) arrived in the United States, they found a physics community that was predominantly experimental. Although doctoral training here was changing, it still emphasized "precision measurement [and] called for instrumental ingenuity, patience and diligence" (Kargon 1982, 41. See also Coben 1979; Forman et al. 1975, 30; Hoch 1983, 235). In the social sciences, as Franz Neumann and others have pointed out, German academics, trained to think historically and theoretically, ran up against their hosts' empirical and ahistorical outlook. The differences between the two traditions had already been clearly identified by Karl Mannheim in 1932. Typically, Mannheim argued, American sociologists began with a practical social problem requiring a solution. But

> if the scholar examining details does not aspire toward a comprehensive view of social reality or shrinks from generalizing hypotheses out of mere caution or due to methical asceticism, then the most excellent work of detail is bound to remain in a vacuum. . . . One almost ventures to say, such works aim in the first place at being exact, and only in the second place at conveying a knowledge of things [Mannheim 1932, 276–77].

Ironically, within a year Mannheim himself had become an emigré, and his own work had to contend with English-speaking audiences. The poor reception of Mannheim's sociology of knowledge by American sociologists perfectly exemplified the clash of traditions he had described.[112] In art history the story was similar. Unlike the Americans, who preferred more narrowly defined problems, the emigrés were typi-

---

112. Rüschemeyer 1981. On the emigré social scientists' experience in the United States see Neumann 1953, esp. p. 24; Adorno 1969, esp. pp. 347, 349.

cally interested in studying art or music as part of the culture or history of a period or place: "European scholars tended to be exploratory on a wider scale without danger of dilettantism. . . . [They were] prepared to go out on a limb more readily than the cautious and accurate, factually oriented scholars in America."[113] And as we saw earlier, German and American biochemistry and psychology differed in similar ways.

It would be quite wrong, however, to attribute such differences to some ahistorical notion of the German versus the American mind. For one thing, neither the German nor the American genetics community was homogeneous (indeed, this very heterogeneity is the central theme for part 2 of this book). For another, there is probably nothing peculiarly German or American about broadly or narrowly focused disciplines. The theory outlined in this chapter suggests that broad conceptions of genetics will occur in any country whose institutional context is comparable to that of interwar Germany. Furthermore, the theory which I have advanced predicts that German-American differences will vary considerably over time and space. First, the interwar differences discussed in this chapter will almost certainly not apply to the later nineteenth century, when visiting American students were often struck by the specialized character of German scholarship (e.g., Thwing 1928, 103, 138, 158, 225). And to the extent that restructuring of the (West) German university since 1945 brought it closer to the American system, national differences will have been eroded. Second, insofar as the departments of the *Technische Hochschulen* were already organized more along American lines during the interwar period (Johnson 1985, 513–14; Riese 1977, 190–92, 223–24), national differences in the applied sciences should also be less pronounced. But third and most important, the theory explains only why national differences should be found in *weakly* institutionalized disciplines, such as art history, sociology, psychology, biochemistry, and genetics. It implies that such differences should be less marked in well-established fields like anatomy or physiology. By implying this, however, the theory creates an anomaly for itself, for as we have just seen, interwar German-American differences seem to have existed even in a well-established discipline such as physics. This raises the possibility that institutional structures may offer an important explanation for the peculiarities of German genetics, but not a sufficient one.

113. Eisler 1969, 613. See also Panofsky 1953. Owsei Temkin had much the same impression of American historians of medicine during the 1930s (Temkin 1977, 22–24).

5 Shifting Focus

I have argued that the different ways in which the boundaries of genetics were drawn in Germany and the United States are attributable to the structure and financing of the institutions in which genetics developed. Let us now explore the limitations of this theory. On closer examination, as we shall see in this chapter, it appears that the differences between German and American geneticists were not confined to their choice of research problems. Their general interests and attitudes also differed in ways which seem at first quite unconnected with their scientific preferences. And this more diffuse contrast of values, unlike their choice of research problems, is difficult to attribute to the kind of institutional constraints discussed in the previous chapter. In order to make sense of these wider differences, one would have to move beyond the structure of scientific institutions to look more generally at the nature of academic roles in German and American society during the interwar period. But wide-ranging cross-national studies of this kind are complex, and the difficulties would be compounded by the fact that neither the German nor the American genetics communities were homogeneous in relevant respects. There were transmission geneticists in Germany as well as developmental geneticists in the United States, even though these constituted minority traditions in each country. It is precisely this heterogeneity, however, which suggests a way to reformulate the problem in such a way as to avoid the methodological problems of cross-national studies. Rather than comparing the research tradition which dominated in the United States with that which was dominant in Germany, why not focus instead upon the corresponding research traditions *within* Germany? Accordingly, my aim in part 2 is to account for the wide range of attitudinal differences between German geneticists of the mainstream, who worked on problems of evolution and development, and those nearer the periphery, who worked in transmission genetics.

Apart from their choice of research problems, were German and American geneticists distinguishable in any other respect? A preliminary answer can be had from a comparative analysis of their obituaries. In table 5.1 one sees that 60 percent of a sample of obituaries of German geneticists noted the breadth of their subjects' biological knowledge,

Table 5.1 Content Analysis of the Obituaries of German and American Geneticists

Attributes	Americans	Germans
Powerful personality	9%	40%
Interested in literature, history, or philosophy	26	67
Interested in fine arts	17	47
Interested in sports	13	13
Interested in politics	13	7
Breadth of biological knowledge	13	60
Unpretentious	22	7
Egalitarian with students	17	13
Hard worker	13	20
Able administrator	30	20
Sense of humor	13	0
Sociable	13	0
"Practical," good with hands	9	0
No mention of personality or extrascientific interests	13	7

while only 13 percent of the obituaries of American geneticists did so.[1] In view of the contrasting structures of the institutions which employed them, this is not particularly surprising. To be considered appointable to chairs of botany or zoology, the Germans had to demonstrate sufficient breadth of knowledge to teach a range of courses in those disci-

---

1. From a sample of about sixty obituaries of American geneticists known to me, I chose a subset at random, omitting very short obituaries (one page or less), those where the subject was foreign-born, and those where two or more obituarists gave different assessments of their subject's qualities. This left the following twenty-three: Emerson, "E. G. Anderson"; Stebbins, "E. B Babcock"; Dunn, "W. E. Castle"; Jenkins, "Roy E. Clausen"; Steiner, "R. E. Cleland"; Johannson, "L. J. Cole"; Plough, "E. G. Conklin"; Snyder, "C. B. Davenport"; Dobzhansky, "L. C. Dunn"; D. F. Jones, "E. M. East" (1939); Beadle, "R. A. Emerson"; Sonneborn, "H. S. Jennings"; Mangelsdorf, "D. F. Jones"; Snell, "C. C. Little"; Sturtevant, "T. H. Morgan"; Wagner, "T. S. Painter"; Painter, "J. T. Patterson"; Jennings, "R. Pearl" (1941); T. F. Anderson, "J. Schultz"; Mangelsdorf, "G. H. Shull"; Rhoades, "L. J. Stadler"; Crow, "W. Stone"; and S. Emerson, "A. H. Sturtevant."

Fifteen obituaries of German geneticists were chosen at random; omitted were those

plines. Besides genetics, for example, Alfred Kühn taught introductory zoology, comparative vertebrate anatomy, invertebrate zoology, animal physiology, embryology, and cytology at various times during his reign at Göttingen. The amazing diversity of subjects on which Richard Goldschmidt published clearly indicates the range of literatures in which he was at home.[2] And Hans Winkler, Friedrich Oehlkers, Fritz von Wettstein, and Otto Renner all taught general botany and plant physiology as well as a variety of other subjects.[3]

On reflection, however, it is evident that many Germans were also "broad" in a more surprising respect. Mastery of the literature of zoology or botany is one thing; a wide range of *research* activities is quite another. In principle it should have been possible for the Germans' research to be just as narrowly focused as their American colleagues', as long as the Germans worked on a classical problem (development or evolution) and were able to teach general zoology or botany. That ought to have been sufficient to secure their credibility as prospective professors of botany or zoology. If most Germans had in fact pursued this career strategy, then in table 4.1 we would have found their research interests concentrated in one of three categories: some in transmission genetics, but most in *either* developmental *or* evolutionary genetics. Instead, the most striking contrast evident in table 4.1 is the much higher proportion of Germans who worked on *both* developmental *and* evolutionary problems. In the case of such men, therefore, even

---

whose author was non-German (e.g., the obituaries of Curt Stern). Where several obituaries were available for a given geneticist, I chose the longest: H. Stubbe, "Erwin Baur" (1959a); Renner, "Carl Correns" (1961b); Klaus Günther and Walter Hirsch, "Hans Nachtsheim zum 80. Geburtstag" (a privately printed biography and bibliography, kindly made available to me by Frau Gisela Eyser); Stern, "K. Belar"; H. Stubbe, "F. von Wettstein"; Henke, "Alfred Kühn" (1955); Stern, "R. B. Goldschmidt" (1967); Mägdefrau, "Otto Renner"; Marquardt, "F. Oehlkers"; Rudolf Haecker, "Valentin Haecker"; Goldschmidt, "Professor Victor Jollos" (typescript, Goldschmidt papers); Schweiger, "J. Hämmerling"; Nachtsheim, "Paula Hertwig"; Kuckuck, "Elisabeth Schiemann"; and Kühn, "F. Süffert."

2. On Kühn's teaching see annual editions of the *Deutsches Hochschulkalender*; on Goldschmidt's range see the bibliography in his 1960. Haecker's range of courses was more restricted but still included general zoology and comparative anatomy in addition to genetics (*Deutsches Hochschulkalender*).

3. During the 1920s Winkler taught evolution. At Freiburg during the 1930s Oehlkers offered plant development as well as a cytology course. At Göttingen and Munich, von Wettstein taught systematics, cytology, and plant development. At Jena, Renner offered cryptogamic botany, systematics, and cytology. Only one of these men (Oehlkers) offered a course in genetics, and it was taught only twice between 1933 and 1939 (*Deutsches Hochschulkalender*).

their research interests might be described as broad. Clearly this was above and beyond what was strictly necessary in career terms; Kühn, Oehlkers, Lehmann, Henke, and Winkler all achieved chairs of zoology or botany despite working on a narrower range of genetic problems (cf. table 4.1).[4] But if so, then what—apart from misguided masochism or the proverbial "teutonic excess"—can have prompted these men to take on *both* of the problems which the Morgan school had shunned as too complex?

The puzzle deepens when we return to the analysis of obituaries (table 5.1), for it appears that the two groups also differed in a variety of nonscientific respects. American obituaries were marginally more likely to cite their subjects' unpretentious manner, sense of humor, sociability, and manual dexterity or practical skill.[5] But other characteristics were singled out in the German obituaries: "strength of personality" and an interest in the humanities or the fine arts. The comparative analysis of obituaries, of course, may be misleading. For the scientific obituary, like any text, embodies conventions. The focus is generally on the scientist's research; his or her popular writings are commonly given subsidiary status, and the scientist's personality or extrascientific interests are usually treated briefly in a concluding paragraph. Discussion of the subject's political views or love life, though sometimes important in understanding his or her work or career, is generally omitted altogether. Thus obituaries may reveal as much about German versus American notions of what is thought to be significant in a scientist's life as about differences in the lives themselves.[6] Such notions are, of course, equally interesting for the comparative analyst, but as it turns out, the contrast between German and American obituaries does not seem to be merely a convention of this kind. In those few cases where American and German obituaries exist for the same person, the portrayals are remarkably similar.[7]

4. For example, Karl Henke was offered the chair in zoology at Göttingen in 1937, despite having concentrated exclusively upon developmental genetics in insects from 1929. And Otto Renner was called to a chair in botany, despite working in transmission genetics, perhaps because he also had a reputation as a plant physiologist (cf. chap. 9).

5. Whether these differences are statistically significant is doubtful, but for the moment their size is less important than their pattern.

6. For a nice example of the revisions which Werner Heisenberg's autobiographical sketches underwent between 1923 and 1975, see Gregor-Dellin 1988. I thank Herbert Mehrtens for drawing my attention to this article.

7. Compare, for example, the obituaries of Renner by Cleland (1966) and Mägdefrau (1961); the obituaries of Bateson by Renner (1961b) and Morgan (1926b); or the obitu-

In any case, we need not depend upon obituaries, for a similar contrast also emerges clearly from records of personal encounters. A particularly rich source of this kind are the letters written home by Germans and other Europeans who were visiting major American genetics establishments, above all the laboratories of T. H. Morgan at Columbia and later at the California Institute of Technology. Because the visitors' judgments were never intended for publication, they are relatively free of the normative constraints to which obituaries are subject. That Richard Goldschmidt locked horns with the American genetics establishment after his emigration to the United States in 1936 is well known. Anticipating that the publication of his *Material Basis of Evolution* would win him few friends, he expressed his hope that the book would be well received in his adopted country by L. C. Dunn, Sewall Wright, "and a few others whose opinion I cherish (including Dobzhansky, J. Huxley, etc.). But I certainly do not mind what the dyed in the wool Drosophilists think, as they are a too narrow-minded and correspondingly overbearing lot."[8] Accordingly, he was pleased with Dobzhansky's appointment to a chair at Columbia: "I am . . . glad that a broad-minded geneticist with a real scientific background (as opposed to the average gene-shuffler) gets into a prominent position and can help to free genetics from the fetters of narrow-mindedness."[9] One might be inclined to dismiss these comments as the sour grapes of an aging emigré who had been forced to leave his mother country, where he was a leading figure in the genetics community, to settle in a land which did not appreciate him.[10] Although there is probably some truth in this, we have already seen (chap. 2) that Goldschmidt's impatience with *Drosophila* geneticists substantially predates his emigration. And in any event, men such as Karl Belar, Theodosius Dobzhansky, and Fritz von Wettstein expressed similar reservations. Above all, it is not just that the visitors found their hosts' view of genetics too restrictive, but that the two groups had little in common even outside the laboratory.

Belar's letters from Pasadena between 1929 and 1931 convey this cultural gulf vividly. From the beginning it seemed clear to Belar that

---

aries of Cleland by Steiner (1982) and W. Stubbe (1976). Similarly, H. J. Muller had little hesitation in paying tribute to "cultivation" in his obituary of E. B. Wilson (1949).

8. Goldschmidt to L. C. Dunn, 27.5.40, Dunn papers.

9. Goldschmidt to Dobzhansky, 19.2.40, Dobzhansky papers.

10. "His reception in America . . . was cold scientifically. The breadth of his biological wisdom was appreciated little, and the iconoclasm of his views was regarded as the irritating fancy of a man of the past" (Stern, "R. B. Goldschmidt" (1958), 307–8).

Figure 5.1 Group portrait in the Kaiser-Wilhelm Institute for Biology during the 1920s; Karl Belar is the last on the right (courtesy: Archiv zur Geschichte der Max-Planck-Gesellschaft).

he would be able to get along with his colleagues in the laboratory but was unlikely to develop any friendships there. Morgan and his wife were the only ones with whom he enjoyed conversation. Bridges was intriguingly eccentric but communicated on another wavelength. Sturtevant, contrary to Belar's expectations, was disappointing and seemed anyway not to like Belar much. Schultz was at least an intellectual—too much so, Belar felt—but unimaginative. E. G. Anderson was nice but boring. Millikan, like many of the people Belar met at Woods Hole, was a "Babbitt": friendly but uninteresting. Many of the people in the laboratory had what Belar regarded as "worthy" hobbies like collecting beetles. But Drosophilists were not the only ones unable to engage Belar's interest. When F. R. Lillie visited Pasadena, Belar found him nice but neither sophisticated nor aesthetically discriminating enough.[11] The fact that Belar was disinclined to talk shop during spare moments in the laboratory, he suspected, cost him credibility among his col-

11. Belar to Curt Stern, 6.10.29, 2.7.29, and 10.3.31, Stern papers.

leagues. Alienated by the general admiration voiced for "hard workers," furthermore, he preferred to spend as much time as possible camping in the desert and mountains. Within three months of arrival, Belar had decided that he could not spend his life in the United States unless he "became a painter and spent three weeks out of four in the desert."[12] Cultural life in Pasadena—even in Los Angeles—was too restricted; Christmas was too commercial, "too Coney Island." Having enjoyed Berlin's lively cultural atmosphere during the early twenties with friends at the Kaiser-Wilhelm Institute for Biology, Belar found America uncultivated.[13]

One of those with whom he spent his free time in Pasadena was a young biochemist, newly arrived in 1931 from Warburg's laboratory at the KWI for Biology: Hans Gaffron. Like Belar, Gaffron found that the only people in the laboratory with whom he could speak his mind were the Morgans; the rest simply misunderstood. This was not just a teething problem; after several years in California he still found that those with whom he could freely talk "are very thin on the ground here."[14]

Although Dobzhansky worked in Morgan's group for a dozen years and built his career there, he seems to have felt himself increasingly the outsider:

> I have definitely left the track of the classical Drosophilaforschung [sic], and becoming [sic] more and more engaged in subjects that are somewhat abhorrent (or at least not very interesting) for the drosophila [sic] fellows here. Fortunately Sturtevant (though sometimes slightly apologetically) is going in the same direction....[15]

By 1937 he was sharply critical of the view that Cal Tech was the "center of the universe in biology."[16] Outside of the laboratory, too, Dobzhansky found that he had little in common with most of the *Drosophila* group. Belar, however, he regarded as both a first-rate cytologist and a "cultured" person with "broad interests." Sharing a fascination with the desert's beauty, the Dobzhanskys and Belars often spent time together.[17] But in 1931 Belar was killed in a car accident, and Dobzhansky confessed to Curt Stern that he was often lonely. Although the

---

12. Belar to Goldschmidt, 11.2.31, Goldschmidt papers.
13. Belar to Stern, 22.12.29 and 7.12.29, Stern papers.
14. Gaffron to Stern, 21.10.31 and 14.10.39, Stern papers.
15. Dobzhansky to Stern, 26.2.36 (cf. 16.9.37), Stern papers.
16. Dobzhansky to Stern, 5.2.37, Stern papers.
17. Dobzhansky, "Reminiscences," 333–34.

United States was on the whole a good place to live, Americans "are a very different kind of peoples [*sic*] from what we were used to [in the USSR]. This difference is somehow a very deep one . . . it is a difference in ideals and in what a person on [*sic*] the bottom of his heart thinks is sacrosanct."[18] Even Sturtevant, whom he liked most of those in the laboratory, seemed different in this respect. Sturtevant was a steady worker, Dobzhansky felt, but devoid of inspiration; he and Bridges were "supreme specialists" rather than "people of culture." Although widely educated and prepared to discuss philosophy, Morgan nevertheless told Dobzhansky that all philosophy was a waste of time, and travel or listening to music were distractions from science.[19] Under the circumstances Dobzhansky was hopeful that Stern would return to Pasadena. In spite of the fact that they did not know one another very well, "it occurs to me that we could understand each other in some respects much better than either of us (or at least myself) could understand these peoples [*sic*]."[20]

That the cultural distance between American and European scientists was not confined to Cal Tech is evident from the diary that Fritz von Wettstein kept during his visit to the United States in 1938.[21] Given the nature of von Wettstein's own research interests (cf. chaps. 2 and 3), his disappointment at the dearth of American work in physiological genetics is hardly surprising. Predictably, his lectures on cytoplasmic inheritance found a cool reception, although the response to his lectures on evolution was more favorable. In his diary von Wettstein made personal as well as professional judgments, and like Belar, he found many American geneticists rather uninteresting. L. J. Cole at Madison was likable "but not very deep." In Pasadena Morgan again won highest marks, but Bridges was "extremely one-sided," and Sturtevant delivered judgments on every topic from a very limited standpoint. E. W. Lindstrom at Iowa City, on the other hand, was scientifically interesting, and his family was "very cultured."[22] Countless entries in the diary indicate that von Wettstein's assessment of his American colleagues was rather similar to his overall reaction to American

18. Dobzhansky to Stern, 9.4.33, Stern papers.
19. Dobzhansky, "Reminiscences," 249, 253–54, 268–70. Dobzhansky reckoned that in fact Morgan enjoyed both of these "distractions."
20. Dobzhansky to Stern, 9.4.33, Stern papers.
21. Henceforth "1938 diary" (kindly made available to me by Diter von Wettstein).
22. On Cole, Morgan et al., and Lindstrom, see von Wettstein's 1938 diary, 27, 55–57, and 34–35, respectively.

society. A devotee of art and music, von Wettstein was delighted with the museums and concerts which he found in New York and Chicago, but as cities he found New York "grotesque," Chicago "monstrous," and various others "soulless." On board ship once again, he was glad to be going home: "It was an interesting country, America, but I don't think in the long term that I could be happy there.... One is drawn to one's own country and to our old culture; it is certainly much more beautiful than America: that colonial country with its incompleteness, its enormous primitiveness, despite its marvelous countryside."[23]

However condescending, these judgments convey a remarkably consistent picture which accords well with the comparative analysis of obituaries. The breadth of biological knowledge and research interests common among German geneticists was associated with considerable respect for "high culture." It will be convenient henceforth to call this dominant research tradition within the German genetics community the "comprehensive" faction. And in view of the grounds on which Morgan et al. declined to work on evolutionary and developmental genetics, it will be convenient to describe the dominant American research tradition as a "pragmatic" one. Thus delineated, is there some logic in each of these complexes of values and attitudes? Studies by Fritz Ringer and Richard Hofstadter suggest that an answer might be found by looking beyond scientific institutions to certain general features of German and American societies at that time. In Germany, academics enjoyed considerable social status during the nineteenth century and saw themselves as the nation's culture-bearers (Ringer 1969). From the late nineteenth century into the 1930s, the professoriate was especially concerned to counteract the evils of specialization in scholarship. Only a unified body of knowledge could secure the unity of German society. Professors were obliged to familiarize themselves with the general features of their colleagues' disciplines and to place their own work in the broadest possible context.[24]

In the United States the status and self-understanding of academics were altogether different. Whether one draws upon the observations of nineteenth-century Americans who had attended German universities or those of emigrés to the United States after 1933, the picture is the same: American academics enjoyed neither the social standing nor the

23. Von Wettstein, 1938 diary, 118.
24. As a visitor to Germany in the late 1860s, William James is said to have been overwhelmed by the breadth of Wilhelm Dilthey's reading, and in this he "spoke for the thousands who came after him" (Shils 1978, 186).

salary of their German counterparts.[25] To Bertrand Russell, American academics seemed to behave more like businessmen than like intellectuals. They lacked the devotion to ideals, whether useful or not, and most seemed to be "vigorous intelligent barbarians" (Feinberg and Kasrils 1973, 45; cf. 44, 355). Theodor Adorno was struck by the fact that "no reverential silence in the presence of everything intellectual prevailed, as it did in Central and Western Europe far beyond the confines of the so-called educated classes" (Adorno 1969, 367). Richard Hofstadter's *Anti-Intellectualism in American Life* (1963) offers an explanation for this pattern. The lack of respect accruing to ideas and intellectuals in the United States, he argues, derives from that country's peculiar history and social structure. It was founded by Europeans in flight from traditional societies in which learning and culture were the prerogatives of an idle aristocracy. The egalitarian ethos of colonial America thus paid tribute to the wisdom of the common man. The lack of an indigenous aristocracy, furthermore, meant that the dominant values in American society by 1900 were overwhelmingly those of an industrial bourgeoisie in especially pure form. In such a society knowledge was generally valued not as a route to wisdom, but as the means to utilitarian ends.[26] The man of knowledge was respected, not as a culture-bearer in the German mold, but as someone whose expertise could solve practical problems. If Hofstadter is correct, breadth of knowledge would have enjoyed no cachet in the American context. Academics would choose problems which either looked easily soluble or promised substantial payoffs, whether theoretical or practical. If specialization were the best means to that end, so be it.

On this analysis, an adequate understanding of the full range of differences between the German and American genetics communities would require exploring the dominant cultural traditions within each country. Unfortunately our understanding of the work and organization of the genetics communities in the two countries is still too primitive to undertake such a formidable study with much prospect of success. Systematic comparative analysis across countries is not only logistically demanding but methodologically complex. There were, af-

25. E.g., Hart 1878; F. Stern 1987, 58; Baumgardt 1965, 242; Chargaff 1978, 195; Bonner 1963, 50; and Veysey 1965, 300–301, 352, 389–90, 442–43.

26. When the German biochemist Hans Gaffron arrived in the United States in 1931, his attention was caught by a piece of folk wisdom framed on the front page of the *Los Angeles Times:* "An education enables you to earn more than an educator" (Gaffron 1969, 30). A generation later the proverbial man in the street would have put it more plainly: "If you're so smart, why ain't you rich?"

ter all, many cultural and social structural differences between early-twentieth-century Germany and the United States, all of which might have been significant. In addition, cross-national comparison is complicated by the fact that neither country's genetics community was homogeneous. As we have seen, Goldschmidt admired the likes of Dunn, Dobzhansky, and Wright.[27] Von Wettstein, similarly, was so impressed by Dobzhansky ("a very fine man" and the cleverest of the Pasadena group) and Wright (an "excellent geneticist, totally at home with evolutionary matters") that he immediately invited both of them to the KWI for Biology.[28] Just as the Germans admired those broader American geneticists, so also did the Americans pay corresponding tribute to the German pragmatic tradition: those working in transmission genetics, such as Baur, Nachtsheim, and Stern.[29] Both genetics communities

27. Although generally unpopular at the Bussey Institution in 1915, Goldschmidt apparently got on well with Wright (Provine 1986, 92).

28. Von Wettstein, 1938 diary, 64 and 30–31; cf. Dobzhansky to Curt Stern, 27.1.39, Stern papers. It is quite possible that the incidence of American biologists with such broad interests was substantially higher before the First World War. In the 1950s Viktor Hamburger wrote from St. Louis: "In the last few years—and decades—I have sorely missed the possibility to discuss things with my German and Swiss friends. I would have given anything to sit and contemplate the situation in biology occasionally with you, Henke or Baltzer. In Europe there is still something of the philosophical frame of mind which has been driven out here by the enthusiasm for progress. Harrison, F. R. Lillie, E. B. Wilson still possessed this frame of mind, and I am glad that I could get to know them" (Hamburger to Alfred Kühn, 28.2.57, Kühn papers, Heidelberg).

It may be significant that Wright, Dobzhansky, and several other American developmental or evolutionary geneticists were very interested in the arts or humanities. Wright had a lifelong interest in philosophy (Provine 1986), and Dobzhansky was interested in a wide range of arts and humanities (Ayala 1985). H. S. Jennings wrote on the philosophy of biology (Burnham). Dunn was described as "a rare combination of scientist, humanist, philosopher and artist" (Gluecksohn-Waelsch 1974). Edmund Sinnott was a gifted painter and sculptor and wrote on the relations between science and philosophy (Whaley 1983). And of Raymond Pearl—who originally intended to study classics at the university—a colleague wrote: "There are two kinds of men of science whose interests and activities greatly contrast. One kind, the orthodox, today very numerous, proceed by a kind of orthogenetical development and do not often step aside from a straight and narrow path. The other kind, rare today, . . . feel that their intense interest in all things—their *philosophical* interest in an older sense of the word . . .—is a safe guide. Such a man was Francis Galton and another . . . was Raymond Pearl" (L. J. Henderson, cited in Jennings 1943, 299).

29. "Baur is the foremost geneticist in Germany and also the ablest," wrote Morgan, and among the younger generation "Nachtsheim is the most intelligent and best balanced . . . the kind of man we should welcome in our lab" (Morgan to Wickliffe Rose, 7.6.25, IEB papers, ser. 1, subser. 4, box 75, folder 1254). Of the six German geneticists whom E. W. Lindstrom singled out as "excellent," five were cytogeneticists or *Drosophila*

were diverse, containing both a dominant and a minority faction. While this diversity makes cross-national analysis more complicated, it also suggests a more promising way forward. For if those differences which distinguished the American and German genetics communities were also present *within* each country, it would be methodologically simpler to study the genesis of these traditions within one country. It is for this reason that the analysis in part 2 shifts to Germany. Chapters 6 and 7 characterize the differences between the dominant comprehensive faction and the minority pragmatic faction in detail, while chapter 8 teases out the connections between these factions and those social changes which transformed German society from the late nineteenth century.

---

geneticists. The sixth—Fritz von Wettstein—was praised not for his work in developmental genetics, but for "first class research in heteroploidy" (Lindstrom, "Notes on Geneticists of Europe and Their Laboratories, 1927–1928," in the logbook of W. E. Tisdale/E. W. Lindstrom/W. J. Robbins, box 37, RG 12.1 [officers' diaries], Rockefeller Foundation papers). H. J. Muller's student Bentley Glass went to Berlin in 1932 in order to work with the men he reckoned were Germany's greatest geneticists: Stern and Timofeeff-Ressovsky (Glass interview). When Stern returned to Germany after two years with Morgan at Columbia, H. J. Muller closed a letter to Stern with greetings to "the geneticists in Berlin," singling out Baur, Nachtsheim, Paula Hertwig, and Emmy Stein (Muller to Stern, 5.10.26, Muller papers). While all four of these geneticists were indeed in Berlin (at the Agricultural College), Muller seems to have overlooked two of Stern's more immediate colleagues, who were very well known and whose work was altogether more typical of German genetics: Correns and Goldschmidt.

Part II STYLES OF THOUGHT WITHIN THE GERMAN GENETICS COMMUNITY

6 Mapping the German Genetics Community

The focus of the analysis now shifts to variation within the German genetics community. If we are to uncover the roots of both pragmatic and comprehensive traditions in Germany, we ought to begin by looking at the institutions in which geneticists in each tradition were employed. As we shall see, German pragmatics and comprehensives were not randomly distributed across the genetics community but clustered to a large extent in different kinds of institutions.

The comprehensive majority was to be found, on the whole, either in the universities or in the KWI for Biology. Hans Winkler at Hamburg and Valentin Haecker at Halle seem to have had few, if any, students who worked in genetics, nor did Correns or Goldschmidt at the KWI for Biology. Otto Renner at Jena and Friedrich Oehlkers at Freiburg did develop schools, but most of Renner's students preferred plant physiology to genetics,[1] and Oehlkers's "Black Forest School" only became sizable from the mid-1930s. Similarly, Timofeeff-Ressovsky's research group at the KWI for Brain Research at Berlin-Buch did not become a separate department until the early 1930s.[2] The largest and most important concentration of comprehensive geneticists during the interwar period was at Göttingen, where Alfred Kühn's and Fritz von Wettstein's shared interest in developmental genetics fostered close relations between the Institutes of Zoology and Botany from 1925 to 1931, a collaboration which was resumed in 1937 at the KWI for Biology. So similar were these two groups that I will treat them for the remainder of this chapter as a single school.

Isolated pragmatics could be found in various institutions; for example, Curt Stern at the KWI for Biology, or Julius Schwemmle at the University of Erlangen. But most of them were associated with Erwin Baur, from 1911 to 1928 at the Berlin Agricultural College and thereafter at the KWI for Breeding Research in Müncheberg outside Berlin. At the Agricultural College, students could work in either plant or ani-

---

1. Renner to Ralph Cleland, 5.1.30, Cleland papers.
2. Eichler's obituary of Timofeeff-Ressovsky (1982, 290) says this occurred in 1931, but the first mention of departmental status in an annual report is in "Tätigkeitsbericht der KWG," *Die Naturwissenschaften* 21 (1933): 426.

mal genetics with Baur and his assistants, Elisabeth Schiemann, Hans Nachtscheim, and Paula Hertwig.

In this chapter I will trace the institutional geography of the interwar German genetics community by comparing the Baur school with that of Kühn and von Wettstein. As we shall see, the two schools differed considerably. While the Kühn/von Wettstein school at the KWI for Biology remained essentially academic in character, Baur's KWI for Breeding Research more closely resembled an industrial laboratory.

As such, the Institutes for Biology and for Breeding Research represented two kinds of institutional solution to the problems facing late-nineteenth-century German universities.[3] On the one hand, rapid industrialization from midcentury had created new research needs which universities were unable to meet. By and large, universities were reluctant to engage in the applied research which industry wanted; those professors who, nonetheless, entered into arrangements with industry—above all, chemists—tried to do so inconspicuously (Manegold 1970; Borscheid 1976). And the *Technische Hochschulen* lacked research facilities until well into the twentieth century. A partial solution for firms was the in-house laboratory, first established by Bayer in 1880 (Beer 1958; Meyer-Thurow 1982). But firms were only happy investing in commercially applicable work; where more fundamental research was needed by an entire industrial sector, firms urged government to shoulder the burden, one result of which was the establishment of the *Physikalisch-Technisches Reichsanstalt* in 1887 (Pfetsch 1970; Cahan 1982). Public-sector laboratories, of course, offered firms no advantages over their competitors; where firms wanted fundamental work done which would not be freely accessible to others, new institutional forms were required. One solution to this problem was Paul Ehrlich's Institute for Serum Testing and Research, established in 1899. Designed to facilitate cooperation among state, industry, and academe, the institute's capital costs were covered jointly by the Reich, the city of Frankfurt, and the dyestuffs firms Hoechst and Casella. Running costs were to be financed from the institute's own serum-testing service and the sale of the new sera which it would develop. The Institute for Experimental Chemotherapy (Georg-Speyer-Haus), established in Frankfurt in 1907 under Ehrlich's direction, was also designed to be self-supporting via the sale of its patents. As we shall see, the structure of the KWI for Breeding Research closely resembled that of Ehrlich's in-

---

3. For a useful survey of the changing institutional landscape of German science at this time, see Beyerchen 1988.

stitutes. This was no accident; Ehrlich was consulted during the formation of the Kaiser-Wilhelm Society, and the KWI for Experimental Therapy (1913) had been modeled on the Georg-Speyer-Haus.[4] Among the first generation of KWIs founded before 1914, some institutes (e.g., the Institute for Physical Chemistry or that for coal research) were set up to conduct applied research; these were predominantly funded by industry, and industrialists were strongly represented in their supervisory committees (*Kuratorien*) (Burchardt 1975).

But others, of which the Institute for Biology is an example, were conceived as sites for basic research and had only slight connections with industry. Since the particular way in which the university system expanded after 1870 was increasingly burdening staff with teaching and administration (cf. chap. 4), the aim of these KWIs was to try to prevent Germany from falling behind in the development of new specialties by using industrial money to establish academically oriented research institutions like the Institute for Biology. Both types of KWI, however, were the brainchildren of a new breed of academic-entrepreneur. Men such as Koch, Ehrlich, Felix Klein, and Emil Fischer (who was instrumental in the founding of the Kaiser-Wilhelm Society [Feldman 1973; Johnson 1990]) were unafraid of contamination from the private sector. Instead they sought new institutional arrangements which would satisfy investors' needs while at the same time expanding the facilities for basic research.

Let us turn now to the two major schools of genetics in interwar Germany, noting the differences between them in research program, organization, strength of ties to academic institutions versus the private sector, patterns of funding, and political fate after 1933.

6.1 Research Programs

In 1934 Richard Goldschmidt regarded Kühn's group as the best school of genetics in Germany.[5] From the late 1920s Kühn's research group at the Institute of Zoology in Göttingen grew rapidly, and by 1931–32 he had fourteen doctoral students.[6] From 1931 many of these

---

4. Lenoir 1988. In the years before World War I the Ministry of Education in Baden sought to encourage similar institutional innovations (Riese 1977, 250–67).

5. W. E. Tisdale to Warren Weaver, 28.8.34 (folder 46: "KWI for Anthropology 1933; KWI for Biology 1929–1935," box 4, ser. 1.1, RG 6.1, 717D, Rockefeller Foundation papers).

6. University Archive Göttingen, AZ: xvi.v.c.v.4.2: Kühn to Kurator, 5.11.32. The number of persons at all levels conducting research in the institute rose from nine in

a

b

Figure 6.1 Alfred Kühn's assistants at the Kaiser-Wilhelm Institute for Biology from 1936: a, Ernst Plagge (left, 1937); b, Erich Becker, ca. 1938; c, Hans Piepho in 1945; d, Viktor Schwartz, ca. 1937 (all courtesy: Prof. Hans Piepho).

c   d

students' dissertations focused on the genetics and development of what had become the institute's favored organism: the flour moth, *Ephestia kühniella Zeller*. Between 1931 and 1936, for example, fourteen of the twenty-nine dissertations which Kühn supervised at Göttingen dealt with *Ephestia,* and four others dealt with wing development or wing pigmentation in other organisms.[7] As we saw in chapter 2, Kühn and Karl Henke had begun in 1924 to study the genetics of *Ephestia*'s wing patterning. And from 1933 Ernst Caspari's methodological breakthrough spawned a series of papers by Kühn's younger students and assistants, Ernst Plagge, Erich Becker, Hans Piepho (1909–), and Viktor Schwartz (1907–).[8]

1925–26 to thirty-three in 1932 and to forty-eight in 1933 (as above, Kühn to Kultusminister, 22.11.32 and 11.7.33).

7. "Von Alfred Kühn angeregte und geleitete Dissertationen," Kühn papers, folder 7, MPG-Archiv.

8. Plagge, Piepho, and Schwartz all received their doctorates with Kühn in 1934–35 and stayed on as assistants. Becker received his Ph.D. in biochemistry from the *Technische Hochschule* Darmstadt in 1937 before joining Kühn's department at the KWI for Biology. Cf. chap. 2.

The degree of collaboration between Göttingen's institutes of zoology and botany during the 1920s was exceptional, helped not only by Kühn's and von Wettstein's common interest in developmental genetics, but also by their close friendship.[9] They each taught the genetics course in alternate years, and their colloquia were held jointly. Von Wettstein's students were obliged to take two laboratory courses in zoology (and Kühn's vice versa), and their doctoral topics ranged from ecology and evolution to cytology and cytoplasmic inheritance (Melchers 1987a). Among von Wettstein's assistants in Göttingen were Gustav Becker (who worked on the effects of polyploidy), L. A. Schlösser (cytoplasmic inheritance), and Georg Melchers, whose genetic and physiological experiments on Alpine plant varieties were designed to illuminate the process of microevolution (Hoffmann 1971; Schlösser 1935; Melchers 1987a). Von Wettstein's call to Munich in 1931 brought this joint teaching venture to an end; although they were reunited at the KWI for Biology from 1937, their teaching opportunities there were restricted.

In Baur's Institute, too, a large number of research students were trained. In 1925 alone there were sixteen doctoral candidates working in the institute, supervised by Baur and his assistants.[10] In the early postwar years, Baur's institute was unique—at least in genetics—in employing exclusively female assistants, all of whom had doctorates. Luise von Graevenitz worked on potato genetics at the institute from 1915 until her early death in 1921 (von Ubisch 1956, 504). Gerda von Ubisch (1882–1965) got her degree in physics before coming to the institute to carry out a Mendelian analysis of barley under Baur's supervision in 1912 and 1913. Employed during the war in the seed-breeding industry, she returned in 1918 as an assistant to work on the genetics of fibrous plants until 1921 (von Ubisch 1956, Deichmann 1991, 354–61). Elisabeth Schiemann (1881–1972) received her doctorate in microbial genetics with Baur in 1912 and stayed on as assistant

---

9. Georg Melchers had experienced nothing comparable in Kiel, Freiburg, Tübingen, or Munich (Melchers 1987a, 376), and the physicist R. W. Pohl regarded it as unique (Pohl to Dean of the Faculty, 11.8.29, University Archive Göttingen, Personalakte F. von Wettstein, AZ: 4Vb 294).

10. Baur to Minister für Landwirtschaft . . . , 14.5.25 (Sign 20282, Rep 87B, ZStA-Merseburg). Some of these students were probably supervised by Paula Hertwig ("Chronik, 1931–1956: Institut für Vererbungs- und Züchtungsforschung [zum 25-jährigen Dienstjubiläums H. Kapperts]," unpublished manuscript, kindly made available to me by Prof. G. Linnert). Elisabeth Schiemann also had Ph.D. students (curriculum vitae dated 2.3.38, Schiemann papers, MPG-Archiv), among them Hermann Kuckuck (Schiemann 1963).

Figure 6.2 Elisabeth Schiemann (in white coat to left of table) outside the Institute for Genetics at Berlin's Agricultural College, demonstrating snapdragon mutants to visitors from the International Genetics Congress in 1927 (courtesy: Prof. Hermann Kuckuck).

from 1914 to 1931, working on various problems in transmission genetics, speciation in the wild strawberry, and the evolution and systematics of domestic varieties of barley and wheat (Kuckuck 1980). Although a *Privatdozent* at the University of Berlin from 1919, Paula Hertwig (1889–1983) occupied one of Baur's assistantships at the Agricultural College from 1921 to 1946, working throughout on mutagenesis and radiation-induced genetic damage in mice.[11] Emmy Stein

11. Kosswig 1959. That Hertwig's interests were primarily in transmission and mutation (rather than evolutionary and developmental) genetics is evident from two monographs. In "Partielle Keimesschädigungen" (1927), she sought to show that while the effects of radiation upon cells had so far attracted primarily embryologists, radium and X-rays had a variety of effects—e.g., upon nondisjunction, chromosomal breaks, sperm fertility, and crossing-over frequency—in ways which could be useful for "geneticists." Although much of the literature which she reviewed in "Artbastarde bei Tieren" (1936) was concerned with developmental genetic problems (e.g., cytoplasmic inheritance), very

(1879–1954), an assistant in the institute from 1918 to 1939, was one of the first geneticists to investigate the effects of radium upon plant tissues.[12] Baur's first male assistants seem to have arrived in 1921.[13] Hans Nachtsheim joined Baur as an assistant and head of the institute's zoological department in 1921. As an assistant in the Institute of Zoology in Munich, he had been studying the cytology of sex determination in bees and annelid worms and doing transmission genetics in *Drosophila*, but at the Agricultural College he worked briefly with pigs before settling upon the rabbit as an ideal experimental organism for studying pathological traits.[14]

After the move to Müncheberg in 1928, most of the research used crosses between domestic and wild species and genera in order to produce either wheat varieties suitable for light German soils or fruit and grain varieties resistant to fungal disease, but several of the assistants also continued Baur's program of mapping in *Antirrhinum*. Along with his work as head of the barley breeding department, for example, Hermann Kuckuck (1903–   ) studied linkage and multiple alleles in *An-*

---

little of the work cited in this section of the book is hers. Though she was familiar with this literature, therefore, these were evidently not the problems which interested her.

12. Schiemann 1955. The effects which she produced behaved much like the dauermodifications studied by Jollos and Woltereck. But whereas the latter geneticists, like various plasmon theorists, were concerned to explore the evolutionary significance (or lack of it) of such phenomena, Stein's aim was merely to identify the cellular or genetic basis of such effects.

13. The relatively large number of women in the Baur school into the mid-1920s is a phenomenon which deserves systematic study. It may have had something to do with Baur's meritocratic inclinations (von Rauch 1944, 89), but the existence of large numbers of female geneticists elsewhere (e.g., with William Bateson [Beatrice Bateson 1928]) suggests that it might have been a fairly general phenomenon, at least during the first generation of Mendelism, deriving from the nature of the labor market. Von Graevenitz, Schiemann, Stein, and von Ubisch all joined him during or immediately after the First World War, when men with doctorates would have been in short supply. Furthermore, Baur obtained some of their labor on the cheap: Both von Graevenitz and Stein evidently came from wealthy families and were prepared to work as unpaid assistants (*Volontär-Assistenten*), in Stein's case from 1918 to 1923 (autobiographical sketch in Stein papers; Schiemann 1955). Since the career situation in genetics remained precarious into the 1920s, Baur may have had little alternative but to employ women. By the late 1920s, however, he had acquired many male students, and when he moved in 1928 to the KWI for Breeding Research, he left most of his female assistants behind. This pattern is consistent with Rossiter's observation that the proportion of women in a field tends to decline as career prospects in that field improve (Rossiter 1982).

14. Plarre 1987, 165. As far as I am aware, the only other male staff member in 1921 was Bernhard Husfeld, an unpaid student assistant (ZStA-Merseburg, Rep 87B, Sign 20281-20283).

a

Figure 6.3 a, Hermann Kuckuck (middle) and Rudolf Schick (right), visiting a plant-breeding station at Quedlinburg in 1947 (courtesy: Prof. Hermann Kuckuck). b, Hans Stubbe in somber mood, ca. 1940 (courtesy: Universitätsarchiv Halle, Rep 40, Nr. I S 132).

b

tirrhinum. Rudolf Schick (1905–1969) collaborated on this linkage work in addition to his responsibilities as head of the potato breeding department. Hans Stubbe (1902–1989) was the only assistant to work exclusively on *Antirrhinum*. His Department of Mutation Research studied the induction of mutations by chemicals and radiation as well as the effects on mutation rate of nutrition, pollen age, and stage of development.[15]

Thus the research programs of these schools were markedly different. Kühn's and von Wettstein's groups conducted basic research, focusing primarily on developmental genetics (and in the latter's institute, on evolution). Baur's school conducted both basic and applied research, the former addressing mainly transmission genetic problems. Only Baur himself worked on evolutionary problems.

6.2 Forms of Organization

Very little information is available on the structures of the institutes at Göttingen and at the Berlin Agricultural College. Much more material of this kind, however, is available once both schools had moved to Kaiser-Wilhelm institutes, and the differences are much in evidence. As table 6.1 indicates, the institute at Müncheberg had many more departments, and they were distinguished not by field or discipline but by experimental organism, reflecting the fact that research at Müncheberg was primarily applied and much narrower in theoretical scope than that at Dahlem. The total numbers of assistants plus department heads, however, were roughly comparable, since the departments at Dahlem each contained several researchers, while most of those at Müncheberg consisted merely of a department head plus technicians and gardening staff. The roles of the institutes' directors were also different. In Müncheberg the institute's division of labor had been conceived by Baur, and he appears to have exercised a certain amount of control over the department's activities. Baur was said to possess a "Führer personality";

15. Cf. *Die Naturwissenschaften*, nos. 17–18 (1934), a special issue dedicated to Baur. Although it has been claimed that Stubbe frequently referred to the evolutionary significance of mutation (P. Hertwig 1962, 2), I find no evidence that Stubbe actually *worked* on problems in evolutionary genetics, at least while he was working in Baur's institute. In a review paper (1934a) Stubbe observed that one of the three central problems in mutation genetics at that time was the significance of micromutations in evolution, but he devoted only two paragraphs of that seven-page paper to the literature on the subject, citing none of his own work. His 400-page monograph on mutation (1938), similarly, discussed the nature of the gene and the process of mutation but not mutations' role in evolution.

Table 6.1 Organization of the Kaiser-Wilhelm Institutes for Biology and for Breeding Research

Department	Head	Assistants
a. KWI for Biology (Berlin-Dahlem) in 1937–38		
Botanical genetics	von Wettstein	Melchers, Stubbe, Pirschle, E. Kuhn
Zoological genetics	Kühn	Schwartz, Piepho, Plagge, E. Becker, Gottschewski (in USA)
Protozoology	Hartmann	Bauer, Hämmerling
Experimental embryology	(closed in 1934)	
b. KWI for Breeding Research (Müncheberg) in 1928		
Director's dept.	Baur	
Mutation genetics	Stubbe	
Potato breeding	Schick	Lehmann, Probach
Barley breeding	Kuckuck	
Rye breeding	Ossent	
Wheat breeding	Oehler	
Feedstuffs breeding	Ufer	
Grape breeding	Husfeld	
Tree breeding	Wolfgang von Wettstein	
Fruit breeding	Nebel	
Berry breeding	Gruber	
Agric. Meteorology	Mäde	
Field-testing methods	Meyle	
Breeding of protein- and oil-containing plants	von Sengbusch	8 technicians

Sources: "Max-Planck Institut für Biologie," *Jahrbuch der Max-Planck Gesellschaft,* Teil II (1962), 111–45; "Tätigkeitsberichte der Kaiser-Wilhelm Gesellschaft," *Die Naturwissenschaften* 26 (1938): 327, 342–43; von Sengbusch papers, Ordner no. 7, MPG-Archiv.

he was the "field officer with a sure grasp of the overall situation who then deploys his troops at the appropriate place" (Schiemann 1934, 97, 100). Exerting such control was made easier, not only by Baur's personal charisma, but also by the fact that on moving in 1928 to the KWI for Breeding Research at Müncheberg, Baur took neither his female assistants nor Nachtsheim with him. Most of the women were nearly Baur's age, and of the five women plus Nachtsheim, only Schiemann had been a student of Baur's.[16] Instead Baur chose to employ a group of much younger men who had recently taken their doctorates with him at the Agricultural College and who were prepared to accept less than an assistant's salary.[17]

If the Institute for Breeding Research was much more hierarchical and bureaucratized than the average university institute, the KWI for Biology was rather less so. In Dahlem the director's function was simply to represent the institute in general administrative dealings with outside bodies such as the Kaiser-Wilhelm Society or various ministries. All of the departments (including the director's) were of equal status. Attempting to describe the institute's structure to an audience more familiar with the traditional "one-man" university institute, Friedrich Glum (general director of the Kaiser-Wilhelm Society) remarked that "the KWI for Biology essentially consists of six little institutes."[18] Some evidently found such an egalitarian structure a bit difficult to imagine; the director of one neighboring institute complained, "One never knows who's boss at that place."[19] Furthermore, even within each de-

16. Perhaps this is why Schiemann was one of the very few women whom Baur had planned to take with him to Müncheberg (cf. "Collegiumsitzung," 28.7.28, Abt 1, Rep 1a, no. 2615, MPG-Archiv). In the event, an argument between the two put an end to this plan (Kuckuck 1980, 522). The other woman was a "Frl. Dr. Becker" who was to work with Stubbe and Baur on mutation genetics, supervise the breeding of vegetable varieties, and look after the institute's library ("Collegiumsitzung," ibid.). There is no record of Becker having made the move, however, and in 1929 she was an *ausserplanmässige* assistant (i.e., on a short-term contract) in Baur's former institute at the Berlin Agricultural College (Sign 20095, Rep 87B, ZStA-Merseburg).

17. On Baur's initial plan for filling the assistantships at Müncheberg, see "Collegiumsitzung," cited in n. 16 above. On salaries see Kuckuck 1988, 21.

18. Glum in Harnack 1928, 19. The sixth "institute" to which Glum referred was a "Guest's Department" for the Danish tissue-culture expert Alfons Fischer which was in operation from 1925 to 1931 (*Jahrbuch der MPG 1961*). In his autobiography Hämmerling confirms the equality of all the departments ("Mein wissenschaftlicher Lebenslauf," 14, Stern papers).

19. ["Man weiss nie, wer an diesem Institut etwas zu sagen hat."] Melchers, "Die Vorgeschichte einer 'Allgemeinen Biologie" in Göttingen . . . ," p. 8, typescript of a talk given at Göttingen on 2.12.83.

partment the assistants seem to have enjoyed rather more autonomy in choosing problems and conducting their research than did department heads in Müncheberg or assistants in universities. This feature certainly impressed Hans Kalmus when he visited the KWI for Biology in the mid-1930s, and it is frequently emphasized by former assistants in their recollections of the period.[20] These assistants' impressions are easily confirmed by surveying the diversity of research problems pursued within each department. In 1928–29, for example, Goldschmidt worked on developmental and evolutionary genetics using the gypsy moth, while his assistants worked on cytogenetics in *Drosophila* (Stern), perceptual physiology in bees (Mathilde Hertz), and various studies using tissue cultures (Tibor Peterfi). In 1931–32 Max Hartmann was working on sex determination in marine algae while his assistants concerned themselves with the effects of nucleus versus cytoplasm on development in *Acetabularia* (Hämmerling) and directed mutation in *Drosophila* (Jollos). In Correns's department, meanwhile, the director studied cytoplasmic inheritance, while his assistants worked on plant physiology (E. Schratz) and the cytogenetics of sex determination (E. Kuhn). In 1937–38 von Wettstein was working on cytoplasmic inheritance in mosses as well as the effects of polyploidy on cell size, but his assistants were free to explore the effects of genes and environment upon flowering (Melchers), mutations and mutagenesis (H. Stubbe from 1936), and mineral metabolism in plants (Pirschle).[21]

Nevertheless, if the working conditions in Dahlem were freer than

---

20. Kalmus interview. Max Hartmann and Carl Correns are said to have allowed their assistants the freedom to choose their own problems (Bauer 1966, 187; Hämmerling, "Mein wissenschaftlicher Lebenslauf," Curt Stern papers). A student of Joachim Hämmerling's told me that the latter's famous experiments on *Acetabularia* were neither proposed nor directed by Hartmann (Schweiger interview). Apart from correcting the proofs of Goldschmidt's 1927 *Physiologische Theorie der Vererbung*, Curt Stern was free to conduct his own research (Caspari, "Curt Stern," *Dictionary of Scientific Biography*, forthcoming). Johannes Holtfreter felt that he had complete freedom in Otto Mangold's Department of Experimental Embryology (Holtfreter to K. Ishihara, 24.6.81; this letter, kindly lent to me by Professor Holtfreter, contains an autobiographical sketch which was published in Japanese in a popular scientific monthly [*Shiz'en*] in January 1982). Melchers enjoyed the same freedom in von Wettstein's department (Melchers interview).

21. Cf. the "Tätigkeitsbericht der KWG," *Die Naturwissenschaften* 17 (1929): 326–27; 20 (1932): 436–37; and 26 (1938): 342–43. This is not to say that *every* department in the KWI for Biology embodied such laissez-faire. Kühn's department, for example, had a more conventional division of labor. Apart from Gottschewski (who had been appointed by Goldschmidt) and Fritz Süffert (who was formally a guest in the department; cf. Kühn 1946), the other four assistants were all working on the developmental genetics of *Ephestia*.

in Müncheberg, the social climate was not. When staff from the Institute for Biology went off to lunch at Harnack House (the Kaiser-Wilhelm Society's social club), strict segregation prevailed, with students and technicians at one table, assistants at another, and department heads elsewhere.[22] Some visitors were unprepared for such formality. When Hans Stubbe and a few colleagues dropped in unannounced at the Dahlem Institute, this "discourtesy" drew a reprimand from Goldschmidt, judging from the obsequious apology which Stubbe later sent.[23] Perhaps Goldschmidt was particularly fussy; von Wettstein seems to have laid rather less emphasis on formality, but anecdotal evidence suggests that Kühn maintained a distinct social distance between himself and most of his assistants, in keeping with academic conventions.[24] In any event, Stubbe et al.'s casual demeanor is hardly surprising given the very different atmosphere in Müncheberg. For although Baur exerted real control over his department heads' activities, his manner was never that of "der Chef," and he never took advantage of his formal status in discussions with students. His open and uncomplicated manner won him much affection from coworkers (Schiemann 1934, 100–101; von Rauch 1944, 87–88; H. Stubbe 1959a, 6). At lunch or over a beer after work he was the "regular guy," talking sports or science with his assistants and amusing them with anecdotes or recollections of boyhood pranks.[25] During the summers in the 'twenties Baur's students were often invited to come out to his farm and help with the harvest. There they went on swimming trips with the family and played with Baur's children; they report having felt very much at home.[26] Moreover, Baur himself was not averse to manual labor, sometimes joining in the harvest at the Agricultural College or transplanting the seedlings for his experiments, and he recommended such work to his assistants as a way to relax (von Rauch 1944, 86–87).

It was not, however, merely the internal structures of these two in-

22. In interviews both Ursula Philip and Charlotte Auerbach mentioned having experienced this as students at the institute in the early 1930s.

23. "Darf ich Sie im Namen aller meiner Kollegen ... bitten, die sehr wenig formeller Art unseres Auftretens zu entschuldigen?" (Stubbe to Goldschmidt, 12.9.30, Goldschmidt papers).

24. Recollections of Hans Piepho, in Grasse 1972, 216; interviews with Hartwig and Straub.

25. Gruber, "Erinnerungen an Müncheberg," typescript, Mappe 2, Baur papers, MPG-Archiv.

26. H. J. Troll to me, 31.3.84; H. Stubbe 1959a, 4; H. Stubbe 1985, 3–4; Kuckuck 1988, 15.

stitutes which differed. The strength of their ties to academic institutions was also distinctive. The Kaiser-Wilhelm Society had originally envisioned close relations between its institutes in Berlin and the university. Department heads would hold guest professorships and be free to offer courses at the university and to take on Ph.D. students. Younger researchers would spend a few years as assistants in a KWI before moving on to an academic career (e.g., Glum; in Harnack 1928). The extent to which this vision was realized at the KWI for Biology varied from one department head to another. Correns and Hartmann taught at the university as guest professors (*Honorar Professoren*), and the latter's course in general biology attracted students such as Curt Stern and Joachim Hämmerling, who later joined the institute as assistants.[27] But Goldschmidt neither taught nor took on doctoral students, and Correns stopped offering courses about 1924. Virtually all of the assistants in genetics became *Privatdozenten* and offered a course at the university.[28] Furthermore, professors of zoology and botany from Berlin and other universities were members of both the institute's supervisory committee (*Kuratorium*) and its academic advisory panel (*wissenschaftlicher Beirat*).[29]

Relations between the University of Berlin and the Institute for Breeding Research, however, were much more tenuous. To begin with, the institute had no academic advisory panel, and the eight academics on its supervisory committee were completely outnumbered by forty-one bankers, estate owners, industrialists, government officials, and representatives of agricultural organizations (Harnack 1928, 210–12). A group of this size was clearly too large to be effective, so a thirteen-man subcommittee (*Verwaltungsausschuss*)—consisting of ministers, estate owners, industrialists, a banker, and a representative of the Kaiser-

---

27. Both von Wettstein and Kühn had (unpaid) personal chairs at the university and offered a course in genetics (cf. volumes of the *Deutsches Hochschulkalender* for 1918–1932 and 1935–1939; Asen 1955).

28. Hans Bauer was one of the very few who declined to habilitate, despite urging by Hartmann and von Wettstein (Bauer to F. Schrader, 19.9.46, Demerec papers). In 1931 Belar, Stern, Jollos, and Hämmerling all offered courses at the university (*Kalender der deutschen Universitäten und Hochschulen,* vols. 109 and 110).

29. In 1928, for example, the KWI for Biology's supervisory committee consisted of the director-general of the Kaiser-Wilhelm Society (Glum), an industrialist (Krupp von Bohlen und Halbach), Dr. Donnevert from the Reich's Ministry of the Interior, the zoologist Karl Heider (representing the Prussian Academy of Sciences), Dr. Richter from the Prussian Ministry of Culture, and Friedrich Schmidt-Ott from the Emergency Council for German Science (*Notgemeinschaft*) as chairman (Harnack 1928, 194–95).

210   Styles of Thought

Wilhelm Society, but no academics—took on this role.[30] Although Müncheberg was commuting distance from Berlin, very few of the institute's assistants habilitated or offered courses at either the Agricultural College or the University of Berlin.[31] They evidently had little interest in a conventional academic career.

One reason for this was probably their educational background. Virtually all of Baur's assistants at Müncheberg had originally come to the Berlin Agricultural College in order to study agricultural sciences. In order to qualify for admission to such a college at this time, the candidate needed to have spent two years doing practical agricultural work (usually on a farm). Thus Baur's students arrived at the college with vocational career aims and a strong commitment to agriculture.[32] In Rudolf Schick's case, for example, his experience of the famine-ridden winter of 1916–17 had convinced him that he should study agriculture in order to contribute to the fight against hunger.[33] When these students decided subsequently to remain at the college and take their doctorates, they were in no sense forsaking the practical in favor of the theoretical, for in Baur's school the two were inseparable.[34] His excellent contacts with landowners and agricultural organizations enabled his students to combine their interests in science and agriculture. These contacts, for example, enabled Schiemann to conduct her experiments on winter and summer wheat varieties during the 1920s at one of the region's seed companies. At one of the institute's annual summer schools for plant breeders, similarly, Kuckuck made contacts with various breeders and arranged several collaborative projects with them.[35] Among Kühn's and von Wettstein's assistants, in

30. Cf. "Satzung," Abt. 1, Rep 1a, no. 2610, MPG-Archiv, and the minutes of various meetings of the subcommittee between 1931 and 1933 (in ibid., no. 2619).

31. Of the thirteen department heads at the Institute for Breeding Research in 1934–35, only Kuckuck and Stubbe habilitated, and this was *after* leaving the Institute (Kuckuck interview; H. Stubbe 1959b, 299; Asen 1955).

32. Hermann Kuckuck, for example, had spent his summer vacations as a schoolboy working on farms, and by the end of his first year at the Berlin Agricultural College, he had resolved to become a plant breeder (Kuckuck 1988, 10–12).

33. Kleinhempel 1985, 34. The same experience persuaded Hans Kappert—Baur's successor at the Agricultural College—to resign his assistantship at the KWI for Biology in 1920 for a research position in the seed-breeding industry (Kappert 1978, 163–64).

34. It was this feature of Baur's approach to genetics which had attracted Hans Nachtsheim away from university zoology (Nachtsheim to Baur, 13.8.23, Sign: 20282, Rep 87B, ZStA-Merseburg).

35. Kuckuck 1988, 14, 24. According to Gruber ("Erinnerungen an Müncheberg," typescript, Mappe 2, Baur papers, MPG-Archiv), these summer schools began in 1929.

contrast, I find only two who were educated in comparable technical institutions.[36]

Given their markedly practical orientation, it is perhaps not surprising that several members of the Baur school—like their "Chef," as we shall see in the next chapter—supported the liberal wing of the German eugenics movement. To be sure, several eminent "comprehensives" such as Carl Correns, Richard Goldschmidt, and Valentin Haecker also published on eugenics topics and were members of various German eugenics organizations (Weindling 1989). If there was a difference, therefore, between pragmatic and comprehensive factions' attitudes toward eugenics—and I would emphasize that this remains to be demonstrated—it may prove less evident among the older members of these factions (Baur, Correns, Haecker, and Goldschmidt were all born between 1864 and 1878) than among their students and assistants (almost all of whom were born between 1890 and 1910). Among the younger members of the Baur school, for example, Hans Nachtsheim endorsed voluntary sterilization on eugenic grounds, not only during the 1920s (Deichmann 1991, 308–9) but also after 1945 (Weingart et al. 1988, chap. 6; Weindling 1989, 566–70). But Hans Stubbe, Emmy Stein, and Hermann Kuckuck also published articles in eugenics journals during the early 1930s.[37] I have not come across a comparable level

36. Kühn's assistant Erich Becker received his doctorate from the *Technische Hochschule* Darmstadt in 1937 (V. Schwartz to me, 1.10.87; H. Piepho to me, 20.10.87 and 7.11.87). Von Wettstein's assistant Karl Pirschle studied four semesters at the *Technische Hochschule* Brunn before taking his degree at the University of Vienna. He then spent six years as head of the plant psychology group at I. G. Farben in Ludwigshafen before joining von Wettstein in 1932 (Berlin Document Center).

Whether geneticists of the "comprehensive" faction were less interested in the application of science is difficult to say, since the evidence is scanty. Consistent with such a contrast, at least, are the claims that Correns, von Wettstein, and Renner (all of them comprehensives, as we shall see) corresponded closely to "the ideal of the German professor, who does science for its own sake, passionately pursuing truth rather than particular purposes" (Baur 1933; cf. Renner 1930).

37. According to Kuckuck, the more liberal Berlin branch of the German Eugenics Society (*Deutsche Gesellschaft für Rassenhygiene*) sought support among Baur's staff for their new journal, *Eugenik* (Kuckuck 1988, 30). Stein, Stubbe, and Kuckuck obliged, to be sure with "technical" articles, discussing genetic phenomena thought to be of relevance to eugenics, rather than overt defenses of eugenic policies (vol. 1 [1930] of *Eugenik*). Stein's and Stubbe's contributions to vol. 2 (1935) of *Der Erbarzt* were similar in character. Paula Hertwig may have been similarly inclined; in an unpublished book manuscript on German geneticists and eugenics, Adela Baer observes that Hertwig, Stein, and Stubbe were the only geneticists not known to be associated with the German Eugenics Society who attended the International Population Congress in Berlin in 1935.

of engagement among young comprehensives. Moreover, the only public critics of the eugenics movement which I have found within the German genetics community were both of "comprehensive" persuasion.³⁸ Tentatively, therefore, I suggest that the Baur school's stance on eugenics may have distinguished it from the comprehensive faction and was in any event consistent with the former's enthusiasm for linking theory with practice.

In their connections to Berlin's prestigious Prussian Academy of Sciences, too, the KWIs for Biology and for Breeding Research were radically different. Adolf von Harnack's original proposal that the Kaiser-Wilhelm Society have close relations to the Academy was largely realized in the case of the Institute for Biology. As we saw in chapter 4, its constitution specified that one of the five members of the supervisory committee and two of the nine members of the academic advisory panel would be drawn from the Prussian Academy.³⁹ The only academic in-

---

38. In his review of a book of essays on eugenics and racial anthropology, Karl Belar was critical of the scientific basis of Alfred Ploetz's contribution, adding that essays on that subject were often excessively speculative (Belar, *Biologisches Zentralblatt* 45 [1925]: 127–28). Fritz von Wettstein's cautious support in 1930 for the gradual introduction of scientifically founded eugenic measures hardly counts as a critique of the movement (von Wettstein 1930b, 405), and three years later he was consulted by those who drafted the Nazis' first eugenic law (Weindling 1989, 506). More remarkable, however, is the manner in which von Wettstein discussed "the genetic bases of racial hygiene" in a collection of essays designed to prepare the medical profession for its role under the new eugenic legislation. At the end of his paper he stressed the complexity of genetic processes in such a way as to undermine facile hopes in the efficacy of negative eugenic measures. Human traits rarely had a simple Mendelian basis; genic interaction meant that it was very difficult to label a given gene as good or bad; most of the alleles which *were* dysgenic, however, were recessive and thus relatively inaccessible; and given how slowly eugenic measures could act, more rapid improvement in the population's mental and physical health could be brought about through better environments for German youth (von Wettstein 1934b, 30–33).

Such cautious support for a moderate eugenics is reminiscent of the "reform eugenists" in other countries. Nevertheless, I have found that, compared with their American or British counterparts, *very* few German geneticists voiced *any* criticism of the eugenics movement during the interwar period, a finding which is confirmed by Adela Baer's much more extensive research on this issue (unpublished book manuscript). Whether this dearth of criticism reflected a high level of eugenic conviction among German geneticists or rather their reluctance to jeopardize the utilitarian image of their fledgling discipline is not clear. I have made a case for the latter hypothesis in Harwood 1989.

39. Of the remaining seven members of the advisory panel, four were to represent other academies plus the University of Berlin, and three would be named by the senate of the Kaiser-Wilhelm Society ("Satzung des KWIs für Biologie," Abt I, Rep 1a, no. 1533, MPG-Archiv). On the response to Harnack's proposal, see Wendel 1975, 116, 125–34.

stitution represented on the supervisory committee of the Institute for Breeding Research, in contrast, was the Berlin Agricultural College.[40] In exchange, several memberships in the Prussian Academy were reserved for directors of KWIs. Correns was elected in 1915, and both Kühn and von Wettstein were elected to membership within a year of their arrival in Berlin. In Baur's case, however, the Academy was evidently in no hurry. It was six years after the establishment of the KWI for Breeding Research before a group within the Academy nominated Baur for membership (to fill the place vacated through Correns's death), and Baur died before the election could take place.[41] Finally, as a glance at their annual reports demonstrates, the two institutes attempted to reach different audiences with their research. Like the group in Dahlem, staff at the Institute for Breeding Research published in a number of standard outlets for genetics and general botany: the *Zeitschrift für induktive Abstammungs- und Vererbungslehre*, the *Handbuch für Vererbungswissenschaft, Forschungen und Fortschritte, Die Naturwissenschaften, Planta*, and the *Zeitschrift für Botanik*. In 1930–31, for example, roughly one-third of their papers appeared in one or another of these journals. Most of their work, however, was aimed at those working in agriculture. For example, another one-third of the group's papers appeared in two journals which seem to have been directed at professional breeders, whether in academe or commerce: the *Zeitschrift für Pflanzenzüchtung* and *Der Züchter*. Both of these were edited by Baur, the latter serving effectively as house journal for the institute at Müncheberg. Most of the remaining third of the papers were published in what appear to be popular agricultural newspapers and magazines (e.g., *Gartenbauwissenschaft, Weinbau- und Kellerwirtschaft, Der Deutsche Forstwirt, Wiener landwirtschaftliche Zeitung, Fortschritte der Landwirtschaft, Mitteilungen der deutschen Landwirtschafts-Gesellschaft, Deutsche landwirtschaftliche Presse, Forstarchiv*, and the *Landbund-Zeitung*.

In terms of both its internal structure and its ties to external organizations, therefore, the KWI for Biology resembled the academic institutions of the day to a far greater extent than did the KWI for Breeding Research.

40. "Satzung des KWIs für Züchtungsforschung ...," Abt I, Rep 1a, no. 2610, MPG-Archiv.
41. Archive of the Preussischen Akademie, Abschn. II, sign II-III, 43, pp. 216–21. Although he was a member of the Bavarian Academy of Sciences and deputy director (like Kühn) of the KWI for Biology from 1919 to 1935, Goldschmidt was nominated but never elected to the Prussian Academy, evidently for internal political reasons (Schlicker et al. 1975, vol. 2).

6.3 Patterns of Funding

As I noted in chapter 4, both Baur's and Kühn's schools attracted research funds from the Rockefeller Foundation. Similarly, the Emergency Council for German Science (*Notgemeinschaft*) supported Kühn and Henke's first work on *Ephestia* as well as Nachtsheim and Hertwig's small animal breeding unit at the Agricultural College, and it "generously equipped" the new laboratories at Müncheberg.[42] Furthermore, the directors of both KWIs received their salaries from the Prussian Ministry of Culture. But there the similarity ends, for, as table 6.2 shows, the running costs for the two institutes were covered by very different sources. While the Institute for Biology relied heavily upon the Kaiser-Wilhelm Society (whose funds were derived after 1911 in varying proportions from the private sector plus various Reich and Prussian ministries) and received only a trivial amount of subsidy directly from industry,[43] the Institute for Breeding Research was funded almost entirely by the Reich's Ministry of Food and Agriculture and by the private sector. In its first year of operation, too, the latter institute's running and capital costs were largely covered by donations from the private sector (see table 6.3).

That both the Ministry of Agriculture and the private sector would play such a large role in the financing of the institute at Müncheberg had been part of Baur's plan from the start. As early as 1917 Baur and the private plant breeder, F. von Lochow, submitted a proposal to the Kaiser-Wilhelm Society for an institute of plant breeding.[44] The German war effort, they argued, would have been unthinkable without recent increases in yield of at least 10 to 15 percent, especially in grain

---

42. Kühn and Henke 1929, 2; ZStA-Merseburg, Rep 87B, Sign 20282; Schiemann 1934, 95. Records of the Biology Panel of the Emergency Council for German Science (*Notgemeinschaft*) indicate further grants in 1933 and 1934 for Kühn, Schiemann, Stubbe, and von Wettstein (file 117, R73, Bundesarchiv). During the 1920s the Emergency Council also had a panel for agriculture, forestry, and veterinary medicine whose members included two professors of animal breeding (R73, file 119). By 1929 this panel had been reorganized into nine subsections, including one each for animal breeding and plant breeding. It would appear, therefore, that the Emergency Council's committee structure provided avenues of support for both basic and applied genetics.

43. The only substantial connection which the Institute for Biology ever enjoyed with agriculture or industry was short-lived. From 1919 Ludwig Armbruster headed a department of bee breeding and biology which was partly supported by the Prussian Ministry of Agriculture. In 1922 it was moved to the Berlin Agricultural College (Abt I, Rep 1a, no. 1553, MPG-Archiv).

44. "Denkschrift über die Gründung eines Forschungsinstituts für Pflanzenzüchtung," folder: "Harnack no. 3," box 19, Emil Fischer papers.

Table 6.2 Funding of the Kaiser-Wilhelm Institutes for Biology and for Breeding Research, 1932–1935

Donor		1932–33[a]	1933–34[b]	1934–35[c]
Kaiser-Wilhelm Gesellschaft	B: BR:	228,000 12,400	237,000 12,400	212,000 12,400
Prussian Ministry of Culture	B: BR:	26,000 14,000	30,000 ?	27,000 12,600
Reich Ministry for Food & Agriculture	B BR:	0 135,000[d] +35,000	0 135,000 +95,000	0 235,000 +92,000
Industry & Commerce	B: BR:	1,200 63,000	1,200 99,000	600 34,000

Note: Sums are in reichsmarks and have been rounded off. Income derived from local government, interest on investments, the institutes' property, or "miscellaneous" sources has been omitted, since it generally constituted a small proportion of overall income. Available data for earlier years reveal the same overall patterns of funding for each institute. For 1928–32 at the Institute for Breeding Research, see MPG-Archiv, Abt I, Rep 1A, nos. 2615, 2616, 2620, and 2665. For 1915–24 and 1929–31 at the Institute for Biology, see nos. 1557, 1558, and 1559.

[a] *Sources:* MPG-Archiv, Rep 8; and Rep 1A, nos. 2624 and 2665.

[b] *Sources:* Correns papers, file 198; MPG-Archiv, Rep 1A, nos. 2665 and 2624.

[c] *Sources:* MPG-Archiv, Rep 8; and Rep 1A, no. 2625.

[d] Where there are two figures listed, the upper one is general support for the institute, and the lower one refers to subsidies for specific projects.

and potatoes, which were attributable to plant breeding. Since Germany would have to reckon with a new set of economic circumstances after the war, it was highly desirable that she reduce her dependence on imports of plant products by developing new domestic varieties, for example, of oil-rich plants. For several reasons, however, private-sector breeders were unlikely to take on this task. First, selection among existing plants had so far been responsible for the improvement of German plant varieties; although it produced no novelty, it was a fast and cheap method with which private breeders could make a profit. But the best varieties of wheat, barley, oats, peas, and beans had now been selected, and no further improvement could be expected with this method. Progress in future would only come from crossing different varieties (e.g., domestic with wild ones) and selecting appropriate recombinants. The trouble with this method, however, was that it was slow and required relatively elaborate facilities. Moreover, there was no patent protection

Table 6.3 Capital and Running Costs of the Kaiser-Wilhelm Institute for Breeding Research, 1928–29

Capital costs

400,000	from Deutsche Rentenbank—Kreditanstalt
50,000	I. G. Farben
25,000	Fritz Behrens Foundation
25,000	Herr Wentzel (private plant breeder)
40,000	Society for the Advancement of German Plant Breeding
5,000	a plant breeders' cooperative in Munich
120,000	several of Berlin's major banks
210,000	Reich Ministry for Food and Agriculture
1,000	an agricultural bank
133,000	Jakob Goldschmidt (Darmstädter und Nationalbank)
21,000	Reichskuratorium [?]
1,600	Notgemeinschaft der Deutschen Wissenschaft
15,000	a fertilizer company
43,400	Reich [?]

Running costs

25,000	various branches of the fertilizer industry
65,000	Reich Ministry for Food and Agriculture
1,000	a Prussian bank
20,000	Fritz Behrens Foundation
10,000	a grain traders' association
10,000	Herr Hagedorn
5,000	Otto Wolf
1,500	German Agricultural Society
3,600	Chamber of Agriculture for Brandenburg
1,000	Kaiser-Wilhelm Society
1,000	local government at Lebus

Sources: MPG-Archiv, Abt I, Rep 1A, no. 2615 (from a meeting on 22.12.27), no. 2620 (list dated 21.9.28), and no. 2620 (list headed "Einnahmen" which was probably the projected budget for 1928–29). Amounts are in reichsmarks.

for new plant varieties, since any farmer could easily grow them and sell the seed himself. Thus no private breeder was likely to risk the necessary investment. Since the gains from this crossing method were of national importance, however, it was essential that the state or other public-sector body establish an institute of plant breeding which would conduct the necessary research and development. The institute should be directed by a geneticist, with each department headed by a plant breeder. Each department would concentrate upon a particular agriculturally significant plant, conducting general research of a kind which might become important to breeding practice, as well as work more closely geared to the expressed needs of the seed industry. Those promising new varieties which bred true would be tested on the institute's experimental farm and thereafter at state agricultural experiment stations. Those which proved successful would be brought back to the institute for replication and finally turned over to private breeders for commercial development.

Raising the necessary funds, Baur and von Lochow thought, should not be difficult; there were wealthy landowners and industrialists who could be tapped, and the Kaiser-Wilhelm Society already had experience with establishing institutes for applied science. Perhaps sensing that finance might not be entirely straightforward, Baur was enthusiastic about the Swedish plant-breeding institute at Svalöf directed by his friend, Hermann Nilsson-Ehle.[45] The institute at Svalöf was built largely with donations from farmers and breeders; its running costs were shared between the Swedish state and a plant-breeding corporation. In exchange for this funding, the corporation received new varieties bred at the institute. When the corporation's profits from the sale of these varieties exceeded 8 percent, a proportion was paid to the state, and this sum was generally large enough to offset the state's contribution to the institute's running costs. Once established, therefore, the institute was essentially self-financing.

For reasons that are not clear, Baur and von Lochow's proposal appears to have been shelved by the Kaiser-Wilhelm Society for nearly a decade. By the beginning of 1927, however, Baur and his organizing committee had secured enough promises of financial backing from interested parties that a similar proposal could be resubmitted. The senate of Kaiser-Wilhelm Society gave its approval, and the new Institute for Breeding Research incorporated some of the features from the Svalöf model.[46] Within three years it had produced its first commercially significant variety: the "sweet lupin." In 1927 one of Baur's assistants at the Agricultural College, Reinhold von Sengbusch (1898–1985), had become interested in the lupin, a plant rich in protein but whose high alkaloid content gave it a bitter taste, thus preventing its use as a forage crop. "Sweet" mutants low in alkaloids were thought to exist in principle, but the problem was how to identify them. Within a year, having devised a simple analytical method which permitted rapid scanning of thousands of seeds, von Sengbusch had selected three such mutants, and by 1930 several varieties of sweet lupin were ready for sale (Hondel-

45. Baur to Dr. Oldenburg, 14.1.18, ZStA-Merseburg, Rep 87B, Sign 13259.
46. Cf. Abt I, Rep 1a, no. 2615, inter alia "Sitzung der KWG" (31.3.27), and minutes of the meeting of the "Engeren Ausschusses" on 12.5.27, MPG-Archiv.

mann 1984). The following year the varieties were sold by the Kaiser-Wilhelm Society to a consortium of German plant-breeding companies for 50,000 reichsmarks plus a percentage of the consortium's profits. Soon Baur was exploring the possibility of taking on a contract from the oil industry to breed an oil-rich lupin. Similarly, in the early 1930s a low-nicotine variety of tobacco, which had been developed in von Sengbusch's department at Müncheberg, was turned over to a government breeding station for development.[47]

To some extent the internal organization of a research institute can probably be ascribed to the aims and personality of its director. Very much the energetic and ebullient academic-entrepreneur, Baur was prepared to *manage* the work conducted by his young staff at Müncheberg so that the Institute of Breeding Research produced results which were—at least in part—economically relevant. Correns, by contrast, was a shy man who disliked administration and was quite happy to let department heads at the Institute for Biology run their own shows. And apart from requiring social deference, most of those department heads were evidently prepared to allow their assistants a great deal of freedom in their research. But there is little doubt that the Institute for Breeding Research's dependence upon agricultural funding had major organizational consequences, both for the nature of the institute's external ties and for its hierarchical division of labor within. It is unlikely that anyone of Correns's disposition would ever have become director of the Institute for Breeding Research; had he done so, by some fluke of circumstance, he would have had little alternative but to organize the institute much as Baur did.

6.4 Institutional Developments after 1933

Thus Baur's school resembled more closely an industrial laboratory than an academic institution. Since the National Socialist German Workers' Party both championed the needs of farmers and sought to gear research more closely to the economic requirements of an expansionist foreign policy, it comes as no surprise that the Party was interested in the work of the Institute for Breeding Research. And in view of the Institute for Biology's concentration upon basic research, one

47. On the sale of the sweet lupin see nos. 2615–17 and 2619, Abt I, Rep 1a, MPG-Archiv. The oil-rich lupin was discussed at a combined meeting of the *Verwaltungsausschuss* and the *Verkaufskommission* on 10.12.32. On the tobacco variety see the minutes of the meeting of these two committees on 12.2.31 (no. 2616, as above).

might have expected it to be subject to rather more pressure than its counterpart in Müncheberg. As David Joravsky has shown, the Soviet disciplines which came under the greatest ideological pressure during Stalin's reign were those with least apparent utility (e.g., philosophy, sociology), while evidently useful fields like chemistry, medicine, and engineering were left alone. Similarly, industrial laboratories in Germany after 1933 experienced far less political intervention than academic ones (Joravsky 1983; Hirsch 1975). In the German genetics community, however, just the opposite occurred. While the Institute for Biology under von Wettstein's directorship became a haven for dissidents, the events at Müncheberg during 1933 are a textbook example of the process of *Gleichschaltung* (forced alignment) in which the National Socialist regime consolidated its control of so many organizations in German society. Tracing the developments at these institutions after the Nazis' takeover will make it apparent that our "map" of the German genetics community cannot be restricted to institutional geography. It must also encompass the ideological landscape.

In May 1933 R. Walther Darré, then president of the German Agricultural Council, was elected a member of the Kaiser-Wilhelm Society's senate. Well informed about the activities of Baur's institute, Darré expressed an interest in becoming chairman of the institute's supervisory committee.[48] The senate, realizing that Darré was about to be named Minister for Food and Agriculture, promptly acceded to his wishes, evidently without consulting Baur.[49] Two weeks later chairman Darré appointed ten new members to the supervisory subcommittee (*Verwaltungsausschuss*), six of whom had joined the Party prior to 1933. Many of these appointees' replies indicate that they had no idea what the institute did nor what their new responsibilities were to be.[50] Since these appointments violated the institute's constitution, the president of the Kaiser-Wilhelm Society (Max Planck) obliged by writing to the senate, recommending that this constitution be amended. The new constitution streamlined the supervisory committee (*Kuratorium*), eliminating representatives from thirteen agricultural, academic, and

48. On Darré's election see no. 2987, Abt I, Rep 1a, MPG-Archiv. In an *Aktennotiz* dated 20.6.33, Glum noted Darré's interest in becoming chairman of the supervisory committee (ibid., no. 2612).
49. Planck to Darré, 30.6.33 (no. 2987), and Baur to Planck, 27.7.33 (no. 2612), Rep 1a, MPG-Archiv.
50. See Darré's letters to the ten, dated 13.7.33 (no. 2612, Rep 1a, MPG-Archiv). Party membership for the ten was established using *Wer Ist's?* (10th ed., 1935).

financial organizations.[51] Planck then wrote to these ousted representatives—some of whom were among the institute's financial patrons, while others had served on its supervisory subcommittee or its sales committee—thanking them for their past service to the institute while expressing regret that those services would no longer be required.[52] Baur's protest at this procedure succeeded only in getting two former members back onto the supervisory committee, and he complained to Planck about not having been consulted about the new membership of the committee. Planck replied meekly that Darré had already eliminated several of the names which Baur had proposed for the new committee, inserting several from "his circle of confidants."[53]

August and September marked a worsening in the relationship between Baur and Darré. A sharp reduction in industrial support for 1933–34 confronted Baur with a deficit of 40,000 marks. When Darré's ministry declared itself unable to cover the deficit due to cuts in its own budget, Baur threatened to fire two hundred employees and close down the institute. As the firings began and reports of the impending closure appeared in the press, Baur reapproached the ministry, armed with a promise of 10,000 marks from a private donor on the condition that the rest be raised elsewhere. Before any more support could be provided, the ministry replied, Baur would have to present his research plan for 1933–34. But this plan, Baur objected, had already been approved by the appropriate body: the supervisory committee on which the ministry was represented. Ignoring Planck's suggestion that he comply with the ministry's request, Baur wrote again to the ministry, restating his financial requirements and announcing that he had so far fired 150 institute personnel.[54] But this kind of brinksmanship—so successful in Baur's earlier attempts to extract funding from the Prussian Ministry of Agriculture and Forestry for his institute at the Agricultural College[55]—failed to persuade. In October Darré informed members of the reconstituted supervisory committee that an investigation of the research activities at the institute was urgently needed, and he re-

51. For Planck's request see Planck to KWG Senators, 19.7.33 (no. 2612, Rep 1a, MPG-Archiv). A typescript of the new constitution, approved by the senate on 19.7.33 and 10.8.33, is in ibid., no. 2610.

52. A master copy of the letter from Planck (dated 26.7.33) with recipients' names penciled in is in no. 2612, Rep 1a, MPG-Archiv.

53. Baur to Planck (27.7.33, 7.8.33, and 16.10.33), Baur to von Cranach (16.10.33), and Planck to Baur (18.10.33), all in no. 2612, Rep 1a, MPG-Archiv.

54. Cf. the correspondence between 31.7.33 and 9.10.33 in no. 2665, Rep 1a, MPG-Archiv.

55. Cf. Sign 20281 and 20282, Rep 87B, ZStA-Merseburg.

quested, in the interest of speed and efficiency, that they transfer the committee's powers temporarily to him as chairman. Although several members of the committee objected to this procedure, Planck agreed, simultaneously assuring Baur that he endorsed (!) the latter's wish for autonomy and suggesting that the "misunderstanding" between Baur and Darré could be cleared up by discussion.[56] In November Darré again wrote to members of the supervisory committee, claiming that his previous letter had been misunderstood and assuring them that he had no intention of radically altering the relationship between his ministry and the Kaiser-Wilhelm Society. This, of course, left open the nature of his intentions with regard to the *institute,* and the ambiguity must have been especially alarming, since in the same letter Darré announced that he had instructed a "trusted expert" to conduct an investigation of the institute's research projects, identifying which of them were economically significant and thus deserving of special funding and which less important ones could be reduced in scale or shut down altogether.[57] Judging from Planck's letter a week later, urging him to sit tight until Darré's representative had visited Müncheberg, Baur must have been worried. And judging from a report to Darré on the institute's financial affairs, Baur had good cause to be worried. Besides noting the cost of the apartment provided for Baur, the purchase of his farm by the institute, and the size of various supplements to his salary, the report also argued that substantial savings could be made by transferring various parts of the institute's research program to university institutes where related work was already going on.[58]

By December 2 Baur was dead of a heart attack, aged fifty-seven. Virtually all of his obituarists attributed this unexpected death in part to the strain of the previous months, during which Baur was allegedly threatened with imprisonment (Kuckuck 1988, 31). Documentary evidence on the causes of the friction between Baur's institute and the

56. Darré to members of the *Kuratorium,* 19.10.33; Planck to Baur, 31.10.33; the members' replies to Planck are dated between 20.10.33 and 26.10.33 (all in no. 2612, Rep 1a, MPG-Archiv).

57. Darré to members of the *Kuratorium,* 17.11.33 (no. 2613). The "expert" was a breeder named Hermann Schneider who was president of the Lower Silesian Chamber of Agriculture and had served as secretary of the Supervisory Committee since his appointment by Darré four months previously. Cf. the *Aktennotiz* by Cranach dated 18.10.33 (no. 2665) and Darré to Schneider, 17.11.33 (no. 2612), all in Rep 1a, MPG-Archiv.

58. Planck to Baur, 24.11.33 (no. 2665). For a copy of this report, which was evidently sent to the Kaiser-Wilhelm Society, see Schneider to Darré, 15 "Nebelung" (November) 1933 (no. 2624), both in Rep 1a, MPG-Archiv.

222   Styles of Thought

Party is difficult to find. According to anecdotal sources, Baur had been interested in National Socialist agrarian policy since the early 1930s and went to Party headquarters about 1932 to learn more. Not at all impressed by Darré, Baur attacked Darré's book *Neuadel aus Blut und Boden* (1930) and was openly critical of the direction which Nazi agrarian policy seemed to be taking early in 1933. Darré complained and eventually retaliated.[59] That this conflict lay between Baur and Darré as individuals, rather than between the institute and the Party, is supported by the fact that the agriculture ministry's appropriations for the institute rapidly increased following Baur's death.[60] And as we shall see in the next chapter, Baur saw considerable promise in the NSDAP's policies on eugenics, agriculture, and possibly other areas as well. That he dared to voice his criticism of Darré so openly probably owes something to Baur's rather blunt and uncompromising manner. But it is also likely that he felt his institute would be safe enough, and his own credibility high enough, under the new regime that he could afford to speak out with some hope of redirecting the Party's agrarian policies. Whence this confidence?

Support for National Socialism among staff at Müncheberg was unusually strong.[61] At least three of the assistants had joined the NSDAP before 1933, and others joined the SS thereafter.[62] Bernhard Husfeld (1900–?), an assistant with Baur on and off since 1921 who served as an acting director of the institute on Baur's death until 1936, probably

---

59. Kuckuck interview; Kuckuck 1988, 31; "Lebenslauf," Ordner no. 7, von Sengbusch papers, MPG-Archiv; H. J. Troll to me, 31.3.84.

60. The ministry's contribution to the institute's recurrent grant increased from RM 135,000 for 1933–34 to RM 235,000 for 1934–35, more than compensating for a decline in private funding (nos. 2624, 2625, and 2665, Rep 1a, MPG-Archiv). Annual reports from the institute between 1934 and 1938 indicate a steady increase in scientific staff (from twenty-six to forty-eight) and new building ("Tätigkeitsbericht der KWG," *Die Naturwissenschaften*, vols. 22 (1934), 23 (1935), 24 (1936), 25 (1937), and 26 (1938).

61. Joachim Hämmerling claimed that Baur's was one of the few KWIs to have been so politicized (letter to Curt Stern, 20.5.46, Stern papers).

62. Klaus von Rosenstiel (1905–?), Baur's son-in-law, joined the Party on 1.6.32 and is described in a recommendation for promotion within the SS as "one of the keenest and most reliable SS men" ("einer der eifrigsten und zuverlässigsten SS-Manner") (Berlin Document Center). Friedrich Gruber (1904–?) joined the Party on 1.4.32 and was in the SS by 1937 (BDC). Joachim Hackbarth (1906–1977) became a Party member on 1.5.32 (BDC). In 1933 Hans Ross (1912–?) joined the SS and Konrad von Rauch (1905–1945) the Party, respectively (BDC). According to Kuckuck (1988, 34), another department head, Meyle, was also in the SS.

joined the NSDAP before 1933 and was a member of the "SS supporters' club."[63] But the staff at Müncheberg were not all pro-Nazi, and life soon became difficult for the institute's social-democratic minority. Early in 1936 Husfeld brought Hans Stubbe, Hermann Kuckuck, and Rudolf Schick before an industrial tribunal for "disturbing the workplace." They had complained in a letter to Husfeld of the way in which Baur's valuable collection of *Antirrhinum* strains was being neglected, and—apparently hoping for support from their colleagues—circulated the letter to other assistants and department heads in the institute. Husfeld replied by rejecting such "strongly liberal methods" of resolving a dispute as "inconsistent with National Socialism" and insisting that Stubbe et al. be disciplined as vigorously as possible.[64] Stubbe et al. appealed in vain to the Kaiser-Wilhelm Society to bring in a neutral referee to settle the matter, but Planck felt the Society could not intervene in a matter which lay within the jurisdiction of the acting director (an argument he had discreetly chosen *not* to employ three years earlier when Darré was crudely overriding Baur's jurisdiction).[65] Appeals for lenience to Wilhelm Rudorf (1891–1969), a former student of Baur's who took over as director at Müncheberg on 1 April 1936, fell on deaf ears, perhaps because Rudorf was himself at least tolerant of National Socialism.[66] On April 24 the Industrial Tribunal ruled against Stubbe

63. Such "Fördernde Mitglieder" paid a subscription to the SS but took no part in SS activities (H. Mehrtens, personal communication). Although documents at the Berlin Document Center date Husfeld's party membership from 1.5.33, several other sources indicate that he may have joined earlier. According to Hans Stubbe, Husfeld is reported to have admitted during denazification inquiries to having joined briefly in 1932 (Stubbe to Klaus Pätau, 12.4.46 and 23.4.46, Pätau correspondence). Furthermore, Georg Melchers recalls Husfeld having worn a Party insignia with a gold rim, indicating that he had joined the NSDAP before the seizure of power and was thus an *alter Kämpfer* (Melchers to me, 2.3.88). In any event, Husfeld's sympathies for the reactionary right were evident as early as 1925, when he joined the Deutsch-Nationale Volkspartei (Berlin Document Center).
64. Letter by fifteen staff members to Husfeld, 12.2.36, no. 2710, Rep 1a, MPG-Archiv.
65. Cf. *Aktennotiz*-Telschow, dated 29.2.36, and Planck to Stubbe, 10.3.36, both in no. 2710, Rep 1a, MPG-Archiv.
66. On the appeals to Rudorf see Glum to Rudorf, 30.3.36 and 3.6.36; Rudorf to Glum, 3.4.36; Stubbe to Rudorf, 5.4.36; all in no. 2711, Rep 1a, MPG-Archiv. According to one observer, Rudorf and Husfeld "were, or at least pretended to be, ardent Nazis" (Lang 1987, 22). After the war Melchers, Hämmerling, and Lang were not amused when the Kaiser-Wilhelm Society allowed Rudorf to remain head of the institute (Melchers to Pätau, 16.11.45, and Lang to Pätau, 12.12.45 and 31.1.46, Pätau correspondence; Hämmerling to Curt Stern, 20.5.46, Stern papers).

et al., and on April 30 Rudorf informed them that they were fired. A few months later Rudorf successfully forced von Sengbusch to resign, allegedly because of political noncompliance.[67] By 1936, therefore, the institute was firmly under Nazi control.

While Kuckuck, Schick, and von Sengbusch found employment in the seed-breeding industry, Stubbe was taken on as an assistant at the KWI for Biology by von Wettstein.[68] Unable to take up several academic job offers because the Reich's Ministry of Education found him politically unreliable, Stubbe regarded the institute in Dahlem as "a spiritual oasis" (Melchers 1972; 1987b, 14; Hagemann 1984, 95–96). Others would have agreed. When Elisabeth Schiemann's right to teach at the University of Berlin was revoked on political grounds in 1940, von Wettstein helped her to obtain a fellowship from the Deutsche Forschungsgemeinschaft so that she could continue her research (Lang 1987, 24–25). One of von Wettstein's assistants, Georg Melchers, had been active in a socialist wing of the youth movement during the 1920s, and another, Karl Pirschle (1900–1945), had lost his right to teach at the University of Berlin in 1936.[69] In this environment Goldschmidt felt reasonably happy despite the abysmal political climate outside. Even the handful of National Socialist sympathizers within the institute were "behaving impeccably."[70] By the end of the year, he was no longer permitted to lecture, publish, or attend colloquia, but Hartmann, Häm-

67. The claim is von Sengbusch's in an autobiographical sketch in Ordner no. 7, von Sengbusch papers, MPG-Archiv. The only published account of this episode so far is by Straub (1986), who observes that although no political motivation is visible in von Sengbusch's removal, his subsequent attempt to habilitate (1940) was declined on grounds of "political unreliability."

68. Ordner no. 7, von Sengbusch papers, MPG-Archiv; Schiemann 1963; Hagemann 1984.

69. Melchers, six-page autobiographical typescript (1985), sketching his involvement with the youth movement, made available to me by the author; Pirschle had joined von Wettstein at the Institute of Botany in Munich in 1932, habilitated in 1933, and moved to Dahlem with von Wettstein in 1934, but for reasons unknown to me, his right to teach was withdrawn in June 1936 (Berlin Document Center). By all accounts, Joachim Hämmerling would also have needed political protection. His wife was a science teacher at Berlin's *Karl Marx Schule* for gifted working-class children, until it was shut down in 1933 (Harris 1982), and in 1939 Hämmerling gave up his job at the KWI for Biology for a position at the German-Italian Institute for Marine Biology in Rovigno because he felt he had no chances of a chair in Germany (Hämmerling to Curt Stern, 20.5.46). It was Ursula Philip's impression (interview) that Hämmerling's politics were socialist.

70. Goldschmidt to Curt Stern, 23.3.34, Stern papers; cf. Goldschmidt to Stern, 2.12.34, and Goldschmidt to Ross Harrison, 26.3.34, Harrison papers.

merling, and others remained friends, and he was optimistic. Ideologically, it would seem, the institutes in Dahlem and Müncheberg were poles apart.

6.5 Conclusion

From the beginning, research in Baur's institute at the Berlin Agricultural College was both narrower in theoretical scope and more applied than at Göttingen. As the schools transferred to Kaiser-Wilhelm institutes, they diverged yet further, in both research program and institutional structure. On moving to Müncheberg in 1928, for example, Baur's group focused heavily (although not exclusively) upon applied research, at least part of which could be sold to private breeders. In addition Baur consolidated his control over staff by employing young men whom he personally had trained as agricultural scientists. In marked contrast, the Institute for Biology in Dahlem was less hierarchically organized than most university institutes. But the two KWIs differed in other respects as well: the strength of their ties to academic institutions and the extent of their financial dependence upon the private sector.

By comparing the institutions in which the Baur and Kühn/von Wettstein schools were located, I hope to have shed further light upon the pragmatic and comprehensive factions which these two schools represent.[71] One would expect that institutions so distinct in aim and char-

71. To use the term *faction* may be somewhat misleading insofar as it implies that members of these two sectors of the genetics community were conscious of their differences. Surprisingly, I have found no signs of group consciousness among either pragmatics or comprehensives, nor signs of tension between them. Just why this was so is not clear, but there are several possible explanations. First, in order to become aware of group differences, one has to *encounter* them, and the facts of institutional segregation may have militated against this. In the winter of 1914–15 Baur's Institute of Genetics acquired temporary accommodation in Potsdam, southwest of Berlin and some fifteen kilometers from the KWI for Biology in Dahlem. By the spring of 1923 his new institute in Dahlem was sufficiently near to completion that some of his staff could move there from Potsdam, but it was early 1925 before Baur himself and the rest of his staff could finally transfer (Schiemann 1934; ZStA-Merseburg, Rep 87B, Sign 20281-20282). Once both institutes were in Dahlem, of course, the likelihood of social and intellectual contact between the two groups was much higher. Elisabeth Schiemann took students with her to seminars in the Institute for Biology, and Hermann Kuckuck recalls the relations between the two groups having been very good (Kuckuck 1988, 13; Kuckuck interview). Within three years, however, the frequency of these contacts must have dropped sharply, since Baur and most of his assistants left for the KWI for Breeding Research in Müncheberg, fifty kilometers east of Berlin. A second consideration is that group consciousness re-

acter would recruit different "kinds" of geneticists, and indeed, Baur's assistants had very different educational backgrounds and career aims than Kühn's or von Wettstein's. More puzzling, perhaps, are the apparently different responses of the two schools to the National Socialist regime after 1933. But what, if anything, have educational and political differences of this kind to do with these schools' different approaches to genetics? If we are to identify the connections among these phenomena, we will need a more detailed inventory of the interests and values of pragmatic and comprehensive geneticists.

---

mained latent because pragmatics and comprehensives did not, by and large, compete for the same pool of resources. Most of Baur's funding came from Reich and Prussian ministries of agriculture and the private sector, while comprehensives looked to various ministries of education and the Kaiser-Wilhelm Society. A third possibility is that the institutional weakness of genetics in Germany meant that recognition of group differences was suppressed whenever the two factions met within such organizations as the German Genetics Society or *ZIAV*, in order that a show of disciplinary solidarity could be maintained. Throughout the interwar period, for example, the editorial board of *ZIAV* included members from Baur's institute as well as from the KWI for Biology, and whenever the president of the Genetics Society was drawn from one of these groups, the secretary was drawn from the other.

7 Imputing Styles of Thought

Is a broad mastery of the literature within one's discipline desirable? Is it important for the geneticist to develop an appreciation of great works outside his or her field, above all in the humanities or fine arts? Is an engagement in party politics compatible with the academic's role? Geneticists of pragmatic and of comprehensive persuasion, as we shall see in this chapter, gave consistently different answers to these questions (table 7.1). What is more, the clusters of interests and values which are revealed in those answers are patterned. In this chapter I will describe these patterned clusters, designating them as pragmatic and comprehensive styles of thought. The recurring theme linking the various sectors of comprehensives' thought is a striving for conceptual synthesis and an integrated worldview. What connects the various facets of the pragmatic style is at first sight not obvious. Concern for synthesis is altogether absent, as is the preoccupation with "high culture," and the aim of science is taken to be simply prediction and control. *Why* these particular attitudes and interests came to be combined into distinctive clusters—and, following Mannheim, what was the "basic intention" underlying them—will only become clear in the next chapter when we examine those styles in sociohistorical context.

Before attempting to extract pattern from among the bewildering array of ideas and attitudes expressed by a group of unique individuals, we had better consider the very real methodological problems which such an enterprise entails. To begin with, there are many individuals whom I have categorized as pragmatic or comprehensive on the basis of their approach to genetics but who are impossible to classify on other dimensions for lack of information. This is a particularly serious problem with junior members of the Baur school and the Kühn/von Wettstein school, for whom detailed biographies are not available.[1] Where obituaries or celebratory biographies exist, moreover, they often omit the kind of information which would be useful: for example, on the geneticist's philosophical or political views. Even where the necessary

---

1. In some cases this is because these individuals died before achieving enough of a reputation to merit an obituary in a professional journal (e.g., Erich Becker, Ernst Plagge, Konrad von Rauch). In a number of cases these individuals are still alive, so that neither personal papers nor biographies/obituaries are yet available, and in my experience interviews are rarely able to elicit the kind of information that is needed.

Table 7.1 Styles of Thought within the German Genetics Community

	Pragmatic	Comprehensive
1. Problem choice in genetics	Transmission genetics	(Transmission) Developmental and/or evolutionary genetics
2. Range of biological knowledge	Specialist	Generalist
3. Interests beyond science	Popular	"High culture"
4. Political stance	Party-political	"Above" politics

information is available, categorization may still be problematic, since an individual's interests and attitudes are not necessarily a fixed characteristic; they respond to altered circumstances.[2] No static classification of the kind proposed here, therefore, will be able to register such change. Some might argue that this methodological complication is essentially self-inflicted. It arises only because I treat "style of thought" as a largely *stable* set of thinking habits which can be assigned to a particular social group. An alternative way to deploy the concept of style of thought is as a *contingent* strategy which arises in particular social situations. According to this latter usage, style does not denote a recurrent mode of thinking but an individual's momentary response to a particular set of circumstances. In this case one does not expect consistency over time in an individual's thought patterns, since they will fluctuate in accord with the particular audiences whom he or she is addressing. Thus an individual might deploy elements of both pragmatic *and* comprehensive styles instrumentally, as needed.[3] I have chosen, nevertheless, to use "style of thought" in the former typological sense, rather than the latter contingent one, not because I regard the latter as implausible, but simply because I find no evidence for it in the German genetics community. Whether geneticists were addressing fellow specialists in technical journals, giving general papers to an edu-

---

2. It is relatively easy to trace a shift in an individual's *research* interests over time; this is clear, for example, in the cases of N. W. Timofeeff-Ressovsky (cf. chaps. 2 and 3) and Georg Melchers (chap. 8). Lack of data, however, generally makes it impossible to document comparable shifts of philosophical, cultural, or political views (but see the case of Peter Michaelis, discussed in chap. 9).

3. A case for this usage is made in Smith and Nicolson 1989, and I am grateful to Malcolm Nicolson for bringing this issue to my attention.

cated lay audience, or writing privately to friends and colleagues, they seem to have been consistent in their patterns of thought. Of course, such consistency might be historically specific. For example, the extreme cultural and political polarization of German society during this period may have effectively forced thinkers to choose, "once and for all," between relatively sharply distinguished styles of thought. In other times or places, one can imagine thinkers enjoying the freedom to deploy both styles, as long as "inconsistency" had no serious consequences.

Arguably, no classification can do complete justice to the objects which it groups. Although styles of thought are the products of shared experience, no two individuals' experience is identical. Differences of biography and personality, therefore, will inject diversity into any human group, no matter how intense the collective experience they have in common. As we shall see (tables 7.2 and 7.3), some individuals display the pragmatic or comprehensive style of thought in very "pure" form (e.g., von Wettstein and Nachtsheim). Others are readily characterizable in terms of their scientific work, but hold anomalous views on another dimension (e.g., Renner, Stubbe, Schwemmle), and in a few cases individuals seem to defy classification altogether. If styles of thought are to be imputed sensibly, therefore, one has to think of individuals lying at various points along a continuum between pure pragmatics at one pole and pure comprehensives at the other, with the unclassifiables dead center. Bearing these cautions in mind, let us turn now to the evidence that pragmatic and comprehensive geneticists differed in style of thought.

7.1 Portraits in Contrast: Alfred Kühn and Erwin Baur

In order to gain a concrete sense of how the various facets of each style of thought were manifest in real individuals, let us begin with Kühn and Baur. These geneticists constitute a useful starting point, but *not* because they embody the two styles of thought in especially perfect form. On the contrary, certain facts about them—that Baur worked in evolutionary genetics, that Kühn joined a political party—distance them from the respective pure polar styles. Nonetheless, I have chosen them as exemplars because of their eminence as geneticists and their centrality as leaders of the two most important schools, and because an abundance of source material makes it relatively easy to reconstruct their views on the issues in question.

The son of a doctor, Alfred Kühn (1885–1968) spent his childhood

230   Styles of Thought

a
b

Figure 7.1   a, Alfred Kühn in 1915; b, Erwin Baur, ca. 1930 (both courtesy: Archiv zur Geschichte der Max-Planck-Gesellschaft).

in various small towns in the province of Baden in southern Germany. His ancestors on both sides of the family for several generations were from the educated middle classes. From 1904 to 1908 he studied zoology at Freiburg with August Weismann and the physiologist Johannes von Kries, serving as assistant under Weismann (and his successor, Franz Doflein) from 1908. In 1918 he went to the Institute of Zoology in Berlin as assistant to Karl Heider, who regarded Kühn as the most gifted of Weismann's students,[4] and in 1920 he was called to the chair of zoology at Göttingen, where he remained until 1937 (Kühn 1959). Erwin Baur (1875–1933) also grew up in a small town in Baden. His brother, father, uncle, and paternal grandfather were pharmacists, while his mother came from a family of craftsmen and innkeepers. From 1894 he studied medicine, taking a year out to work as an assistant in botany at Kiel, before qualifying as a physician in 1900. The following year he remained at Kiel as assistant in marine

---

4. Karl Heider to Geheimer Rat [Krüss], 11.3.19 (ZStA-Merseburg, rep 76, Vc, Sekt 2, Tit 23, Litt A, Nr. 112, Bd II.).

bacteriology. Completing his military service as a navy physician in 1902, he spent a further year in medical practice at a psychiatric clinic and mental asylum before finally returning to botany. He received his doctorate at Freiburg in 1903 and went to Berlin as first assistant in the botanical institute, where he habilitated in bacterial physiology. Failing to be called to several university chairs of botany in 1910, he felt handicapped by his family's lack of connections to academically influential circles.[5] But in 1911 he accepted the chair of botany at the Berlin Agricultural College on the understanding that a genetics institute would be built for him. From 1914 until 1929 he was professor of genetics at the college.[6] Thus Baur came from a more modest class background than Kühn, and his career took a number of unconventional twists before settling down in genetics at an institution which did not enjoy university status.

As we have seen, the two men's research interests were very different. Baur was an early advocate of the chromosome theory, demonstrating its validity in the snapdragon (*Antirrhinum*). His studies of natural populations and his consistent defense of natural selection have earned him a reputation as one of the contributors to the evolutionary synthesis (cf. chap. 3). While Baur worked in transmission and evolutionary genetics, Kühn was primarily interested in developmental genetics. As he confided to Karl Henke, "Your group's developmental-genetic studies [on *Drosophila*] are very promising. This work will certainly advance our understanding of gene-action. *Drosophila* genetics—which has frankly become rather boring (don't mention this to Timofeeff!)—thus acquires a new appeal."[7] Furthermore, the role which genetic research played in each man's career was different. Although Baur researched and lectured in bacteriology and mycology during his period as an assistant in Berlin (1903–1911), he became interested in genetics soon after the rediscovery of Mendel's laws, holding his first course on genetics in 1905 and writing one of the earliest German textbooks on the subject in 1911 (von Rauch 1944, 85). Thereafter genetics commanded virtually all of his attention; he had very little interest, for example, in general botany (von Ubisch 1956, 421). Weighing up the attractions of a chair in botany at a university versus one at the Berlin Agricultural College, he felt that the latter was more appropriate, given the nature of his research interests (von Rauch 1944, 94). Kühn, on the other

---

5. Baur to Correns, 21.11.10, cited in von Rauch 1944, 92.
6. Schiemann 1934; autobiographical sketch, Baur correspondence, SBPK.
7. Kühn to Karl Henke, 12.1.45, Kühn papers, Heidelberg.

hand, came to genetics only in the early 1920s, once he had taken up his chair at Göttingen,[8] and it never displaced his other biological interests. The first twenty years of his career were devoted to a variety of cytological and physiological problems, including cell division in amoebas and color perception in bees, the latter work culminating in an influential little book on spatial orientation in animals[9] and the founding (with Karl von Frisch) of the *Zeitschrift für vergleichende Physiologie* in 1924. On repeated visits to the Naples Zoological Station during the interwar period, Kühn worked on the development, physiology, and systematics of hydroids, but occasionally, too, on color perception in the octopus, confessing to the station's director that "the problem of [animals' spatial] orientation remains my old love."[10]

This difference in the range of Baur's and Kühn's biological interests is also evident in their other academic activities. The dissertations which Baur supervised at the agricultural college were usually on linkage and mapping in *Antirrhinum* (von Rauch 1944, 295; Kuckuck 1988, 26). At a time when Kühn's research was focused on developmental genetics, however, he continued to supervise a wide variety of zoological topics. Of the thirty-three doctoral dissertations completed under his supervision between 1930 and 1936, only half concerned developmental genetics.[11] Both men wrote successful textbooks, but on different subjects. Baur wrote an introductory genetics text and an introduction to the principles of plant breeding for students of agriculture and forestry, and was coauthor of the first textbook on human genetics, which soon became the standard such text during the interwar period.[12] Kühn also published a genetics textbook, but it never acquired the significance of either his text on development or his well-known introduction to

8. In a notebook spanning the years 1916–44 (henceforth "Kühn notebook," kindly lent to me by Professor A. Egelhaaf), Kühn recorded selected quotes from his scientific and nonscientific reading. His notebook entries (pp. 67–72) indicate that he was reading the second edition of Johannsen's genetics textbook about 1921; cf. Kühn to Doflein, 1921 (probably August or September), Kühn papers, MPG-Archiv.

9. Kühn 1919. In his Nobel Prize acceptance speech, Konrad Lorenz attributed the concept of "taxes" to Kühn (Burkhardt 1981, 74), and Kühn's analysis of different modes of orientation was said to have borne fruit in the ethological work of Otto Koehler and his students (Kühn 1959, 277).

10. Kühn to Reinhard Dohrn, 18.6.35, Naples Zoological Station Archive (ASZN: Be, 1935, K).

11. Cf. list of dissertations supervised by Kühn, folder 7, Kühn papers, MPG-Archiv.

12. Baur 1911 (4th ed. 1930); Baur 1921 (5th ed. 1924; Spanish translation by 1933); Baur, Fischer, and Lenz 1921 (4th ed. 1932, English and Swedish translations by 1931).

zoology, which remained the major text of its kind in West Germany well into the postwar era.[13] Both men's editorial activities reflect the same pattern. Baur was not only cofounder and editor of the world's first genetics journal in 1908 (*Zeitschrift für induktive Abstammungs- und Vererbungslehre* [*ZIAV*]), but he also founded and edited two other genetics journals and one for plant breeding.[14] Kühn was on the editorial board of *ZIAV* from 1932, but the journals which he founded were in other areas: comparative physiology and general science (*Zeitschrift für Naturforschung*, established in 1946).

Kühn's broad mastery of zoology and the integral organization of his textbooks were widely acclaimed by colleagues.[15] Richard Goldschmidt, whose own range within zoology was considerable, was not alone in thinking that Kühn was exceptional. Jane Oppenheimer was enthralled by Kühn's textbook on embryology:

He is unique in having at his command an exhaustive knowledge of the development of a wide variety of organisms that is unequalled in scope by that of any other investigator who currently concerns himself with developmental problems.... To his familiarity with morphogenetic phenomena he adds an equally profound understanding not only of the heredity of organisms, but also of their characters of life and habit, and he can therefore evaluate their developmental traits in terms of their broadest possible biological significance [Oppenheimer 1956, 32].

Of the three zoologists generally reckoned as most important in Germany during the thirties—Hans Spemann, von Frisch, and Kühn— Kühn was the "all-around man."[16] Maintaining this broad perspective

13. Respectively, Kühn 1934e (4th ed. 1965); Kühn 1955 (2d ed. 1965); Kühn 1922 (17th ed. 1968).

14. In 1928 he and Max Hartmann founded *Handbuch der Vererbungswissenschaft*. In addition Baur founded *Bibliotheca Genetica* (1917) and *Der Züchter* (1929). On Carl Fruwirth's death in 1930, Baur also took over editing the *Zeitschrift für Pflanzenzüchtung* (H. Stubbe 1959a, 5).

15. Cf. Autrum 1965, 173; Richard Goldschmidt to Kühn, 8.8.55 and 15.4.55; Richard Harder to Kühn, 23.3.53; and H. Stubbe to Kühn, 22.4.55, all in Kühn papers, Heidelberg.

16. This view was frequently voiced about 1930 (interview with H. Hartwig; interview with Ernst Caspari; Spemann to F. R. Lillie, 10.11.32 [Lillie papers]) as well as during the postwar period [Oehlkers to Karl Jaspers, 4.6.62 (in the possession of Prof. G. Linnert, Freie Universität, Berlin)]; Viktor Hamburger to me, 23.9.85; Otto Koehler to Otto Hahn, 7.6.56 [Koehler papers]).

When von Frisch was being considered for a chair of zoology in the early 1920s, there were evidently vague misgivings about the narrowness of his research (Kühn to F. Doflein, 2.4.21, Kühn papers, MPG-Archiv). In 1929 the range of research in von Frisch's

took up a great deal of time which Kühn could have devoted to his research. Since he often claimed that research was his first priority and that he did not enjoy updating textbooks,[17] Kühn apparently felt some sense of obligation to maintain a broad perspective despite the demands which it imposed.

Underlying this sense of obligation was Kühn's concern with specialization. During the nineteenth century, he felt, scientific polymaths began to be replaced by discipline-based specialists who lacked any grasp of the interrelation among disciplines.[18] In his book on the work of Anton Dohrn, for example, Kühn drew attention to the nineteenth-century divorce of comparative morphology from physiology (Kühn 1950). Haeckel and Weismann had brought an evolutionary perspective to morphology, but physiology remained medically oriented and ahistorical. It had been Dohrn's hope that the Naples Zoological Station would encourage a synthesis of evolutionary theory with physiology by enabling biologists to study the evolving physiology of organisms in their natural habitats. What instead emerged late in the century was a comparative embryology (*Entwicklungsgeschichte*) inspired by Haeckel which was evolutionary but descriptive, and an experimental embryology in reaction to Haeckel and Weismann which, while physiological, had abandoned evolutionary questions. Dohrn's call for an evolutionary physiology—which Kühn felt was still relevant in the 1940s—had fallen on deaf ears (Kühn 1950, 18–19, 33, 151, 160–61, 171–72, 179). When he discussed the new experimentalists at the station in the late nineteenth century, Kühn's sympathies were again apparent. In his rejection of both the problem of evolution and the hegemony of comparative anatomy, Hans Driesch had abandoned Dohrn's hopes for the station, and Kühn was not impressed (Kühn

---

institute was somewhat more focused than in Kühn's: Of nineteen projects, about three-quarters dealt with bees and other related work (International Education Board papers, ser. 1, subser. 2, box 35, folder 491). By 1933 he had twenty doctoral students, "nearly all" working on the physiology of perception in fish and bees (Rockefeller Foundation papers, RG 1.1, ser. 717D, box 14, folder 134).

Hermann Hartwig's experience in the laboratories of both Kühn and Spemann confirms this picture. Virtually all of Spemann's Ph.D. students worked on amphibian development, and his introductory zoology course focused heavily on the hydra as an illustrative organism. Anecdotes circulated in the laboratory about Spemann's ignorance of systematics (and genetics) and his inclination to remain "immer an der Urmundlippe" (Hartwig interview).

17. Kühn to Koehler, 20.11.47 (Koehler papers); Kühn to R. Dohrn, 14.10.29, 8.2.30, and 7.12.28 (Naples Zoological Station, Dohrn Archives, sections A and B).

18. Kühn notebook, pp. 105–6, 111, 153–54.

1950, 175–77). But not all of those at the station made such a sharp break with the older morphological tradition. Kühn admired Theodor Boveri, for example, because the latter was just as strict an experimentalist as Driesch, while refusing to discard evolutionary questions: "For the biologist who genuinely notices the diversity of organisms in nature, the question of their transformation is simply inescapable.... The description and comparison of forms deserves a place alongside experimental work" (Kühn 1950, 178).

The trend toward specialization accelerated in the twentieth century, but since it was sometimes fruitful—and in any event impossible to reverse—Kühn saw teamwork as one way to counteract its fragmenting effects: "*Universitas* will be an empty word if the various 'disciplines' together constitute merely a sum...."[19] Creating a community of related specialists under one roof was, as he saw it, the organizational genius of the KWI for Biology. And his keenness to collaborate with others at Göttingen, such as the physicist Robert Pohl or the biochemist Adolf Butenandt, made Kühn particularly attractive to the Rockefeller Foundation's program in molecular biology.[20] Nevertheless, the establishment of a team did not absolve its members of responsibility for grasping the overall problem with which the team was concerned. Collaboration in multidisciplinary teams demanded much more broadly educated researchers than did research pursued by isolated specialists.[21] But despite the inexorability of the trend toward specialization and the growth of teamwork, Kühn's admiration was reserved for those heroic individuals who swam against the tide: "Goldschmidt... has never been a pure geneticist. He has been active in many fields: morphology, development, histology and cytology. And only a *whole biologist*, who always keeps the diversity of the natural world in view, could achieve such grand syntheses" (Kühn 1948a, 242 [emphasis in original]). On the death of the Swiss zoologist Jean Strohl, his close friend since their days as students in Weismann's institute, Kühn reflected:

> Strohl is irreplaceable and not just for us, his friends. For our times produce ever more specialists. Not only is the broad humanistic overview of all areas of knowledge disappearing, but even in our own field one finds a narrowing not only in the empirical scope of work but in the perception of problems. This

19. Kühn to H. Spemann, 7.6.41 (Spemann papers).
20. Rockefeller Foundation papers, RG 1.1, ser. 717D, box 13, folder 123, memo from W. E. Tisdale, dated 18.9.34.
21. Kühn 1939a, viii–x. History of science, Kühn believed, provided an important means of broadening the perspective of specialists, and he called for the creation of more posts in that field.

phenomenon has to occur with the growing complexity of work in each area and sometimes yields depth of understanding, but it has to be compensated by others who place individual results in an overall problem-structure as well as in their historical context. Strohl was particularly gifted in this respect....[22]

The prospect of a single mind achieving intellectual synthesis remained his ideal.

For Erwin Baur, in contrast, specialization was not proceeding rapidly enough. Agricultural and medical education at various universities was too general to be of real value, and Baur was skeptical of the traditional value placed upon all-around learning:

> My proposals [for colleges to concentrate their resources upon a few subjects in which they are strong] will be criticized on grounds that they destroy the old ideal of *Universitas*—but is there any German university or college today which really embodies this ideal? To do so, it would need to have its own faculties for technology, agricultural sciences, veterinary medicine, etc. These and many other sciences belong to a genuine *Universitas* just as much as a medical and a law faculty do.[23]

Rather than trying to integrate genetics within a wider intellectual context, Baur sought to link it to agricultural and medical practice. Irritated by what he saw as the failure of government agrarian policy during the 1914–1918 war, which had brought unnecessary civilian hunger and malnutrition, Baur was a tireless champion of genetics' agricultural importance. At the first public meeting of the Emergency Council for German Science in 1922, in the presence of government ministers and high officials, Baur argued that such hunger could have been avoided had Germany applied genetics to plant breeding in the way that Sweden had (Schiemann 1934, 82). As others have noted, Baur's vision of the economic and political gains to be had from large-scale funding of both strategic and applied research, guided by a coherent agrarian policy within a planned economy, was extraordinarily modern (Zirnstein 1972). Already before 1914 Baur had been a member of the most important German agricultural associations: the Association for Applied Botany, the German Agricultural Society, and the Society for the Advancement of Plant Breeding. He attended their conferences regularly

---

22. Kühn to H. Fischer, 18.11.42, Kühn papers, Heidelberg; cf. Kühn to R. Dohrn, 13.10.42, Naples Zoological Station Archive.

23. Baur (1932b), "Müssen wir wirklich Forschungsanstalten und Hochschulen schliessen?" *Deutsche Allgemeine Zeitung*, 10.2.32; several professors' responses to these proposals, followed by Baur's reply, were published under the same title in the 27.2.32 and 18.3.32 issues of this paper.

and was active in various of their committees, receiving the Agricultural Society's silver medal in 1925 (Schiemann 1934, 91–93). He frequently gave lectures to these organizations, as well as to regional chambers of agriculture, on various aspects of plant breeding. These were not simply lectures on recent developments in basic genetics, packaged so as to look attractive to breeders. In a refresher course for animal breeders, for example, he pointed out which of their traditional practices were out of line with genetic principles and thus ineffective.[24] One of Baur's major concerns, furthermore, was to persuade private breeders to coordinate their research efforts in order to avoid duplication, and he had achieved some success with fruit breeders along these lines by the early 1930s.[25] And at the Fifth International Congress of Genetics in 1927 he gave a paper stressing the need for patent rights to be extended to new plant varieties in order that the work of private plant-breeding firms would continue (Baur 1928, 399–400).

As an ardent eugenicist, moreover, he was equally concerned that genetics should inform the state's population policies (Schiemann 1934, 105–7). He joined the Berlin branch of the German Society for Racial Hygiene soon after its founding in 1907, serving as chairman from 1917 to 1919 and as a member of the editorial board of the society's journal (the *Archiv für Rassen- und Gesellschafts- Biologie*) in 1933.[26] With other figures from Berlin's genetics community he advised the Prussian Health Council's Standing Committee on Population and Race Hygiene. In 1920, for example, he prepared for the committee a summary of the eugenic laws and guidelines adopted by various states in the United States.[27] He was also a member of the council's committee to explore the feasibility of an imperial institute for race biology and race hygiene, and accordingly played a major role in founding and staffing the KWI for Anthropology, Human Genetics, and Eugenics (established 1927).[28] Race hygiene, in Baur's view, was "the prototype of a prudent rational management of human life."[29] A declining birthrate

24. Baur, "Neuzeitliche Vererbungserkenntnis," eleven-page typescript of a lecture on 7.10.27 to the fourth Fortbildungslehrgang für Tierzuchtbeamte in Schleswig-Holstein (kindly provided by Frau Helga Stromeyer-Baur).
25. Zirnstein 1972, 63; Baur, "Die volkswirtschaftliche Auswirkung der Pflanzenzüchtung . . . ," typescript dated 6.5.33, Bundesarchiv, R73, Nr. 28.
26. *Archiv für Rassen- und Gesellschafts-Biologie,* vol. 27; the society also awarded him its Rudolf Virchow medal (E. Fischer 1933, 392).
27. Baur to C. B. Davenport, 24.11.20, Davenport papers.
28. Baur to von Luschan, 1.3.22 (von Luschan papers); Schiemann 1934, 106.
29. Form letter for the Berlin Society for Racial Hygiene, dated 18.12.17, folder 90, R86, Bundesarchiv, cited by Weiss 1987, 151.

overall, combined with the higher fertility of genetically inferior individuals, he felt, was hastening the biological degeneration of modern civilized societies (Baur 1932c; 1932d). With advancing civilization came the growth of cities, which robbed the best human types from rural areas. Urban populations had especially low fertility, however, since neither gifted strata nor the working classes could afford to have adequate numbers of children. Baur's solution was threefold: (1) change the pattern of taxation so that all strata could afford at least three to four children; (2) reduce the fertility of the unfit by sterilizing hereditary criminals and confining antisocial individuals in humane asylums; and (3) take measures to curtail the flight from the land. As things stood, Baur argued, government policy was destroying German agriculture because the free market in agricultural produce favored cheap foreign imports with which the German farmer could hardly compete. A radical change in agrarian policy could save the farmers while simultaneously reducing unemployment and freeing Germany from dependence upon foreign suppliers, which would be undesirable in the event of another war (Schiemann 1934, 91–97). To that end, the government would have to start devoting more attention to plant breeding, expand the amount of acreage under cultivation, and better coordinate the agrarian policies of provincial governments and Reich.[30] Thus eugenic policy and agrarian policy were closely linked in Baur's technocratic vision of the integration of science in a planned economy.

His excellent agricultural connections, organizational talents, and sheer energy made Baur by far the most successful institution builder in the German genetics community. Having persuaded the Berlin Agricultural College that it needed a separate chair of genetics, Baur eventually managed to get the Prussian Ministry of Agriculture to build him a decently equipped institute with good facilities for training students (von Rauch 1944, 98–124). But it had been a constant struggle to secure the necessary funding, aggravated by the war and the postwar inflationary crisis. And by the time the institute was finally completed in 1925, the grounds in which it was situated were simply too small for plant-breeding research on the scale which Baur envisioned, and he had lost enthusiasm for the Sisyphean task of extracting funds from the ministry.[31] By 1927 he had persuaded both the Kaiser-Wilhelm Society and

30. Baur, "Die volkswirtschaftliche Auswirkung der Pflanzenzüchtung . . . ," typescript, 6.5.33, Bundesarchiv, R73, Nr. 28; Zirnstein 1972.

31. Furious at the reduced budget he was allowed in the winter of 1923–24, Baur threatened to close down all of his plant-breeding program as well as the breeding of large animals. Only by firing most of his gardening staff and selling off the animal breed-

private plant-breeding firms to endow a new KWI for Breeding Research at Müncheberg near Berlin, insisting that within a few years an institute with sufficiently large-scale facilities would be able to support itself with contracts from the seed industry and breeders' associations.[32]

In the geneticists' professional associations, too, Baur was a moving force. At the Fourth International Congress of Genetics at Paris in 1911 he had undertaken to host a fifth congress in Germany in 1916. Since the war forced postponement, and German participation in international scientific organizations after 1918 was subject to boycotts, Baur (with Correns and Goldschmidt) was instrumental in organizing the German Genetics Society in order to facilitate contact with foreign geneticists. He hosted its first meeting in Berlin in 1921 (Nachtsheim 1921), and the second meeting at Vienna the following year attracted seven hundred participants from virtually all countries but France (von Rauch 1944, 262–63). By 1927 he succeeded in bringing the Fifth International Congress to Berlin. One of the first such congresses to be held in Germany following suspension of the boycott, it coincided with the opening of his own KWI for Breeding Research, as well as the KWI for Anthropology, Human Genetics, and Eugenics. Baur's importance to the genetics community as an institution builder and his extensive contacts within the agricultural community were widely acknowledged by his colleagues, and his sudden death in 1933 was deeply felt. Otto Renner called it the worst disaster which could have befallen German biology. Though Richard Goldschmidt was critical of Baur in several respects, he regarded him nonetheless as "utterly irreplaceable." Germany had lost more than just an outstanding scientist, Max Hartmann emphasized, for such scientists rarely possessed Baur's vision of the potential practical applications of science.[33]

---

ing stock, he insisted, would he be able to make the financial savings required of him by the ministry. Cleverly recruiting Schmidt-Ott from the *Notgemeinschaft* as an ally, Baur was soon able to pressure the ministry to fund him at the level he had requested (see ZStA-Merseburg, Rep 87B, Sign 20282, correspondence among Baur, Schmidt-Ott, and the ministry between December 1923 and February 1924).

32. The institute at Müncheberg—with twelve departments rather than two—was far grander than that at the Agricultural College. Von Sengbusch's department at Müncheberg had between ten and twenty technicians during the period 1928 to 1931; at the Institute of Genetics at the Agricultural College in 1926–27 he had two (von Sengbusch papers, Ordner 8, "Chronik"). In 1933 the institute at Müncheberg employed two hundred workers, in addition to the assistants and technicians (Nr. 2665, Abt I, Rep 1a, MPG-Archiv).

33. Otto Renner to Ralph Cleland, 12.12.33 (Cleland papers); Richard Goldschmidt to Curt Stern, 4.12.33 (Stern papers); Goldschmidt 1960, 272; Hartmann 1934a, 259–

If Kühn's appreciation of the many facets of each biological problem was truly comprehensive, Baur's conception of genetics as a key element in a rationally planned society was altogether pragmatic. Their differences of perspective, however, were not confined to the biological sciences. For in the value which they placed upon nonscientific forms of knowledge, as in their conceptions of the academic's role, Kühn and Baur were once again radically different. Kühn embraced the ideal of well-rounded learning. As students in Freiburg, he and Jean Strohl together read not only the classics of nineteenth-century biology, but also Gottfried Keller, Hermann Hesse, and the cultural historian Jacob Burckhardt.[34] In later years, too, his reading encompassed contemporary psychology (Hugo Münster, A. Messer), a variety of historians and philosophers (Simmel, Schopenhauer, Plutarch, Alloys Riehl, F. Hebbel, Droysen, Ranke, Dilthey), and—of course—Goethe and Schiller.[35] Their works provided him with rather more than just a pleasant intellectual diversion for evenings and vacations. When Reinhard Dohrn invited him to contribute an essay on the work of Anton Dohrn to a special issue of the Naples Zoological Station's journal, Kühn spent the better part of 1941 and 1942 reading late-nineteenth-century evolutionary and morphological literature in order to place Dohrn's work in intellectual context.[36] From various entries in his notebook it is clear that he took seriously this task of biographer and historian of ideas. To understand fully how new discoveries were made, he noted, one had to take into account the personal qualities of the discoverer; but the latter's concerns were also shaped by his or her epoch.[37] Under the circumstances, it is hardly surprising that the commissioned "essay" grew into a two-hundred-page book: *Anton Dohrn und die Zoologie seiner Zeit.* Similarly, in preparing his essays on Goethe's significance for biology (of which more later), Kühn went to the Goethe-Schiller-Archiv as well as to the Goethe-Museum rather than simply relying on their published works (Henke 1955, 197). Nor were Kühn's interests outside science confined to history and philosophy. Possessed of a strongly visual aesthetic sense, he was attracted not only to the variety of animal forms

---

60; Richard Hertwig expressed the same view in a letter to Goldschmidt, 11.12.33 (Goldschmidt papers), as did Paula Hertwig (1956, 129).

34. Kühn to H. Fischer, 18.11.42, Kühn letters, Heidelberg.

35. Kühn notebook.

36. Kühn to Dohrn, 23.1.41, 4.4.41, and 9.8.42, Naples Zoological Station Archive.

37. Kühn notebook, pp. 87, 139–40, 165–66, 188, 107–8, 154, 77–78, 178, 182–83.

but also to architecture, Italian Renaissance art, and expressionism.[38] An involvement with art, he believed, lent breadth and perspective: "It is an observation which I have found confirmed through much experience: that the most insightful scholars are those who have also been interested in art."[39]

Underlying the enormous breadth of Kühn's biological knowledge, his interest in the arts and humanities, and his critique of specialization was a particular form of holism. From the entries in his notebook, it appears that he endorsed Goethe's prediction that mechanistic-atomistic thought would eventually be replaced by a dynamic-chemical perspective.[40] Just as the biographer had to pay close attention to the historical context in which his or her subject moved, so also was the part shaped by the whole in the biological process of development: "The comparative analysis of development shows that the organism dominates cell-formation by employing one, several or many cells for the same purpose, accumulating cell material and directing its movements, and forming its organs as if the cells themselves did not exist or as if they only existed in complete subordination to the organism's will."[41] To justify his rejection of mechanism, Kühn drew upon the arguments of the Machian positivist Joseph Petzoldt: that the idea of material substances in science had to be replaced by the concept of forces, that atomism was merely a convention, and that the classical notion of causality had to give way to "functional relations" or the "conditioning" of events.[42] Knowledge did not arise through passive contemplation of nature but through interrogation by interested persons. Fantasy and presupposition thus had a legitimate role to play in this process,[43] and theories rose and declined due as much to changing assumptions as to their explanatory strengths or weaknesses.[44] As Kühn

38. See the documents and recollections by Kühn's former colleagues in Grasse 1972, 50, 220, 246; Egelhaaf interview.
39. Kühn notebook, p. 174, quoting Lichtenberg.
40. Kühn notebook, p. 187.
41. Kühn notebook, p. 26, transcribed in 1917 from Oskar Hertwig's *Das Werden der Organismen*. On page 86, similarly, he transcribed a quote from Droysen, who noted the parallel between the individual in history and the cell in the organism but insisted that neither history nor the organic world could be accounted for solely in terms of their parts.
42. Kühn notebook, 52–65.
43. Kühn notebook, 144–45, 150–52, 33–34, 87, 135, 172–73.
44. In Kühn's copy of E. Radl's *Geschichte der biologischen Theorien* (Leipzig: W. Engelmann, I. Teil, 1905; II. Teil, 1909), kindly lent to me by Prof. H. Querner, he

explained to a student, doing science was like getting across a lake whose layer of ice was beginning to break up in early spring. One could get across it without sinking as long as one did not remain too long on any one ice fragment. Each fragment was like a scientific concept; it should be used instrumentally without putting it under too much strain or insisting upon its precise definition.[45]

But if mechanism was inadequate, vitalism was no better. Kühn was critical of those whom he regarded as mystics (*Schwärmgeister*), like Adolf Meyer-Abich or Pascual Jordan.[46] Human beings can only understand those living phenomena which are amenable to physical-chemical explanation, and there were no issues which were in principle unresearchable.[47] The only tenable position was thus an intermediate one, combining a materialist conception of nature with the epistemological insights from idealism. This, Kühn reckoned, was Kant's seminal contribution.[48] The acquisition of knowledge, similarly, demanded a balance between scrupulous observation and analysis on the one hand and imagination on the other. The former was absolutely essential; despite the emphasis which he placed on theory, Kühn always described himself as an empiricist and was critical of those colleagues who indulged in undisciplined speculation.[49] Kühn greatly respected Boveri for his unusual combination of observational acuity, analytical rigor, and imagination.[50]

The same insistence on balance is evident in Kühn's assessment of the Romantic tradition in biology. While Goethe's scientific work won his admiration, many *Naturphilosophen* did not. Kühn was attracted to the holism of the Romantics, their search for intellectual synthesis and

---

underlined various passages in which Radl argues this conventionalist position (e.g., Teil I, 2–3, 8; Teil II, 579). See also Kühn notebook, 87–88; Kühn to Dr. Rupp, 12.2.47 (Kühn letters, Heidelberg); Kühn to Hans Spemann, 26.12.40 (Spemann papers).

45. Conversation with Bernhard Hassenstein, 30.5.84.

46. Kühn to Erwin Bünning, 18.3.45 (Kühn letters, Heidelberg).

47. Kühn notebook, 60–61, 162–63, 186–87; Kühn 1932–33, 69.

48. Kühn notebook, 51, 83, 180–81, 46, 92, 150–52, 85, 3–15. Kühn regarded himself as a Kantian (Kühn to Koehler, 9.9.47, Koehler papers).

49. Max Hartmann came in for criticism on this score (Kühn to H. Spemann, 30.12.21, Spemann papers), but Kühn also had reservations about giving Goldschmidt's *Physiologische Theorie der Vererbung* to students, despite his respect for Goldschmidt's work in general (Viktor Schwartz to me, 26.10.84). On Kühn's self-understanding as an empiricist, see the anecdote related by Autrum 1969, 264.

50. Kühn 1938b, i. Weismann was credited with essentially the same gifts (Kühn 1957b, 194–95). On the need for balance between observation and imagination, see also Kühn notebook, pp. 33–34 and 93.

Imputing Styles of Thought 243

their critique of naive empiricist theories of knowledge. But he condemned those guilty of unbridled speculation and the overenthusiastic search for analogies in nature. High marks were awarded instead to C. F. Kielmeyer for avoiding speculation and to Franz Unger for being an excellent observer who disavowed Lorenz Oken's excessive fantasizing. Goethe represented the best of the Romantic tradition not only because of his remarkable polymathy but also because he embodied Kühn's methodological ideal: the combination of ambitious theorizing and disciplined observation. Some of Goethe's theories, of course, turned out to be wrong, occasionally because he failed to ground them empirically. But in Kühn's judgment, a holistic perspective helped Goethe to anticipate the central problems of what would later be called perceptual physiology, ecology, and, above all, embryology (Kühn 1932–33, 60–62, 66–68; 1948b, 217, 219, 222–23, 234).

To describe Kühn as a comprehensive, therefore, is to denote not only his approach to genetics or his resistance to specialization in science. It is also to convey his view that the search for universal understanding is central to the scholar's role. An emphasis upon the practical value of knowledge was utterly foreign to him:

The final goal of every science is to provide answers to the ultimate questions: What are we in the world? Where did we come from, and where are we going? Even if the scholar may occasionally lose sight of these questions, they will continue to be raised, and the results of specialist research will ultimately be judged on the extent to which it can answer them. Even if short-sighted individuals may tend to judge inventions and discoveries by the degree to which they advance material culture, these discoveries will nevertheless remain barren if they do not also yield spiritual values.[51]

Like the arts, the sciences were built upon a philosophical core, and it was important that both scientists and philosophers recognize this.[52] Hans Spemann most perfectly embodied Kühn's ideal of the scholar. His life's work was the closest approximation to a work of art which Kühn could imagine. For Spemann, experiment was not a matter of applying "levers and screws" so as to force nature to respond, but of "conducting a conversation with a living partner." The value of zoology, in Spemann's view, was educative rather than immediately applicable, and Spemann's broad humanistic education prevented him from

51. Kühn notebook, pp. 90–92, quoting H. Kienle; cf. pp. 106–7. In a letter to Spemann (7.6.41, Spemann papers), he regretted the tendency to judge scientific achievements "by results" rather than by the intellectual vision that spawned them.
52. Kühn notebook, pp. 98–99; Kühn to Pater Silvester, 3.3.49 (Kühn letters, Heidelberg).

244   Styles of Thought

falling prey to a naturalistic worldview (Kühn 1942, 102). "Not only your achievement," Kühn wrote to Spemann, "but above all your whole person represents for me that which one should strive for...."[53]

Erwin Baur's world, however, was a very different place. At school he had performed much better in the sciences than in the humanities, and later in life he took little time for theater, concerts, or reading outside his field.[54] This posed difficulties for his friendship with that quintessential comprehensive thinker, William Bateson. The two men had got on well together during Bateson's visits to Berlin and at international congresses before the First World War:

He is so fresh and bright. Perfectly unaffected, doesn't mind going first through a door, and altogether a natural, clever and humorous man.[55]

... the best Mendelist *ausserhalb* England, I think.[56]

Nonetheless Bateson had slight misgivings:

He is a splendid companion ... but so utterly immersed in his notebooks that I think his work may suffer a little from want of breadth.[57]

On subsequent visits Bateson became increasingly disenchanted:

He is easier to travel with than I expected, but he is also duller. The end of his ideas and experiences is very soon reached. The whole range not only of the arts (including music) is sealed to him, but all literature, too. He knows not even Goethe or Schiller, and has in fact read literally nothing in any language, outside science ... a simple soul, you see. He will last me well for a week or so, but I would rather have someone more complicated, if we were to walk for a fortnight.[58]

Nor did philosophy rate highly in Baur's books; it was suspect because speculative. Far from scientists needing an understanding of philosophy, it was politicians, historians, lawyers, and sociologists who were

---

53. Kühn to Spemann, 9.7.29 (Spemann papers); see also Kühn 1939b, 426; Kühn 1941, 371.

54. Schiemann 1934, 53; von Rauch 1944, 68–69. The only sign that Baur had any interest in music is that he seems to have owned some jazz records ("Tagebuch einer Anatolienreise," 1926, typescript, Baur papers/MPG-Archiv). At that time jazz was presumably not regarded in educated middle-class circles as a form of high culture.

55. W. Bateson to B. Bateson, 25.9.10, Bateson papers.

56. W. Bateson to B. Bateson, 18.12.09; cf. Bateson to L. J. Cole, 8.4.25, and to C. C. Hurst, 17.1.10, all in the Bateson papers.

57. W. Bateson to B. Bateson, 28.9.10, Bateson papers.

58. W. Bateson to B. Bateson, 1.10.12; cf. W. Bateson to B. Bateson, 1.1.14, both in the Bateson papers.

ignorant: namely, of biology. His few remarks on scientific method stressed the importance of experiment over theorizing,[59] and he appears to have endorsed a reductionist research strategy. Plants, Baur wrote, were his "reagents":[60] "I can work with genes and the genetic combinations of my plants exactly like the chemist with his [atoms] and molecules and with his formulae" (quoted in von Rauch 1944, 285). Mendel's laws had the same significance for biology as Dalton's laws for chemistry.

What free time Baur had was spent outdoors: in gardening, hunting, and sports. From his student days he had been a keen mountain climber, skier, rower, and sailor.[61] With a playful sense of humor, an aversion to formality, and abundant enthusiasm, Baur had an engaging manner which won him the affection of colleagues and students.[62] In addition, his persuasive skill as well as his ability to express complex ideas very simply made him a highly effective speaker and teacher (e.g., Lang 1987; Tschermak 1933; von Rauch 1944, 81; Kuckuck 1988, 11, 15). But despite his intellectual prowess and forceful personality, Baur was often described by colleagues as an open, uncomplicated man who radiated health and solidity.[63] In keeping with his affection for rural life, he was buried on his farm in Müncheberg, adjacent to the institute he had founded (E. Fischer 1933).

7.2 Imputing Styles of Thought

Clearly Kühn and Baur held contrasting views on a wide variety of issues, but to what extent were these views typical of the genetics community? Sociologically speaking, a style of thought is a cluster of related ideas and attitudes which is associated with a particular group. At this point, therefore, we must establish which features of Baur's and Kühn's thought were merely idiosyncratic and which were shared, not only with other members of their schools, but also with other geneticists

---

59. On the lamentable ignorance of biology, see Baur 1932c, passim. On Baur's preference for experiment over theory, see Schiemann 1934, 66; Baur 1911 [1919], 348; von Rauch 1944, 103, 234.

60. von Rauch 1944, 285–86; Baur to Steinbrinck, 22.11.08, Steinbrinck papers.

61. See the obituaries of Baur by Stubbe (1934b, 1959a); von Rauch 1944, 26–34; Schiemann 1934, 58–59.

62. See, for example, E. Fischer 1933, 391–92; Kuckuck 1980, 522; Lang 1987; Stubbe 1959a; Gruber, "Festrede zum 100. Geburtstags Erwin Baurs am 16.4.76 . . . ," typescript, Mappe 2, Baur papers, MPG-Archiv.

63. E.g., Stubbe 1934b; H. J. Troll to me, 31.3.84.

whose choice of research problem was recognizably pragmatic or comprehensive, respectively (tables 7.2 and 7.3).[64]

Let us begin by considering the breadth of these individuals' biological knowledge and interests. If one looks at the judgments made by colleagues in obituaries and other biographical sketches, the picture is remarkably consistent. While those whose approach to genetics is identifiably comprehensive were invariably praised for "never losing sight of the whole,"[65] pragmatics never were. Such praise, as we saw in chapter 5, usefully tells us something about the values taken for granted by comprehensives' biographers, but once again, evidence from more objective sources confirms that the comprehensives were indeed more familiar with the entire range of biological problems and literatures than were the pragmatics.[66] Consider individual geneticists. Friedrich Brie-

---

64. This sample of forty-five individuals includes all of the major figures in the interwar German genetics community, plus a number of younger ones who had achieved at least the status of assistant by 1933.

65. The quote is from a Correns obituary (von Wettstein 1938, 148), but virtually the same phrase was used in praise of Hans Winkler, "who never gets lost in details but always keeps his eye on the broader problem [*stets den Blick auf's Allgemeine richtet*]" (Dekan to Minister für Wissenschaft . . . , 1.6.23, ZStA-Merseburg, Rep 76, Va, Sekt 2, Tit IV, Sign 68D, Bd II). The currency which this methodological injunction enjoyed among substantial sections of the German scientific community is evident in Kühn's 1938 presidential address to the *Gesellschaft Deutscher Naturforscher und Ärzte*. Despite the unavoidable need to specialize, he said, "the value of an individual fact is merely that of a part in the *whole of our knowledge of nature*." The "real danger" of "losing our way in the search for a unitary picture of nature" must be countered through teamwork in research, which could produce cognitive "syntheses bridging many specialties" (Kühn 1939a, viii–ix; emphasis in original).

66. In a few cases, to be sure, the existing evidence makes it difficult to ascertain the extent of particular geneticists' biological knowledge and interests. For example, much of Karl Belar's reputation was based on his book on the cytology of mitosis, but he also worked on the physiology of sexual and asexual reproduction, and taught a course on evolution (Caspari, "Karl Belar," *Dictionary of Scientific Biography*, forthcoming). On the other hand, neither Caspari nor Belar's other obituarists go so far as to claim that he possessed—or aspired to—particularly broad biological knowledge (cf. Goldschmidt 1956, 142–48; Bauer 1963; Stern 1931).

Similarly, neither Nachtsheim (1959) nor Kosswig (1959) mentions the extent of Paula Hertwig's interests in biology, although she offered courses at the University of Berlin during the 1920s, not only in genetics, but also in evolution, embryology, and general biology (cf. *Deutsches Hochschulkalender,* 1920–1927).

Julius Schwemmle is another such case. Neither Ziegler (1980) nor Arnold (1981) claims that his interests were broad, but he was appointed to Erlangen's chair of botany at a relatively young age (thirty-six) and taught genetics, general botany, plant systematics, and pharmacognosy (*Deutsches Hochschulkalender,* 1930–1939) while supervising dissertations in morphology, physiology, development, and genetics (Ziegler 1980).

Table 7.2 Comparing Styles of Thought: The Baur School versus those of Kühn and von Wettstein

	Broad knowl. of biology?	Cultural interests?	Holist?	Politically active?[a]
The Baur school				
Baur	−	−	−	radical rt.
Nachtsheim	−	−	−	?
Kuckuck	?[b]	−	?	SOCIAL-DEM.
Stubbe	?	+	?	SOCIAL-DEM.
Schick	−	−	?	SOCIAL-DEM.
v. Sengbusch	−	−	?	?
Hertwig	+/−[c]	?	−	LEFT-LIB.
von Ubisch	+?	?	?	LEFT-LIB.
Stein	?	+	?	?
Schiemann	+	+	?	conserv.
Michaelis	+/−	+	?	conserv.?
von Rauch	?	?	?	?
Husfeld	?	?	?	CONS/NSDAP
Gruber	?	?	?	NSDAP
v. Rosenstiel	?	?	?	NSDAP
Hackbarth	?	?	?	NSDAP
Kühn et al.				
Kühn	+	+	+	LEFT-LIB.
Henke	?	+	+?	left-lib?
Caspari	?	+	?	?
Plagge	?	?	?	?
Piepho	?	?	?	?
Schwartz	?	?	?	?
E. Becker	−?	?	?	?
von Wettstein et al.				
von Wettstein	+	+	+	conserv.
Pirschle	?	+	?	?
Melchers	−?	?	−	SOCIAL-DEM.
G. Becker	?	+	?	?
Schlösser	?	?	?	?
Knapp (1934–36)	−	−?	?	?
Barthelmess (1936–38)	?	?	?	NSDAP
E. Kuhn (1934–38)	?	?	?	?

[a] Under this heading, capitals denote party membership or activism prior to 1933; lowercase conveys the apparent party sympathies of those who were not politically active. Key: radical rt. = right-wing radical; social-dem = social-democratic (SPD); left-lib = left-liberal (DDP); conserv = conservative-nationalist (DNVP); NSDAP = National Socialist.

[b] ? = evidence lacking.

[c] +/− = evidence ambiguous.

Table 7.3 Comparing Styles of Thought: Geneticists outside the Schools of Baur, Kühn, and von Wettstein

	Broad knowl. of biology?	Cultural interests?	Holist?	Politically active?[a]
Pragmatics				
Bauer	?[b]	?	?	?
Grüneberg	−	+	?	?
Kappert	−	−	?	liberal?
Schwemmle	+/−[c]	+	?	?
Stern	−	+/−	−	liberal?
Comprehensives				
Belar	+/−	+	?	?
Brieger	+	+	?	?
Correns	+	+	?	?
Goldschmidt	+	+	+	conserv.
Haecker	+	+	?	?
Jollos	+	+	?	?
Oehlkers	+	+	+	liberal?
Renner	+	+	−	conserv.
Timofeeff-Ress.	+	?	?	?
Winkler	+	?	?	?
Unclassifiable				
Hämmerling	−	+	−	social-dem
Lehmann	?	?	−	RADICAL-RT.

[a] See table 7.2, note *a*.
[b] ? = evidence lacking.
[c] +/− = evidence ambiguous.

ger's obituarist stresses the range of his knowledge, a claim borne out by the fact that as a *Privatdozent* Brieger taught not only genetics but plant development, cryptogamic botany, bacteriology, and ecology. The polymath J. B. S. Haldane regarded Brieger as first-rate, noting that he had "a more physiological outlook than most geneticists."[67] Correns's extensive knowledge of systematics and his early morphological and subsequent plant-physiological interests are well documented (see, e.g., von Wettstein 1938; Sierp 1924; Hartmann 1934b; Renner 1961b). Goldschmidt's pride in the range of subjects on which

---

67. See the curriculum vitae and a reference from J. B. S. Haldane in Brieger's file, Society for the Protection of Science and Learning papers. Cf. Linskens 1986.

he published is obvious in his autobiography (Goldschmidt 1960). In addition to his work in developmental and evolutionary genetics, plus several papers and a book on the inheritance of musical talent, Valentin Haecker also published in cytology and ornithology. He was, as one of his students put it, "not just a geneticist... but a zoologist in the true sense of the word" (Heberer 1964; cf. R. Haecker 1965). Several of Victor Jollos's colleagues remarked on the "unusually broad and profound" range of his biological knowledge.[68] Critical of those biologists whom he regarded as narrow specialists, Friedrich Oehlkers encouraged his students and *Privatdozenten* to acquire broad knowledge in botany.[69] Otto Renner enjoyed a reputation not only as a geneticist but also as a plant physiologist, and all of his biographers stress this scientific breadth.[70] The extent of N. W. Timofeeff-Ressovsky's knowledge of natural populations surprised H. J. Muller, and various biographers have noted his "enormous breadth and erudition."[71] When Richard von Wettstein died in 1931, he had only begun to rework the fourth edition of his *Handbuch der systematischen Botanik*. His son Fritz, however, took over this task, not merely out of filial piety but because he was genuinely interested in the subject matter:

> Probably no experimental botanist could have done this but von Wettstein who had been steeped in systematics since his youth and for whom the greatest biological problem had always been the variety of form in nature, since this problem contained all others.... [In Berlin] the universality of von Wettstein's interests flourished and he could pursue genetics as the science of genesis in the widest sense [Renner 1944–48, 263–64; cf. Kühn 1947].

Of Hans Winkler one referee wrote that "his work is free of one-sidedness and indicates a broad perspective."[72]

In contrast, I have yet to find a biography of a pragmatic geneticist

68. The phrase is from Brink 1941, but Goldschmidt's unpublished obituary concurs (Goldschmidt papers).
69. Oehlkers to Prof. Lehmann, 18.4.58 (Kühn letters, Heidelberg); Oehlkers to Karl Jaspers, 4.4.62 (in the possession of Prof. G. Linnert, Freie Universität Berlin); cf. chaps. 4 and 9.
70. See the biographies of Renner by von Frisch (1960–61); H. Stubbe (1962); Mägdefrau (1961); Kühn (1960a); Cleland (1966); Bünning (1944, ix); cf. Renner 1930, 4–5.
71. Medvedev 1982, 5; Adams 1980a, 269; cf. obituaries of Timofeeff-Ressovsky by Eichler (1982) and Zimmer (1982).
72. [*lassen auf universelle Ausbildung schliessen*] Prof. Kolkwitz to Dekan der Philosophischen Fakultät, 29.1.23 (ZStA-Merseburg, Rep 76, Va, Sekt 2, tit IV, Sign 68D, Bd II).

which celebrates its subject's breadth.[73] Some of them, in fact, stress the pragmatics' specialized interests. For example, although trained in botany, Hans Kappert was said to have "specialized from the beginning as a geneticist" (Hertwig 1950, 133). Although Hans Nachtsheim habilitated in zoology and comparative anatomy at Munich in 1919 and could have made his career in that discipline,[74] he accepted Baur's offer of a post at the Berlin Agricultural College. There, as Germany's first *Privatdozent* for genetics, he began to focus his research and teaching wholly upon the new discipline.[75] Of Reinhold von Sengbusch a colleague wrote: "Rarely has an individual researcher . . . been specialized to such a high degree" (Stubbe 1958, 2). As a student at the University of Berlin in the early 1920s, Curt Stern sought out Max Hartmann's course in modern biology because he was bored with general zoology, and neither of his biographers so far has attributed particular breadth

---

73. Neither Stein ("Hans Bauer," four-page typescript, Emmy Stein papers) nor Bauer (1966) refers to Hans Bauer's breadth of biological knowledge or interests. Nor do Abel (1980) and Butterfass (1980) make this claim for Edgar Knapp. The same applies to Fischbeck's (1979) and Schiemann's (1963) tributes to Hermann Kuckuck, as well as to biographical sketches of Rudolf Schick by Kleinhempel (1985), Stubbe (1985), and Kuckuck and Stubbe (1970). None of the biographies of Hans Stubbe makes such claims: Hagemann 1984; Rübensam 1982; Melchers 1972; and Stubbe 1959b. The same is true for the only biography of Hans Grüneberg which I have found (Lewis and Hunt 1984). According to Hans Kalmus, Grüneberg's colleague of many years at University College London, Grüneberg's specialized approach to research was closer in character to the Anglo-Saxon mode than to the continental tradition whence they both came. In Kalmus's judgment, the fact that Grüneberg knew little physiology, botany, or ecology was a serious disadvantage in his last research project on polymorphism in Sri Lankan snails (Kalmus interview).

74. There was certainly no dearth of soon-to-be vacated chairs in zoology at that time. Correspondence between Richard Hertwig and Richard Goldschmidt (in the latter's paper) indicates that between 1921 and 1925 chairs in zoology were being filled at Berlin, Breslau, Tübingen, Vienna, Rostock, Munich, and Königsberg.

75. For Nachtsheim's willingness to concentrate upon genetics, see his autobiographical sketch (1966). His decision to alter his habilitation at the college from zoology to genetics seems to have been taken freely; his fellow assistants there—Paula Hertwig and Elisabeth Schiemann—elected to habilitate in zoology and botany, respectively (Rektor to Minister für Landwirtschaft, 20.5.21, Sign 20101, R87B, ZStA-Merseburg). Most of Nachtsheim's biographers make no claims for the breadth of his biological interests (Günther and Hirsch, "Hans Nachtsheim zum 80. Geburtstag," privately printed, 1970; Vogel 1980; Friedrich Vogel, "Hans Nachtsheim: Persönlichkeit und wissenschaftliches Werk," typescript, Gedenkvortrag zum 90. Geburtstag, Berlin 1980). Ulrich's tribute to his "Weite und Vielseitigkeit" appears to refer to the diversity of problems in *genetics* which Nachtsheim addressed: his work on *Drosophila*, the domestication of animals, and congenital abnormalities in rabbits and man, and his attacks on neo-Lamarckism (Ulrich 1950, 9).

of knowledge to him.[76] As with Kühn and Baur, therefore, the pragmatics' range of biological knowledge and interests was more restricted than the comprehensives'. (This is not to say that the pragmatics' range of knowledge *in general* was necessarily narrower. As we saw in section 1, Baur may have been uninterested in general botany or in relating genetics to the wider body of, say, developmental biology, but he was very knowledgeable about both agriculture and eugenics and keen to relate genetic knowledge to such practical matters.)

Marked differences are also evident when we survey both groups' attitudes toward "high culture." Several of the most prominent comprehensives shared Kühn's interest in history, philosophy, and literature.[77] Correns was one of the few scientist members of Berlin's elite *Mittwochs-Gesellschaft*[78] and was fond of eighteenth-century English literature.[79] Otto Renner was said to be "at home in cultural history to an extent rarely seen among scientists" (Mägdefrau 1961, 106; see also Cleland 1966; Melchers 1961). His knowledge of the classics, too, was "profound," and on graduating from school he had been undecided

---

76. On Stern's student days, see Hämmerling, "Mein wissenschaftlicher Lebenslauf," Stern papers. Of Stern's biographies (Neel 1987 and Caspari, "Curt Stern," *Dictionary of Scientific Biography,* forthcoming), the latter is particularly interesting in this regard, since Caspari does *not* credit Stern with the wide knowledge of the biological literature which he earlier recognized in Richard Goldschmidt (Caspari 1960). According to his student Ursula Philip, Stern was not interested in evolution and became interested in development only after emigrating to the United States in 1933 (Philip interview). Nevertheless, Stern's case is not entirely straightforward. In two general lectures on science which he gave in the 1940s he endorsed specialization as the only realistic research strategy, at the same time urging his audience not to lose sight of the general issues without which specialized research becomes meaningless (notes for "On Being a Scientist" [n.d.] and "Research and Science" [1944], box CS, Stern papers). What is not clear, however, is whether Stern would have expressed this view prior to 1933 or whether he developed it only in response to the high levels of specialization which he encountered among his American colleagues.

77. On Victor Jollos see the unpublished obituary by Richard Goldschmidt (Goldschmidt papers). On Gustav Becker see Hoffmann 1971. On Elisabeth Schiemann see Kuckuck 1980. Ernst Caspari reckoned that an ability to discuss Aristotle with the president of Wesleyan University in Connecticut (a philosopher by training) helped him to obtain a post there in 1946 (Eicher 1987). Karl Henke was associated in Göttingen with the "Nohl-Kreis," a circle of intellectuals interested in the philosophical and pedagogical ideas of Hermann Nohl (Hartwig interview; cf. Caspari 1957), and he contributed an essay on pedagogy in biology to one of Nohl's journals (Henke 1930).

78. A discussion group consisting mostly of academics but including a few top civil servants, officers, businessmen, and politicians (see Scholder 1982).

79. The papers which Correns gave to the *Gesellschaft* (on biological topics) are in R106/17, Bundesarchiv. His interest in English literature comes across in letters to Carl Steinbrinck (14.9.23, 18.1.27, 26.1.27, all in the Steinbrinck papers).

whether to study classics or botany. Like Kühn, both Renner and Valentin Haecker published essays on Goethe's biological studies. Although Richard Goldschmidt claimed neither to understand nor to appreciate philosophy, he did not deny his admiration for Goethe.[80] Friedrich Oehlkers shared an interest in the philosophy of biology with his friend and colleague in Freiburg, Karl Jaspers, and his wife had studied philosophy and was particularly interested in Heidegger.[81]

Establishing just what kind of philosophy attracted the comprehensives is difficult, since few geneticists articulated—or perhaps recognized—the metaphysical assumptions underlying their work. Obituaries are of little use for this purpose, since, having been written in most cases since 1945, they reflect the conventional wisdom that reference to philosophical premises is irrelevant or even indiscreet.[82] Nevertheless, the comprehensives' writings on genetics and other subjects occasionally allow tentative inferences as to their philosophical premises. While Oehlkers seems not to have shared Richard Woltereck's antimaterialist tendencies, he respected the latter's attempt to develop a philosophically informed approach to biology. Oehlkers's holistic sympathies, for example, are visible in his analysis of different traditions in genetics. The old Mendelism, he argued, was 'analytical' in approach, while the newer developmental genetics was 'synthetic.' The former "breaks apart [*zerlegt*] the totality of the organism into the sum of its traits"; in consequence, such "atomistic conceptions" fail to explain the emergence of the unique individual in its phenotypic "totality." Developmental genetics, though no less causal in outlook, approached this problem in a holistic manner.[83] Richard Goldschmidt's preference for a holistic model of the chromosome rather than the Morgan school's atomistic one is well known and seems to have been grounded in some kind of systems theory in which the properties at each level were more than the sum of those at a lower level (Allen 1974; Goldschmidt 1954, esp. p. 709). Similarly, a central feature of von Wettstein's plasmon theory of cytoplasmic inheritance was its holistic structure (cf. chap. 9).[84]

80. Goldschmidt 1960, 29, 32 (confirmed by Witschi 1959). The works on Goethe referred to are Renner 1949 and Haecker 1927.

81. Interviews with Marquardt, Linnert, and Harte.

82. The difficulty is compounded, I suspect, by the prewar associations between National Socialism and particular philosophical positions, notably holistic ones.

83. The quotes are from Oehlkers 1927, 2–7. For his appreciation of Woltereck's work see Oehlkers 1933–34. For his views on biology's methodological relations to other sciences, see Oehlkers 1930.

84. The evidence for Karl Henke's sympathy for holism is more suggestive. In an essay defending expressionist painting, he argued that impressionism was associated with

Imputing Styles of Thought   253

When we turn to the pragmatic faction, the picture is very different. Of those pragmatics in the Baur school, plus those listed in table 7.3, only four seem to have shared the comprehensives' interest in high culture.[85] Others' interests lay elsewhere. For von Sengbusch this meant sports, photography, and car repairs; for Schick, gardening. The only avocation mentioned by Hans Kappert's biographers is mountain climbing.[86] One student of Nachtsheim's noted that he had few interests outside of science and had probably never been to a concert. "The pragmatic, not the philosophical," another noted, "is the essence of this great geneticist."[87] Given their limited interest in philosophy, the pragmatics' ontological assumptions are difficult to uncover. Anecdotal evidence suggests that Hertwig, Nachtsheim, and Stern may have been skeptical of holistic concepts in genetics.[88] The pragmatics' conceptions

---

individualism and a mechanistic worldview, while expressionism gave precedence to the community. Hans von Marées' (1837–1887) early work had enjoyed little impact, since "he was too strongly oriented toward the totality to have achieved success in an age so obsessed with specialization" (Henke 1928, 239; cf. 224, 229–30). The philosophical assumptions which seem to underlie Henke's ideas about pedagogy in biology were antipositivist in several respects, and perhaps antireductionist as well (cf. Henke 1930, 384–85).

85. Schwemmle was a keen pianist and was interested in architecture (Ziegler; interview with Berthold Schwemmle). Stubbe was interested in painting, sculpture, and literature (Hagemann 1984). Grüneberg was interested in Indian architecture and Chinese painting (Lewis and Hunt 1984).

Stern is an ambiguous case. Caspari (entry on Stern, *Dictionary of Scientific Biography*, forthcoming) says that Stern was interested in philosophy but does not elaborate. When I asked Bentley Glass to compare the interests and personalities of Stern and Goldschmidt, he remarked that Stern was not as "cultured" as Goldschmidt (interview). By all accounts, Stern was much happier in Morgan's laboratory than was Belar, and seems to have been better suited to the American academic climate (see chap. 8, section 4). While the extent of Stern's interest in high culture is unclear, therefore, his overall classification as a pragmatic is defensible.

86. On von Sengbusch see Ordner 4, von Sengbusch papers, and Kuckuck and Stubbe 1970. On Kappert see Rudorf 1955b.

87. The first student was Friedrich Vogel (interview). The quote is from Günther and Hirsch, "Hans Nachtsheim zum 80. Geburtstag," p. ii; privately printed, 1970). Perhaps Paula Hertwig might have shared Nachtsheim's priorities. In a letter congratulating Hertwig on her seventy-fifth birthday, her lifelong friend Elisabeth Schiemann reminisces about their years as postgraduate students at the University of Berlin. While Schiemann enjoyed preparing for the oral examination in philosophy as part of her doctoral examinations, Hertwig evidently did not; as Schiemann wrote her, "Ich glaube, Sie haben es mehr als Last empfunden. . . ." (item no. 117, Schiemann papers).

88. Interviews with F. Vogel and U. Philip. In her 1934 review paper Hertwig was openly skeptical about claims for the holistic nature of cytoplasmic heredity (cf. chap. 9). It is perhaps significant that she originally gave this paper at a meeting of the *Gesellschaft für empirische Philosophie,* Berlin's counterpart to the Vienna Circle.

254   Styles of Thought

of scientific method, on the other hand, are rather more visible. In contrast with Kühn's emphasis upon imagination and presupposition or Goldschmidt's bold theorizing, the pragmatics were much more empiricist. On Nachtsheim's seventieth birthday, for example, his students and friends presented him with a medal inscribed with the motto, "What one can see, hear, experience: That is what I prefer." Accordingly, Nachtsheim regarded Goldschmidt as "ingenious but vacuous."[89] Edgar Knapp is said to have been an unemotional man with a "strict and logical" mind who was quick to attack propositions which he regarded as unclear or speculative. (Abel 1980; Butterfass 1980). Julius Schwemmle's views on scientific method, too, were the very model of sobriety. The qualities which made for scientific excellence, he maintained, were the ability to choose an appropriate organism, exhaustive experimentation, correct interpretation of data, and the capacity for self-criticism. Theoretical imagination, it seems, was not among them. Schwemmle's papers sometimes begin not with an introductory section in which the problem is outlined and the literature surveyed, but, abruptly, with his results. And in the conclusion, where one might expect to find a discussion of the theoretical implications of the author's data, one finds simply a summary of his findings.[90] In other papers he seems reluctant even to generalize. When confronted with conflicting evidence, he did not attempt to reconcile the differences through developing a unifying theory or treating the deviant instance as a special case or artifact. Instead—reveling, it would seem, in the diversity of nature—he was content to allow contradictory cases to stand as is.[91] Methodologically cautious in the extreme, Schwemmle held a view of scientific method worlds apart from the comprehensives' search for synthesis.

Perhaps the most striking—and ultimately illuminating—difference

---

89. [ein geistreicher Schwätzer] Vogel interview; the motto is mentioned in Günther and Hirsch, "Hans Nachtsheim zum 80. Geburtstag," p. ii (privately printed, 1970).

90. E.g., Schwemmle and Zintl 1939; Schwemmle 1941b, 321–35. At the end of his paper on chloroplast mutations, for example, one looks in vain for any discussion of the evolutionary or developmental significance of such mutants (Schwemmle 1941a, 171–87). The qualities which he identified with scientific excellence are found in Schwemmle 1934.

91. Responding to the findings of Renner and Kupper (1921), for example, Schwemmle first declared his argument that *some* traits' maternal inheritance was due solely to the maternal cytoplasm. With *other* maternally inherited abnormalities, however, his explanation was Mendelian: The cytoplasm merely modified the degree of expression. But with these latter traits he was cautious: "I do not wish to imply that [such] abnormalities in other cases could not be due to the cytoplasm alone" (1927, 48). For other examples see chap. 9.

Imputing Styles of Thought 255

between the two factions, however, is the depth of the comprehensives' interest in the fine arts. Some of them were simply devotees of the concert and art museum (e.g., Jollos, Haecker, and Renner).[92] On his lecture tour of the United States in 1938, von Wettstein took every opportunity to visit major art museums, and his diary of the trip is filled with details of the various collections he saw as well as judgments of their quality.[93] Apart from visiting museums and trying his hand at musical composition, Karl Pirschle began to learn Egyptian toward the end of the war so that he could better appreciate Egyptian art (Stubbe 1955). Goldschmidt's collection of oriental porcelain contained two hundred pieces by 1939, some of which had been displayed in the international exhibition of Chinese art in Berlin as well as at various exhibitions and art museums in San Francisco.[94] And Friedrich Brieger helped to establish a school of music—in Brazil, to which he had emigrated in 1936 (Linskens 1986). Other comprehensives were actively engaged in performance. Peter Michaelis was a gifted pianist,[95] and, as we have seen, Belar derived his greatest pleasures in Pasadena from sketching and painting. The son of a painter, Correns evidently possessed sufficient talent to have become one himself, had he so chosen.[96] On leaving school, Joachim Hämmerling hesitated between the study of music and that of biology, and a student of Oehlkers's (Hans Marquardt) obtained doctorates in both musicology and botany before settling down as an assistant in genetics.[97] Henke's knowledge of contemporary painting was sufficiently extensive that he was asked to contribute a lengthy essay on the subject to a volume on modernism in various fields (Henke 1928).

But what, exactly, is the relationship between comprehensives' science and their love of the arts? It would be wrong, I think, to draw a

92. Goldschmidt, "Professor Victor Jollos," typescript, Goldschmidt papers; R. Haecker 1965; Mägdefrau 1961; von Frisch 1960–61.
93. From his diary of a tour of the Southwest and Mexico at the end of his 1938 visit, as well as from the unpublished recollections of his military service in World War I (both kindly lent to me by Diter von Wettstein), it is obvious that von Wettstein was very sensitive to nature's beauty. See also his obituaries by H. Stubbe (1951, 4) and Kühn (1947, 6).
94. Goldschmidt to F. R. Lillie, 18.10.39, Lillie papers.
95. Interviews with Kuckuck, Rothe, and W. Stubbe.
96. Correns's mother encouraged him in this direction (von Wettstein 1938, 141; Saha 1984, 104–5), but his father blocked an artistic career (autobiographical sketch, Correns papers).
97. Hämmerling, "Mein wissenschaftlicher Lebenslauf," Stern papers; Marquardt interview.

sharp boundary between their aesthetic sensibility and their research, treating the former as recreation and the latter as profession. For art's significance to them was by no means so neatly circumscribed. To call a scientist's achievement "artistic" was the highest form of tribute, and not just among geneticists.[98] Goldschmidt's autobiography is full of admiration for those whom he described as cultured, refined, or artistic (e.g., Goldschmidt 1960, 106–9). For von Wettstein, Correns's research was a "work of art from the realm of ideas"; his life was "a marvelous work of art" (von Wettstein 1938, 151–52, 154). One of Michaelis's technicians described him as typical of an earlier, less specialized generation: "less a scientist than an artist."[99] Renner honored Julius Sachs as the greatest botanist, "both as researcher and as artist," and a decade later the same phrase was applied to Alfred Kühn.[100] The meaning of this tribute for geneticists of comprehensive persuasion, as for Kühn himself, was in part an epistemological one. Scholarly excellence required a delicate balance between intellect and imagination. Renner's "clear and undistorted understanding of nature derives as much from analytical thought as from aesthetic and religious feelings."[101] "Scientific imagination in a great scientist," Goldschmidt argued, "has always in it an ingredient of artistic feeling" (1956, 123). Theodor Boveri was a perfect example; his brilliant contributions to biology bore the marks of his personality, as philosophical and artistic as they were insightful. He was "not only a great scholar but a noble, harmonious man" (Goldschmidt 1916). Alfred Kühn, similarly, was admired for "the beauty, harmony and force of his character."[102] The enormous respect for art among comprehensives, therefore, reflected an epistemological ideal in which insight demanded not only analytical skill but an artist's intuition.[103]

98. Cf. Heilbron (1986, 52) on Max Planck and other physicists. The botanist Gottfried Haberlandt was "not only a botanist . . . he is at heart [*im tiefsten Wesen*] an artist" (Jost 1933–34, 172). The applied entomologist K. Escherich modestly staked his claim to cultivation, saying, "The heavy emphasis upon art in my life and that of my family protected us from one-sidedness, narrowness and the most dangerous enemy of mankind: philistinism" (Escherich 1944, 28; cf. 209).

99. Rothe interview.

100. Renner 1949, 108; Otto Koehler, "Die Fakultät ehrt . . ." (undated speech recommending Kühn for an honorary doctorate), Koehler papers.

101. Bünning 1944, viii; virtually the same was said of the zoologist Karl Heider (Ulrich 1960, 873).

102. [*geschlossene Kraft seines Wesens*] Otto Koehler to Otto Hahn, 7.6.56 (Koehler papers).

103. When Belar dismissed Jack Schultz as "too intellectual" (cf. chap. 5), I suggest, it was with reference to this ideal of the harmonious person.

But the connection between art and biology also had an ontological dimension. Common to both realms, I suggest, was an interest in *form*. The aesthetic appeal of biological form, combined with the intellectual puzzle as to how it is generated in development and modified in the evolutionary process, made biology especially attractive to this generation. Kühn was most certainly not the only *Augenmensch* among the comprehensives. Hämmerling felt that many biologists of his day were attracted to the discipline because of their visual sensibility, and Max Hartmann agreed.[104] The beauty that Kühn and Henke found in a moth's wing pattern, Belar evidently found under a microscope: "The constant contact with form and its aesthetic appeal" was "one of the most attractive aspects of biology" (Belar 1924, 375). Goldschmidt believed that the high incidence of artistic ability among biologists of his generation—he cited Otto Bütschli, Ernst Haeckel, Carl Chun, and many others—was no accident: "This ability can be understood as a part of the sense of form which makes a good naturalist and morphologist" (Goldschmidt 1956, 62–63).

The comprehensives' admiration for art and the artist thus constitutes a kind of nodal point in their style of thought. It bears common features not only with their decision to study biology—and especially development—but also with their striving for breadth of perspective beyond their discipline. Indeed, their interest in form points toward a deeper thread which runs though all of the facets of comprehensive thought. Thirty years ago Jane Oppenheimer asked herself why so many of the greatest embryologists had been German. Her answer:

[They were] in a sense in love with the embryo, and thus able to fathom some of its secrets by processes of understanding that transcend the usual . . . procedures of scientific thinking. And why? Because they could see the embryo whole. . . . They were the flower of the . . . romantic movement which could enable them to comprehend the wholeness of the embryo as part of the wholeness of nature . . . [Oppenheimer 1956, 33].

Her answer has since been echoed by others who knew personally some of the early pioneers in embryology: Their fascination with form was a fascination with wholes (Hamburger 1980b, 303–8; Holtfreter 1968, xii). On the aesthetic level it is the integration of shape and color in evocative proportion which is effective in visual art, just as in music it is the successful integration of sound and silence which gives rise to emergent properties like harmony and rhythm. And on the analytical level the fascination with form lies in understanding the creative pro-

---

104. Hämmerling, "Mein wissenschaftlicher Lebenslauf," 4 (Stern papers).

cess—be it in the studio or in nature—whereby integral forms somehow emerge out of constituent elements assembled in sequence. Whatever comprehensive geneticists may have thought of "the romantic movement," Oppenheimer is right that an ontological holism is the recurring element common to all sectors of their thought. In their insistence that genetic theory encompass not only transmission genetics but also development and evolution; in their resistance to specialization; in the all-around cultivation to which they aspired—in each of these arenas the comprehensives sought a breadth of perspective which could capture what they took to be the wholeness of nature. And on those rare occasions when metaphysical assumptions surfaced in their writing, a preference for holism is evident.

In contrast, very few pragmatics were preoccupied with the arts or other forms of "cultivation." And in those instances where their ontological or epistemological premises can be inferred, the pragmatics displayed neither the comprehensives' holism nor their enthusiasm for adventurous theorizing and conceptual synthesis. Furthermore, as we shall now see, pragmatic geneticists' views were distinct from those of the comprehensive faction on other issues than science and culture.

7.3 Differences of Political Outlook

In many respects the views of the Baur school and other pragmatics seem recognizably "modern." In their attitudes toward specialization and in their casual relation to high culture, they resemble most scientists today. The outlook of comprehensive geneticists, on the other hand, corresponds much more closely to the usual stereotype of the traditional German professor. When it comes to politics, therefore, one might assume that the pragmatics would be generally more sympathetic to democratic ideas than would the comprehensives. Significantly, this was not the case; both factions were highly diverse. Among the comprehensives were geneticists of conservative-nationalist persuasion as well as some sympathetic to social democracy and to the German Democratic Party, while the pragmatics were about equally split between liberals, social democrats, and Nazis. It is thus impossible to array pragmatics and comprehensives along a conventional left-right political spectrum. What distinguishes these factions, nonetheless, is their attitude toward party-political *activity*. By "activity" I do not mean simply voting or expressing political views; rather, pragmatics were more likely to join a party or other political organization (of whatever stripe) than were comprehensives. And the latter believed that engagement in party politics undermined national unity and was thus incompatible with the academic's proper role.

It should be noted that the categorizations of party-political activity presented in tables 7.2 and 7.3 refer to the period *prior* to 1933. Trying to infer political attitudes after 1933 is notoriously difficult for a variety of reasons. Almost all of the assistants in both schools joined some kind of National Socialist organization, but the motivations for joining varied enormously. Admittedly, in a few cases the particular National Socialist organization which a geneticist joined does correlate with independent indicators of his/her political outlook. For example, some geneticists seem to have made only minimal concessions to Nazi pressure, joining only the relatively harmless *Nationalsozialistische Volkswohlfahrt* (a welfare organization) or the near-mandatory teachers' league (*Nationalsozialistischer-Lehrerbund* or *-Dozentenbund*),[105] while some who joined the SS, SA, or party were plainly enthusiastic about NSDAP policies.[106] But many either were opportunists who hoped membership would advance their careers, or joined in response to pressure of varying degrees.[107] But the extent of pressure which local Party organizations were able to exert upon academics varied considerably from one university to another in ways that are not yet well mapped, making it hazardous to draw conclusions about an individual's ideological stance from the Nazi organizations of which he or she was a member.[108] The point is nicely summed up by the geneticist Klaus Pätau:

105. E.g., Kuckuck and others in the Baur school to be discussed below.
106. For example, Kühn's long-standing assistant at the Institute of Zoology in Göttingen, Otto Kuhn, put pressure on doctoral students to join the SA and had considerable influence over academic appointments in the university after 1933 (Schappacher 1987, 352; Eichler interview; Herter 1979, 182).
107. Returning to Berlin from abroad in the summer of 1933, Hans Bauer was asked by his college fraternity to join one of the Nazi organizations or forfeit his fraternity membership. Loathe to leave the latter because of several long-standing friendships, and impressed by the support which the Party seemed to enjoy from all social classes, he joined the SS (until 1935) because its Berlin branch was more moderate than the SA (Bauer to Franz Schrader, 19.9.46, Demerec papers). Hans Winkler, on the other hand, faced virtual blackmail. When the *NS-Dozentenbund* at Hamburg learned in 1937 that Winkler had not disclosed the fact that one of his assistants was Jewish, Winkler was threatened with firing unless he joined the Party (Höfler 1949, 249). At Göttingen, socialist students were allowed to resume their studies if they joined the SA (Dahms 1987a, 31).
108. I am indebted to Herbert Mehrtens for this point and, more generally, for helping me to understand the complexity of National Socialist influence within the universities.

260   Styles of Thought

It is evidently extremely difficult for our liberators [a reference to the Nuremberg war-crimes trials] to believe that certain organizations which gave every appearance of being highly committed, actually contained a host of members whom no one would have regarded as Nazis. This was at least the case with the SA, and even few Germans probably realize that completely harmless people like B. [*sic*] preferred the SS to other Nazi organizations for reasons which had a lot to do with beer-drinking and nothing to do with politics.[109]

In order to assess political outlook, therefore, it is necessary to focus upon the *pre*-1933 period. And in so doing, the sharpest differences between pragmatics and comprehensives emerge from a comparison of the schools of Baur and Kühn/von Wettstein.

As we saw in the last chapter, many of Baur's assistants displayed a high level of enthusiasm for National Socialism. But the Baur school was by no means politically homogeneous. Several of the assistants whom Baur had left behind at the Berlin Agricultural College made only minimal concessions to Nazism. After 1933, for example, Emmy Stein joined neither the Party nor even the *Nationalsozialistischer-Lehrerbund*, and Elisabeth Schiemann's outspoken opposition to anti-Semitism cost her the right to teach at the University of Berlin from 1940.[110] Hans Nachtsheim was not a member of any political party prior to 1933 but was concerned enough about his future under the Nazi regime to explore the possibilities of emigration to the United States. In the event, he remained in Germany throughout the National Socialist period, becoming in 1941 head of the Department of Genetic Diseases (*Experimentelle Erbpathologie*) at the KWI for Anthropology, Eugenics, and Human Genetics. According to his fellow department head, the psychologist Kurt Gottschaldt, he and Nachtsheim were the only non–Party members in that institute at a time when approximately one-third of the staff were in the SS, and more were in the SA.[111] Others in Baur's school were politically active before 1933. Gerda von Ubisch was a member of the left-liberal German Democratic Party be-

---

109. Pätau to "Antoscha" (Anton Lang), 2.2.46, Pätau papers. Pätau (1908–1975) himself was no apologist for National Socialism. A student of Max Hartmann's from 1933–34 who later became an assistant in the department, Pätau emerges from his correspondence over the period 1942 through 1947 as an anti-Nazi of socialist persuasion (cf. Pätau to Stubbe, 6.2.46), and he was active in the Social Democratic Party after 1945 (Anton Lang to me, 17.9.87).

110. Stein file, Berlin Document Center; on Schiemann see Lang 1980, 8; Kuckuck 1980.

111. Telephone conversation with Gottschaldt, 2.3.88. Nachtsheim appears to have joined only the *NS-Lehrerbund* (his file, Berlin Document Center). On his worries in 1933, see Nachtsheim to W. Landauer, 24.10.33, Dunn papers.

fore fleeing to Brazil in 1935 (von Ubisch 1956, 504). Paula Hertwig was a member of the liberal Weimar Circle and stood as a parliamentary candidate for the German Democratic Party in late 1932 or early 1933.[112] I have no evidence of party membership for Stubbe, Schick, and Kuckuck, who were ousted from the Institute for Breeding Research in a political purge in 1936 (cf. chap. 6), but there are several indications that they were sympathetic to the Social Democratic Party.[113] Still others were generally conservative, though not associated with a particular party. Peter Michaelis, a latecomer whom Baur brought to the Institute for Breeding Research in 1932, seems to have avoided becoming entangled with either the Nazi or socialist factions there. All that is known of his political inclinations is that he was strongly anticommunist (W. Stubbe 1987). Elisabeth Schiemann appears to have shared her father's conservative nationalism.[114] In party-political terms, therefore, the Baur school was a motley bunch.

Baur's own political outlook has sometimes been described by his students as "liberal-democratic" (Kuckuck 1988, 20; Schiemann 1934, 100; cf. Deichmann 1991, 410, 414). More commonly they suggest that his death was caused in part by the strain of defending his institute against destruction by the Minister of Agriculture, Walther Darré, leaving the implication that Baur was not trusted by the Nazis.[115] This,

112. Hertwig interview with Paul Weindling (Wellcome Unit for the History of Medicine, Oxford). On her connections with the *Weimarer Kreis,* see Döring 1975, 123; and Baumgardt 1965, 257.

113. Stubbe and Kuckuck have recalled the countless discussions which they and Schick had from the late twenties on social and political issues, eventually culminating in their clash with the Nazi group at Müncheberg (Kuckuck and Stubbe 1970, 3). Furthermore, in his autobiography Kuckuck makes no attempt to deny the political charges which were levied against the three in 1936: among other things, "associating with Jews and socialists, reading left-oriented newspapers and books, behavior on the first of May 1932" (Kuckuck 1988, 34; Straub 1986, 21). Significantly, all three of them chose to remain in the Soviet-occupied zone after the war. Kuckuck joined the East German Communist Party (SED) in 1946 but soon came under political pressure because of his public criticism of Lysenko. In 1949 he left the Party in protest, and the following year he resigned his directorship (of the former KWI for Breeding Research at Müncheberg) in order to seek a position in West Berlin (Kuckuck 1988, 69–75). Stubbe's political credentials were evidently sound enough for him to be appointed in 1945 to a committee in charge of land reform, and the regional Soviet commandant also entrusted him with the direction of three local seed companies (Stubbe 1980, 478–79).

114. Kuckuck 1980, 519. Schiemann's father, Professor Theodor Schiemann, was associated with conservative groups before 1914 and an advocate of German annexationist policies during the First World War (Schwabe 1969, 91, 98, 193).

115. E.g., Stein 1952; von Sengbusch, "Lebenslauf ab 1927," Ordner no. 7, von Sengbusch papers. In an interview, Hermann Kuckuck said that Baur's death was so

however, is to attribute far too much significance to Baur's clash with Darré. Sniping and rivalries among power blocs *within* the NSDAP were commonplace, and the reckless manner in which Baur could voice criticism readily made him enemies.[116] What Baur's students meant when they described him as a democrat, I suggest, was his unpretentious and anti-elitist manner, so unusual among professors of his generation. If Baur had been a genuine democrat, it is unlikely that the National Socialist newspaper, *Der Völkische Beobachter*, would have paid such tribute to him on his death, that he would have visited Party headquarters in 1932 to learn more about their agrarian policies, or that Goldschmidt would have described him as "a Nazi killed by super-Nazism."[117] In fact, Baur's political views prior to 1933 demonstrate a number of affinities to National Socialism. His outspoken nationalism, directed principally against the British and French, was evident from at least 1911, and in 1918 he was already referring to the likelihood of another war to exact revenge (von Rauch 1944, 45, 48). In 1925 the University of Wisconsin decided not to appoint him the Carl Schurz Visiting Lecturer for fear that he would use that platform to denounce the American role in the First World War.[118] Economically he was strongly opposed to laissez-faire policies, not least because he thought they would destroy German agriculture (von Rauch 1944, 47; Baur 1932d, 3). The "unhealthy influence of capitalism" had to be curbed. A population of industrial workers eking out an existence in huge urban tenements would never be able to reproduce itself; only a rural population living in decent conditions (or industrial workers with their own gardens) could do this.[119] A strong nation with the necessary autarky could only be achieved through "state socialism" (which he distinguished sharply from the "idiotic egalitarianism" of social democracy).[120] Accordingly, Baur seems to have sympathized with Kapp's

---

unexpected that rumors were circulating in the institute that Nazis might have murdered him.

116. Stubbe 1959b. Baur's sharp tongue won him the nickname "*der Baur mit dem Löwenmaul*" (literally, "the farmer with the lion's maw," a double pun referring both to Baur's outspoken manner and to his work with the snapdragon, *Löwenmäulchen*).

117. *Völkischer Beobachter*, 6.12.33; Schlösser 1933; Goldschmidt 1960, 272.

118. L. J. Cole to William Bateson, 26.3.25; cf. Bateson to Michael Pease, 18.11.22, Bateson papers.

119. Von Rauch 1944, 46–47; Baur 1932d, 2. For Baur's hope that the "repellent luxury" and "disgusting hedonism" would disappear under the pressure of postwar poverty, see his "Eugenik im neuen Deutschland" (accompanying his letter of 12.10.21 to A. F. Blakeslee, Blakeslee papers).

120. Von Rauch 1944, 60; Baur to Lotsy, 29.3.15, reprinted in von Rauch 1944, 47.

right-wing attempt to overthrow the Weimar government in 1920.[121] The centralized administration of higher education and research under one man and one ministry was essential; Baur was irritated by the duplication and inefficiency of a system which distributed responsibility for science among sixteen ministries of culture, the *Deutsche Forschungsgemeinschaft,* the Kaiser-Wilhelm Society, and Prussian and Reich ministries of agriculture (von Rauch 1944, 62–63; Baur 1932b). These state socialist predilections, along with Baur's evident anticlericalism,[122] place him closer to radical right movements like the NSDAP than to the traditional conservatism of the Deutsch-Nationale Volkspartei. Although he never belonged to a political party (von Rauch 1944, 45), the vigor with which Baur campaigned for eugenic and agrarian reforms, plus his affinities with the radical right, justify classing him as a political activist in table 7.2.

The political complexion of the Kühn school is more difficult to establish (for lack of information), but seems to have ranged less widely. Enthusiasm for National Socialism, though present, was less prominent than in Baur's group. Ernst Caspari recalls many of the doctoral students in Göttingen—like German students in most disciplines at that time—having been pro-Nazi, but only one of the assistants (Otto Kuhn) was ideologically committed to the Party.[123] In Karl Henke's case, it is difficult to judge whether he was politically active prior to 1933 and, if so, how to characterize his inclinations.[124] Kühn's own political leanings were clear; although he displayed no resistance to the regime, the Nazis were suspicious because of his membership of the

---

121. In a letter of 14.3.20 to Bateson (Bateson papers), he comments: "Seit gestern sind wir wieder in der Revolution. Es war aber auch die Corruption in der alten Regierung so gross—Erzberger ist nur ein Beispiel, dass diese Reaktion kommen musste."

122. "Reise nach Spanien mit Professor Baur," unpublished manuscript (1928), Baur papers, MPG-Archiv.

123. Caspari interview.

124. During the political screening process prior to his appointment as professor of zoology at Göttingen in 1937, the Party had no doubts about Henke's political reliability, perhaps because he had served in the Freikorps after 1918 (Henke file, Berlin Document Center; cf. Deichmann 1991, 395). On the other hand, to judge from the volume in which he published his 1928 defense of expressionist painting, Henke was then sympathetic to those new developments in architecture and pedagogy which had a populist character (Henke 1928). The political coloration of the Nohl-Kreis, with which Henke was associated in Göttingen, is a matter of dispute. What seems clear is that Nohl's populist/egalitarian pedagogical views found little support from either traditional conservatives or Nazis. But his pedagogy is difficult to link with any political party because one of its central aims was the constitution of a new *Volksgemeinschaft* which would transcend sectional interests (Ratzke 1987, 204, 207).

German Democratic Party, his Jewish friends, and his association in Berlin with a circle of "elitist communists" (*Edelkommunisten*).[125] He seems, indeed, to have got on well with his famous Jewish colleagues in the science faculty at Göttingen—Max Born, Richard Courant, and James Franck—but once they and his close friend, the geologist Viktor Goldschmidt, had emigrated, life at the university became increasingly difficult for him. Among other things, he evidently felt it necessary to keep a low profile so as not to draw attention to the fact that Caspari remained as an assistant, and Werner Braun as a doctoral student, in the institute until emigrating in 1935 and 1936, respectively.[126] The political pressures upon Kühn, however, mounted steadily, for in 1936 he sought to be released from his post in Göttingen with no new job offer yet in hand. Had the Ministry of Education forbidden him to move to the KWI for Biology—which von Wettstein was trying to arrange—Kühn was prepared to accept an offer from the United States.[127]

Von Wettstein's group encompassed the same diversity of political outlooks as Baur's. Although it had its share of Nazi sympathizers, only one had joined the Party prior to 1933, and by 1938—whether freely or with the aid of a little push from von Wettstein—all of these assistants had left the KWI for Biology.[128] This left von Wettstein freer

125. See letters of 29.5.35 and 14.1.37 from the Party's *Kreispersonalamtleiter*, Kühn file, Berlin Document Center. The communist group in question would appear to be the Internationale Sozialistische Kampfbund, a splinter group of the Social Democratic Party, whose socialism was of an ethical and non-Marxist variety. Its many sympathizers in the science faculty at Göttingen included the mathematicians David Hilbert and Richard Courant (Dahms 1987a).

126. Kühn to Otto Koehler, 16.1.48, Koehler papers.

127. Kühn to O. Koehler, 20.11.47, Koehler papers; Kühn to Carlo Schmid (n.d., probably late 1940s), and von Wettstein to Herr Geheimrat, 15.9.36, Kühn papers, Heidelberg.

128. Most of these men's associations with von Wettstein after 1933 were relatively brief. Alfred Barthelmess (1910–   ) joined the Party in 1930, and was a guest in von Wettstein's department from 1936 to 1938 (Barthelmess file, Berlin Document Center). Eckhard Kuhn (1904–1942) was said to be "enthusiastic" about the Nazi takeover (Knapp 1943, 346), and Goldschmidt regarded him as an "enthusiastic SA Man" (Goldschmidt to Curt Stern, 4.12.33, Stern papers). On taking over Correns's department in 1934, von Wettstein had inherited Kuhn, and in 1938 it was arranged that Kuhn would "transfer" to the Institute of Botany in Hamburg (Hünemörder 1987). L. A. Schlösser (1906–1973), who had been a student and assistant of von Wettstein's at Göttingen and Munich, was *Schulungsleiter* in the SS's *Rasse- und Siedlungsamt* in 1933 (Schlösser 1933), but seems not to have moved to Berlin with von Wettstein in 1934 (judging from annual reports of the Kaiser-Wilhelm Society from 1934 through 1938). Edgar Knapp's Party membership (referred to in a letter from Melchers to Pätau, 14.3.45, Pätau papers) prob-

to offer help and refuge to social democrats (Stubbe, Melchers, and Pätau), the liberal Kühn, and the conservative Schiemann. The SS's suspicions of the KWI for Biology as a haven for dissidents were, therefore, justified, and its characterization of the institute's directors as "crypto-communists, pro-Jewish and pro-democratic" was, to some extent at least, applicable to Kühn.[129] But it was ludicrously off-target for von Wettstein. Altogether unsympathetic to socialism, von Wettstein welcomed some features of the Hitler regime—e.g., the *Anschluss* in 1938—and on his visit to the United States that year he was at pains to correct the "false impressions" about Germany which he felt some emigrés (e.g., Thomas Mann) had purveyed.[130] Nor could he be portrayed as a liberal. He showed no enthusiasm for the formation of an Austrian republic in 1918 and confessed half-seriously to Melchers that he was at heart "an advocate of enlightened absolutism."[131] Though conservative, however, von Wettstein displayed a good deal more *Zivilcourage* than most of the professoriate. He repeatedly defended the autonomy of German science against National Socialist interference, thus

---

ably helped him to secure the newly established chair of genetics at the University of Strassburg in 1941 (Deichmann 1991, 330; cf. chap. 9). Knapp, too, had originally been hired by Correns, and in 1936 von Wettstein arranged for him to switch to the institute at Müncheberg in exchange for Stubbe ("Tätigkeitsbericht der KWG," *Die Naturwissenschaften* 25 [1937]: 376). In Georg Melchers's opinion, Kühn's and Knapp's involvement with the Party was essentially opportunistic (Melchers to me, 2.3.88, and interview).

129. *SS-Ahnenerbe* to *Reichssicherheitshauptamt* (1942), cited by Epstein 1960, 84. Another Nazi agency also expressed reservations about von Wettstein's "democratic-liberalizing views" (Grau, Schlicker, and Zeil 1979, 115).

130. See von Wettstein's 1938 diary (pp. 36, 58–59, and 98) and his recollections of military service in World War I (which Diter von Wettstein kindly lent me). In the latter he vented his anger at the fact that his company had lost a battle due to an ammunition workers' strike: "Ich glaube jeder von uns . . . hätte in die Arbeiterbande hineingeknallt, die wochenlang wegen Lohnstreitigkeiten gestreikt hatten. . . . Ist es da ein Wunder, dass wir auf diese rote Bande noch heute eine Wut ohne gleichen haben?" (p. 67; according to Diter von Wettstein, his father recorded these recollections in 1929). From 1931 von Wettstein had been a member of the Deutsch-Osterreichische Arbeitsgemeinschaft, an all-party organization of intellectuals founded in 1925—by, among others, his father, Richard von Wettstein, professor of botany at the University of Vienna—which sought to promote stronger cultural, legal, and economic relations between Germany and Austria. (His membership is recorded in a questionnaire in Rep 76, Vc, Sekt 2, Tit 23, Litt A, Nr. 112, Bd III, ZStA-Merseburg. On the organization itself, see Ball 1937, 58, 85.)

131. Melchers 1987b, 399. In his diary of the First World War, von Wettstein wrote: "Die Republik kam, lustlos aufgenommen, man wusste nicht wozu. Wir mussten uns einleben . . . wir begannen zu studieren" (pp. 73–74).

incurring personal risks which the more liberal Kühn was probably not prepared to take.[132] At Munich in 1934, when a number of professors were becoming annoyed at the increasing proportion of their students' time which had to be spent on Nazi political and sporting activities, von Wettstein—then dean of the science faculty—protested more bluntly to the Ministry of Education than did any of the other deans (Roegele 1966, 162–63). Six months later, as the Kaiser-Wilhelm Society's memorial ceremony in Berlin for Fritz Haber approached, von Wettstein informed staff of the KWI for Biology that since attendance at the ceremony would be regarded by Nazi authorities as a sign of defiance,[133] each individual would have to decide for him- or herself whether to take the risk; he, however, would be attending. On the day, almost all institute staff, including the members of Nazi organizations, chose to attend.[134] Consistent with this portrayal of him as a conservative anti-Nazi is the fact that some of the persons involved with the plot to assassinate Hitler on 20 July 1944 had held meetings at von Wettstein's house.[135] Politically, therefore, von Wettstein was the very embodiment of the conservative nationalist professor who resisted political intrusions upon scholarship and sought to remain "above" party politics (cf. chap. 8).

Turning from the Baur and Kühn/von Wettstein schools to the remainder of the genetics community, the evidence on political attitudes is much weaker but again suggests diversity within each camp (table 7.3). Among the pragmatics, Hans Kappert may have voted for the (middle-of-the-road) Catholic Center Party.[136] Which parties Bauer, Grüneberg, and Schwemmle voted for prior to 1933 is unknown. At a guess, Stern was some kind of left-liberal.[137] Among the comprehen-

---

132. This comparison was drawn by W.-D. Eichler (interview), among others, who studied with both Kühn and von Wettstein after 1933.

133. Formerly director of the KWI for Chemistry and a Jew, Haber had emigrated in 1933 and died six months later in exile.

134. Klaus Pätau to Dr. Havemann, 1.2.46, Pätau papers. According to his son, von Wettstein had told his wife that he had to attend the Haber ceremony, even if it meant the family would have to emigrate (Diter von Wettstein to me, 26.6.85).

135. Diter von Wettstein to me, 26.6.85.

136. Questionnaire issued by the American Military Government, dated 31.3.48, box 1, Kappert papers.

137. Stern saved an excerpt from Ignazio Silone's "Farewell to Moscow" (*Harper's*, November 1949) in which Silone declares his faith in "socialist values" such as brotherhood rather than in any particular doctrine (Stern papers, "Memorabilia and Miscellaneous"), and in 1950 he was evidently opposed to the University of California's request that staff sign loyalty oaths, even though he felt that he personally was not at risk (Stern to Ernst Caspari, 4.7.50, Caspari papers). That Stern would not have regarded himself

sives I have found next to nothing about the pre-1933 political views of Belar, Brieger, Correns, Haecker, Jollos, Timofeeff-Ressovsky, or Winkler, but Friedrich Oehlker's political sympathies were evidently middle-of-the-road.[138] Goldschmidt, like von Wettstein, is best described as a conservative nationalist.[139] Otto Renner, similarly, was no friend of the Weimar Republic. In 1926 police in Berlin uncovered plans for a putsch by the right-wing *Vaterländische Verbände* with which Professor Ludwig Bernhard was associated. When police searched Bernhard's house, thirty-one professors at the University of Jena, including Renner (Fliess 1959, 300–302), signed a public letter of sympathy for Bernhard, criticizing "the treasonous founders of the Republic" (*Hochverräter von 1918*) for bothering "a patriotic man." Four years later, as president of the German Botanical Society, Renner opened the annual meeting by declaring: "I extend particular greetings to those colleagues who have come from outside the Reich's borders, and I may as well say that we regard the Germans among them as fellow countrymen, even if the momentary political boundaries separate them from us" (Renner 1930, 2). But like many traditional conservatives, Renner was ambivalent about the Hitler regime. He welcomed the new emphasis upon eugenics but was no anti-Semite (Renner 1935,

---

as a socialist is implied in a letter from his friend Karl Belar, mentioning that he had met a man in Pasadena who was interesting "although a socialist" (Belar to Stern, 5.8.30, Stern papers). One of Stern's students in the early 1930s—herself a member of a social democratic students' association—described him without hesitation as a liberal (Ursula Philip interview).

138. I have no evidence that Oehlkers was politically active before 1933, and one of his students believes that Oehlkers would never have considered joining a party or other political organization (Hans Marquardt to me, 13.9.89). Cornelia Harte's impression that his sympathies lay with the "liberale Mitte" (letter to me, 24.6.88) is borne out by occasional remarks in his correspondence. In 1930 and 1931 he was concerned that nationalists were exploiting the high level of unemployment; Germany, he feared, was heading for political and economic disaster (Oehlkers to Ralph Cleland, 30.8.30 and 2.1.31, Cleland papers).

139. Goldschmidt enjoyed his life in imperial Germany and was then "something of a monarchist . . . like most unpolitical Germans" (Goldschmidt 1956, 91). An ardent German patriot in World War I, he found the Treaty of Versailles stupid (Goldschmidt 1960, 47, 152–57, 269). Even late in 1933, disillusioned as he was with the anti-Semitism of "uncultured Nazis," he found National Socialism "distinctly preferable to . . . Bolshevism or Communism which might have been, and may still be, the alternative" (Diary of HMM, dated 13.12.33, folder 109, box 13, ser. 717D, Rockefeller Foundation papers). In his autobiography he does, to be sure, pay various compliments to governments of the Weimar Republic (1960, 187–88, 194, 268), but these judgments may have been colored by hindsight. Ursula Philip (interview) recalls him as "very much *deutsch-national*."

1977–79). In May 1933 he publicly defended German Jewish scientists, among them his assistant Leo Brauner, who had been banned from the Botanical Institute by Nazi students.[140] The price of such defiance, however, was that Renner was rejected by the Ministry of Education for the chair of botany at Munich in 1934 (although ranked first by the faculty) in favor of a Nazi whom the faculty did not want.[141] Like von Wettstein, Renner was strongly opposed to the imposition of political criteria in academic appointments, and he fought successfully against the appointment of the holistic philosopher Adolf Meyer to a chair in "theoretical biology" at Jena.[142] But Renner regretted having to spend his time on politics: "My *Oenothera* work could have been finished long ago if only the ugly noise of an agitated world, with which I want nothing whatever to do, didn't keep forcing its way into my study [*in meine Klause dränge*]."[143] Like von Wettstein and Goldschmidt, therefore, Renner resisted National Socialism from an "unpolitical" conservative-nationalist standpoint.

What, then, can be said about the pre-1933 political predilections of pragmatics and comprehensives? To begin with, it is important to recognize that neither faction was politically homogeneous. The pragmatics, in particular, embraced nearly the entire spectrum of Weimar's political parties, from social democrats to Nazis as well as various points in between. As a result, it is impossible to distinguish pragmatics from comprehensives in the conventional language of "left" and "right." This is not to say, however, that the two factions possessed identical political profiles. The higher proportion of "early Nazis" and social democrats among pragmatics is particularly noticeable, as is the number of conservative nationalists among comprehensives (a contrast to which I will return in the next chapter). What does emerge from tables 7.2 and 7.3, however, is a difference in the level of party-political *activity*. While eleven of the twenty-one pragmatics were engaged in some form of

---

140. Bünning 1977, 13; for Georg Melchers (a social democrat), this made Renner "der mutigste aller mir überhaupt bekannten Genetik-Ordinarien" (Melchers to Klaus Pätau, 31.3.47, Pätau papers). Renner publicly denounced the attacks upon Jews and ransacking of synagogues in November 1938 ("Reichskristallnacht") (Schmidt et al. 1983, 290).

141. Renner to Ralph Cleland, 25.12.34, Cleland papers.

142. Renner to Ralph Cleland, 28.12.36, Cleland papers; Renner to Goldschmidt 26.12.37, Goldschmidt papers; Renner to Gratzl, 29.3.36 and 3.1.37, Renner papers.

143. Renner to Ralph Cleland, 19.1.36, Cleland papers. Perhaps Correns would have sympathized. Writing from a spa town late in 1932, he complained, "Die Politik ist übler als je, und ich bin froh, es hier wie der Vogel Strauss machen zu dürfen und den Kopf in den Sand stecken zu können" (Correns to C. Steinbrinck, 29.9.32, Steinbrinck papers).

party-political activity before 1933, only two of the twenty-three comprehensives were so engaged.[144] Clearly we are *not* entitled to conclude from this contrast that comprehensives were, on the whole, more politically neutral than pragmatics; von Wettstein, Renner, and Goldschmidt are examples to the contrary. Instead, as we shall see in the next chapter, this is a contrast between two conceptions of the academic's relation to the political arena. Pragmatics regarded party membership as entirely compatible with scholarship, while comprehensives were more likely to claim for themselves an "unpolitical" higher ground, above the petty squabbling of political parties.

7.4 Conclusion

Can geneticists' thought be classified as "pragmatic" or "comprehensive"? Bearing certain provisos in mind—that relevant information on many individuals is lacking, that some individuals defy classification,[145] and that one must allow for diversity within both pragmatic and comprehensive factions—the answer seems to be yes. The successful imputation of styles of thought to different sectors of the genetics community, however, still leaves us with a major problem. To identify a style of thought is not merely to catalog a group's attitudes on various issues; one must also demonstrate some *coherence* among those attitudes. A style of thought is not simply an aggregate; it is structured. As we have seen, a certain degree of coherence is evident within compre-

---

144. In producing these figures I have treated Hämmerling and Lehmann as unclassifiable (cf. section 4). Notice, too, that neither the Baur school nor the Kühn/von Wettstein school maps cleanly onto the pragmatic and comprehensive factions, respectively. Michaelis, for example, is classed as a comprehensive (albeit a problematic one, as we shall see in chap. 9). Elisabeth Schiemann, too, is better described as a comprehensive, while Knapp and Melchers are best classified as pragmatics (see chap. 8, section 4).

145. Joachim Hämmerling is a case in point. His research in developmental genetics, as well as the importance to him of fine arts (cf. section 2), suggests that he might be classified as a comprehensive, but in other respects he more closely resembles the pragmatics. In his unpublished autobiography ("Mein wissenschaftlicher Lebenslauf," Curt Stern papers), for example, he notes that he was never particularly interested in systematics and found general zoology boring. This, combined with the fact that his entire career was devoted to *Acetabularia,* may have prevented him from getting a university chair (cf. chap. 4). Finally, Hämmerling was no holist; he seems to have been convinced from an early stage that all biological phenomena must be reducible to chemical and physical ones (cf. his autobiography, pp. 15 and 16).

On the basis of his work in physiological genetics, Ernst Lehmann might also have been classified as a comprehensive, but his views on a variety of other matters were more typical of pragmatics (cf. chap. 9).

270   Styles of Thought

Table 7.4 Universities Where Comprehensives and Pragmatics Received Their Doctorates, Worked as Assistants, and Acquired Their First Posts

	Doctorate	Assistantship	First post
Comprehensives			
von Wettstein	Vienna 1915	KWIB '19–24 (Berlin '23)	Göttingen 1925
Goldschmidt	Heidelberg, 1902	Munich '03–14 (1904)	KWIB 1914[a]
Kühn	*Freiburg '08*	*Freiburg '10–18* (1910)	Göttingen 1920
Correns	*Munich '89*	*Tübingen '92–02* (1892)	Leipzig 1902
Renner	*Munich 1906*	*Munich '07–11* (1911)	*Munich 1913*
Haecker	*Tübingen '89*	Freiburg '90–00 (1892)	*T. H. Stuttgart* (1900)
Oehlkers	*Munich 1917*	*Ag. College '20–22, Tübingen '22–25* (1922)	*Tübingen 1925*
Belar	Vienna 1919	KWIB '19–29 (Berlin 1924)	Berlin 1930
Henke	*Göttingen '23*	*Göttingen '24–30* (1929)	*Göttingen 1937*

hensives' thought: in their broad approach to the problems of genetics, their attitudes toward breadth of biological knowledge, and their cultivation of artistic sensibility, the recurring theme is a striving for an all-embracing conceptual synthesis, occasionally manifest in sympathies for holism. What distinguishes the pragmatics' views is a kind of metaphysical indifference, manifest on two axes. First, the aim of science for pragmatics was to develop not a unified picture of nature, but rather a tool for prediction and control. And second, unlike the comprehensives', the pragmatics' conception of science is not tightly linked to their views on culture or politics by any recurring ontological theme. On inspection, that is, the various facets of the pragmatic style appear to

Table 7.4 *continued*

	Doctorate	Assistantship	First post
Pragmatics			
Baur	*Freiburg '03*	Berlin '03–11 (1904)	*Berlin Ag. Coll.*[b] 1911
Nachtsheim	*Munich 1914*	*Freiburg/Munich* 1914–1921 (1919)	*Berlin Ag. Coll.* 1923
v. Sengbusch	*Halle 1924*	*seed company* 1924–26	KWIBR 1927[c]
Kappert	Münster 1914	KWIB 1914–20 (*Berlin Ag. Coll.* 1929)	*Berlin Ag. Coll.* 1931
Kuckuck	*Berlin Ag. Coll.*, 1929	(*Berlin 1942*)	KWIBR 1929

Note: The individuals listed in this table are those pragmatics and comprehensives from tables 7.2 and 7.3 for whom I have information on the nature of their extrascientific interests.

By "first post" I mean a salaried position higher in status than the ordinary assistant or *Privatdozent*, such as *Extraordinarius* or *Abteilungsleiter*.

The date in brackets under the column labeled "assistant" designates the date (and sometimes place) of *Habilitation*.

The names of institutions have been italicized if, during the period in question, they possessed a separate science faculty, or if their philosophical faculty was split into philological and scientific sections, or if the institution was not a university.

[a] KWIB = KWI for Biology.
[b] Ag. Coll. = Agricultural College.
[c] KWIBR = KWI for Breeding Research.

bear no relation to one another. But if that is so, how can one claim any coherence for the pragmatic style?

Let us consider what the coherence of a style of thought might mean. It is obviously not logical in character. There is no reason in principle, for example, why a transmission geneticist should be any less broadly informed in biology or more skeptical about holistic assumptions than a developmental or evolutionary geneticist. Nor is there any logical reason why a narrowly specialized scientist need be any less interested in high culture than a generalist; numerous counterexamples can be found on any campus today. And it is not at all evident why broad synthetic thinkers should be less likely to join political parties than are narrowly analytical ones. Thus even if a recurring "synthetic" theme runs through most sectors of comprehensive thought, the resulting congruence is clearly a contingent outcome, not a necessary one. The mere existence of a recurring theme, however, seems a pretty weak cement for bonding otherwise diverse sectors of thought into a robust style. What gives a style its stability, its capacity to persist over the course of a generation? The answer, according to Mannheim, is to locate the "basic intention" of a style of thought. One cannot rest content with the analysis of disembodied thought patterns; one has to analyze those patterns in action, looking at how they are used to advance the carrier group's aims. We have to ask, in other words, *why* comprehensives and pragmatics advocated such distinctive conceptions of science and of the academic's role. Only by returning the comprehensive and pragmatic styles to the historical context whence they came can we make those basic intentions visible.

But what kinds of collectivity and what kinds of social circumstances can generate a style of thought? Given that the Kühn/von Wettstein and Baur schools, as loci of the comprehensive and pragmatic styles, respectively, were organized in such distinctive ways, the explanation for these styles might once again be thought to lie in the structure of scientific institutions. It is not obvious, of course, how the contrasting political perspectives of these two schools could derive from their institutional locations. But it might seem less farfetched to claim a link between the character of each school's institutional situation and its cultural interests. For throughout the interwar period the natural sciences at many German universities were located within the faculty of philosophy. At these universities doctoral candidates had to pass an oral examination in philosophy, and professors in humanistic subjects were entitled to participate in *Habilitations* procedures, appointments, and policy decisions affecting the sciences. This would have placed scientists under a certain amount of pressure to pay at least lip service—and perhaps more—to the traditional ideals of broad knowledge and cultivation.[146] Since the humanities were not taught at agricultural colleges, however, Baur and his colleagues in Berlin would have been free of such pressure. One difficulty with this hypothesis is the fact that in several cases, as we have seen in this chapter, comprehensives' interests in the fine arts or classics were already well developed before they entered the university. Nevertheless, the hypothesis is at least testable: From an early date *some* universities possessed separate science faculties or had

---

146. I am grateful to Harald Weinrich for drawing this possibility to my attention.

split their philosophy faculty into an arts and a science section which functioned to some extent independently of one another.[147] This enables us to ask whether there is any correlation between geneticists' styles of thought and the faculty structure of the university or college where they spent various phases of their careers. Table 7.4 presents the dates and universities where a sample of comprehensives and pragmatics received their doctorates, served as assistants, and were first appointed to tenured positions. It is evident that all of the comprehensive geneticists (like their pragmatic counterparts) spent one or more phases of their careers in universities which had *separate* science faculties or sections: for example, von Wettstein, Kühn, and Henke at Göttingen; Correns, Haecker, and Oehlkers at Tübingen; and Goldschmidt and Renner at Munich.[148] Faculty structure thus provides no explanation for the differing cultural interests of pragmatics and comprehensives, nor is it clear that *any* institutional structure could account either for these interests or for the two factions' different political perspectives. In order to resolve this problem, we will need to turn our attention away from the structure of scientific institutions toward the macrosocial changes which were transforming German society from the late nineteenth century.

147. Separate science faculties were established at Tübingen (1863), Heidelberg (1890), Strassburg, 1871), Frankfurt and Hamburg from their foundings in 1914 and 1919, respectively, and Göttingen, Halle, and Jena in the early 1920s. At Munich (1870), Freiburg (by 1900), and several other universities during the 1920s, the philosophical faculty was divided into arts and sciences sections.

148. Should Kühn's, Correns's, and Belar's first posts appear to support an institutional explanation, it should be noted that (1) Göttingen's philosophical faculty split within two years of Kühn's arrival; (2) Correns spent the bulk of his career (1914–1933) at the KWI for Biology; and (3) Belar's "first post" at the University of Berlin was merely a professorial title. His entire postdoctoral career was spent at the KWI for Biology and (from 1929 to 1931) at the California Institute of Technology.

8 Mandarins Confront Modernization

Both institutionally and ideologically, the German genetics community was segmented. In their approaches to genetics, as in their attitudes to a variety of issues having apparently nothing to do with science, the comprehensive and pragmatic factions were distinct. In order to understand the genesis and meaning of the styles of thought associated with these factions, one must look beyond the confines of the genetics community and its institutions. For similar phenomena are to be found in many sectors of the German professoriate from the late nineteenth century, as indeed within the educated middle class (*Bildungsbürgertum*) of which academics were the most articulate spokesmen. Since styles of thought are fashioned by groups in action, my first task in this chapter is to outline the social context in which the German professoriate found itself at that time. For it is against the backdrop of the professoriate's struggle to recapture what it took to be its proper role in Wilhelmian society—a role threatened by that constellation of social changes best described as "modernization"—that we can begin to understand the basic intention of the comprehensive style of thought. Conversely, if we are to find any coherence among the otherwise disparate elements of the pragmatic style, we must recognize that the professoriate did not react to modernization with one voice.

Although controversial in some quarters, the concept of modernization is indispensable for understanding German history during the Wilhelmian period (1871–1918). I use the term as shorthand for the complex of social changes—urbanization, social differentiation, political liberalization—accompanying the industrialization of the German states from the mid-nineteenth century and accelerating rapidly after formation of the *Reich* in 1871. More generally, of course, social transformations of this kind, in Germany as elsewhere during the nineteenth century, were the stuff of classical sociological theory. Modernization in this usage is not an ahistorical process with some kind of inbuilt dynamic which eventually restructures all societies (social evolution in all but name), but a contingent historical fact: That in the aftermath of the late-eighteenth-century political and industrial revolutions in France and England, many nations have undergone a series of similar social changes (Bendix 1967; Wehler 1975; cf. Lepsius 1977). Nor is modernization intended to imply a balanced process of simultaneous and coordinated changes in all sectors of a society; tradition and modernity can coexist stably in the same society, and economic development is no guarantor of democratization. Wilhelmian Germany was "partially modernized" in just this sense.[1]

But modernization entails more than simply social structural change. Identifying the *cultural* dimensions of this process was of prime interest to German sociologists at that time. For Max Weber the most fundamental of these was secularization (or 'rationalization'): the displacement of overarching meaning systems—be they religious or otherwise metaphysical—by naturalistic forms of knowledge whose intent was instrumental rather than expressive or interpretive.[2] Once freed of the obligation to lend meaning, Nietzsche argued, such naturalistic knowledge was incapable of creating a unitary worldview: "The sciences, pursued without any restraint and in a spirit of the blindest laissez-faire, are shattering and dissolving all firmly held belief."[3] Accordingly, the social role of the man of knowledge was in transition. The traditional aim of education for leadership, Weber argued, was to cultivate a personality of the desired kind, such as the literary mandarin of the ancient Chinese bureaucracy, or the English gentleman. Modernization, however, created a need for specialized training, and the new specialists were struggling to assert themselves against their "cultivated" predecessors, a "fight [which] intrudes into all intimate cultural questions."[4] Indeed, Weber himself joined this fight, chiding his more traditional colleagues for failing to recognize that the specialized knowledge appropriate to modern academe offered no license for prophecy. The true, the good, and the beautiful were not to be conflated.[5]

As I will argue below, it was precisely this fight which separated comprehensives from pragmatics in the German genetics community.

1. On "partial modernization" in general, see Bendix 1967 and Rüschemeyer 1976. On Wilhelmian Germany see Dahrendorf 1967 and Wehler 1973.
2. Cf. Lash and Whimster 1987; Robin Horton finds the same difference between Western science and African traditional thought in his 1971.
3. Nietzsche, *Unzeitgemässe Betrachtungen,* cited by Frisby 1985, 31. Many inhabitants of the modern age, Georg Simmel remarked, perceived art as a meaningful refuge from the fragmented and restless character of everyday life (Frisby 1985, 44–45).
4. Weber, "Bureaucracy," 243 (cf. Weber, "The Chinese Literati," 426), in Gerth and Mills 1970, 196–266, 416–44.
5. Weber, "Science as a Vocation," in Gerth and Mills 1970, 129–56. Max Scheler's sympathies were evidently with the other side: "The most *profound* perversion of the hierarchy of values," he wrote, "is the *subordination of vital values to utility values,* which gains force as modern morality develops." This he called "the ethos of industrialism" (cited in Stikkers's introduction to Scheler 1980, 21).

Throughout most of the nineteenth century, the educated middle class had seen themselves as occupying a privileged status as culture-bearers whose creation of a unitary scholarship would point the way toward unification of the German states. By the late nineteenth century, however, the reality was very different. Disappointment with Bismarck's "small Germany"—which failed to include German-speaking parts of Austria-Hungary—was compounded by the loss of the educated middle class's political influence to a rapidly growing bourgeoisie as well as by the social strife and instability of the young nation. The solution to this crisis, academics argued, lay in *their* hands, as mandarins whose loyalty to the nation was 'above politics' and whose wisdom would guide the ship of state through troubled waters. Orientation was to be provided by the intellectual syntheses so valued by the academic community at that time (and so characteristic of comprehensive thought). But the professoriate was unable to speak with one voice, for modernization was taking its toll even within the professoriate's own ranks. As academics were increasingly recruited from industrial and commercial strata, the traditional ideological cohesion of this class was progressively eroded. With their pragmatic style of thought, these newcomers to academe—who tended to be educated and employed in the newer sectors of an educational system in transition—were at once more sympathetic to the aims of an industrializing society and altogether happier to embrace the role of specialized expert than were their comprehensive colleagues. Both comprehensive and pragmatic styles of thought, therefore, were forged in the heat of modernization, albeit by different strata with different aims.

8.1 *Bildung* as Ideology

The key to the comprehensive style of thought lies in the concept of *Bildung*. Derived from the verb *bilden* ("to form"), *Bildung* is generally translated as "cultivation." It was an educational ideal which emphasized not simply the nurture of intellect, but the development of the whole person. By engaging with the highest cultural values—those of ancient Greece—the developing person would become aesthetically and morally aware as well as learned (Vierhaus 1972; Bruford 1975). A well-proportioned personality was commonly admired as a "harmonious work of art" (Thomas 1976–77). Once the entire range of human capacities was perfected, one acquired balanced judgment and perspective, making it possible to apprehend the whole truth. "The best thing a man can do with his life," Wilhelm von Humboldt wrote, "is to

[form] a living picture of the world, properly unified." A century later Max Scheler echoed the same view; the aim of *Bildung* was to acquire a "*complete* view of the world," to grasp the "*essence* of the whole world's structure."[6]

The concept of *Bildung* is associated with the emergence of the educated middle class during the latter part of the eighteenth century (Elias 1978; cf. Vierhaus 1972). From the start the concept expressed membership of both class and nation. On the one hand *Bildung* served to distinguish the middle class from the aristocracy. Against the aristocracy's obsession with form, gentility, and birth, *Bildung* represented achievement. While eighteenth-century German aristocrats were self-consciously cosmopolitan—they spoke French, adopted the manners of their French counterparts, and admired the French literature and drama of the period—the emergent middle class spoke German. Although thinly dispersed across the multitude of German states, these new *Bildungsbürger* acquired a sense of their cultural identity through the now classic *German* literature and philosophy of the late eighteenth century: Lessing, Herder, Schiller, and Goethe expressed the ideals and frustrations of this new stratum becoming aware of itself. Though as yet lacking political coherence, the new Germany, its middle-class champions argued, would be a *Kulturnation,* distinguished from the French nation by its *Bildung*. Circumstances in the preindustrial German principalities, however, made it difficult for such a dispersed class to acquire political self-consciousness. Compared with its more powerful counterparts in France or Britain, therefore, the nascent German middle class was much less concerned with achievement in economic or political realms than in learning or art: Literature was the medium of its protest. This inward-looking and "unpolitical" characteristic of *Bildung* was evident very early in Wilhelm von Humboldt's remark that Goethe had done more for the world by developing his personality than by deliberately working for the good of society (Bruford 1975, 17).

From the end of the eighteenth century members of the educated middle class had been insisting that the participation of cultivated persons in the state was essential for a healthy and progressive society. Following French military victories early in the new century, they renewed their advocacy of *Bildung*. Reforms were necessary in which the middle class acquired political rights, and the state recognized its obligation to support *Bildung* at school and university level. The one-sided

---

6. The two quotes are taken, respectively, from Bruford 1975, 24 and Scheler 1980, 96 (emphasis in the original).

emphasis in French education upon intellect was said to have spawned the revolution, but *Bildung* made for balanced judgment and political moderation. By endowing *Bildung* the German states would secure middle-class loyalty, and the cultural unity conferred by *Bildung* would pave the way for eventual political unification. In the midst of the Napoleonic Wars the German states were persuaded. Prussia took the initiative by giving *Bildung* pride of place in its university reforms; the new universities were to demonstrate Prussia's cultural prowess to an admiring world as well as turning out well-educated graduates who would loyally serve the state (Turner 1971; McClelland 1980). Professors were to be free to teach (and students, to study) what they wanted, because the aim of *Bildung*—objective knowledge of the world— required pure contemplation, freed from base motives or interests.[7] Thus reformers like Wilhelm von Humboldt had insisted from the beginning that the state must not interfere with education. *Bildung* (with all of its benefits to the state) would thrive only in institutions free of one-sided vocational or political aims. Unlike French educational institutions, German secondary schools and universities were to be strongholds of pure scholarship.

At the same time that *Bildung* distinguished the German states from their powerful neighbor, it also served as an entrée to ruling circles in early-nineteenth-century German society. *Bildung*, its champions insisted, distinguished the elite from the masses. Thus top civil servants were often ennobled, and the aristocracy itself attended the new educational institutions in order to become cultivated. The newer commercial and industrial middle classes, in contrast, could not be admitted to this ruling stratum because the utilitarian and specialized forms of education which they preferred did not confer *Bildung*.[8] By the mid-nineteenth century the university-educated middle class had grown in size, in social esteem, and in political importance as the ideological vanguard of the movement for national unity. Since the industrialization of the German economy was state-led, the educated middle class's domination of the highest positions in the civil service gave it a central role in this process. While it would be wrong to describe this as a "rul-

---

7. In a presidential address to the *Gesellschaft Deutscher Naturforscher und Ärzte*, Alfred Kühn contrasted pure and applied research in the same manner, referring to the tension between "apprehending nature as a whole versus problem-oriented research, between totally unhampered research which develops in all directions versus the necessity of concentrating upon particular urgent tasks" (Kühn 1938b, vi).

8. O'Boyle 1968. In Friedrich Paulsen's view, the educated middle class was still deploying much the same argument at the end of the century (Paulsen 1921, 686–87).

ing class" throughout this period, there is no doubt that its members *saw* themselves as part of a ruling class.⁹ And throughout the nineteenth century *Bildung* retained a peculiar appeal for those members of the bourgeoisie and other newcomers who sought social recognition from ruling circles.

As Fritz Ringer has shown, however, during the latter half of the century the social changes wrought by rapid industrialization began to undermine the privileged position of the educated middle class (Ringer 1969). By the 1930s, in one German sociologist's view, that displacement was undeniable: "If *one* sector of [nineteenth-century] society can complain that its rank and significance in the social structure have disappeared, it is the educated middle class [*Stand der Gebildeten*], to a much greater extent than the old property-owning classes" (Geiger 1932, 100). The bourgeoisie was not merely far wealthier than the educated middle class; it was also able to exert influence upon government policies via its representation in the *Reichstag*. And while the proportion of the *Reichstag* deputies who were industrialists rose sharply after about 1880, the proportion of academics and top civil servants declined. The facts of urban social pathology and the growing politicization of an industrial working class, German academics declared, all pointed to the fact that society was in crisis.¹⁰ Although in retrospect it seems obvious that the roots of this crisis were economic and social, German academics diagnosed a crisis of *values*. Since the early nineteenth century they had seen themselves as a kind of idealist vanguard of the movement for German unification; a unitary German scholarship would pave the way for political unification. That the young German nation should now be in such fragmented disarray, therefore, seemed a natural consequence of the fact that academics were being displaced from the corridors of power. They no longer occupied a social position from which they could dispense a worldview capable of forging the nation's warring factions into a solidaristic whole. As a result, academics complained, genuine *Kultur* was steadily losing ground to a complex of decidedly inferior values, those of *Zivilisation*.

Kultur was that blend of learning, art, and morality characteristic of Germany, while *Zivilisation* was the legacy of Western Europe. The latter dealt with means rather than ends; it referred to progressive and

9. Vierhaus 1972, 543. Wehler (1973) regards the upper reaches of the Prussian civil service as having achieved membership of the ruling stratum for nearly two decades after the Napoleonic Wars.

10. From his content analysis of Wilhelmian intellectuals' autobiographies, Doerry finds them preoccupied with notions of social harmony (Doerry 1986, 165–70).

practical ways of organizing society in the realms of economics (including technology) and politics. *Kultur* designated the spiritual and artistic values around which a society was (or should be) organized. As with the defense of *Bildung* in the eighteenth century, this distinction was dual-purpose: It served to demarcate the newly unified German nation from its international competitors as well as to identify the enemy within. Since the United States embodied all of the social trends most feared by German academics, *Amerikanismus* and *Amerikanisierung* ("Americanism," "Americanization") became synonyms for *Zivilisation*. For some it meant the creation of dangerous ties between the university and industry (e.g., Manegold 1970, 114; Wendel 1975, 324; Schönemann 1930, 59); for others, the introduction of parliamentary democracy (Schwabe 1969, 154). For the biologist Jakob von Uexküll (1933, 60) it meant the use of wealth as the sole measure of social status. Martin Heidegger dismissed America as a land without history in which quantity prevailed over quality, and a "bourgeois-democratic spirit" reigned (Tertulian 1988, 14). To the biologist Theodor Boveri, Americanization meant the domination of the practical over the cultural; through American influence Germany was in danger of drifting away from Hellenic ideals (Wien 1918, 158). For the writer Ludwig Klages it was tantamount to industrialization, through which the world "is gradually being transformed to look like one great Chicago, interspersed with agriculture" (cited in Thomas 1985, 84). For the right-wing radicals associated with the magazine *Die Tat*, Americanism represented mass production and the rationalization of industry (Herf 1984, 42). In nearly all of these cases the attack upon Americanization was also a rejection of the ugly face of German modernization (Trommler 1985). In this way Americanism (or *Zivilisation*) provided an ideological weapon with which the educated middle class sought to fend off its class rival.

As the academics saw it, late-nineteenth-century German society lacked its former cultural vitality. Life was increasingly shallow and devoid of meaning. Although antimodernist critiques were common in several European countries toward the end of the nineteenth century, the movement was especially vocal in Germany (Stern 1961; Wohl 1980). The crisis of values was felt even within academe itself. The unchecked spread of *Zivilisation* had brought about a "crisis of learning" (*Krise der Wissenschaft*). Suspicion fell upon the growth of the natural sciences and focused in part upon questions of method. From the 1890s scholars in the humanities had tried to block the advance of naive realism and empiricism from the natural sciences into the humanities and social sciences (Hughes 1959; Ringer 1969, chap. 6). But the problem was more serious than infatuation with an inappropriate epistemology. It seemed as though a fundamental shift was occurring in the *kind* of knowledge cultivated in the universities. As Friedrich Paulsen put it:

> Scientific research does not seem to redeem its promise to supply a complete and certain theory of the universe and a practical world-wisdom grounded in the very necessity of thought. Former generations had been supplied with such conceptions by religion or [philosophy].... Then a new generation ... turned to science with the expectation that exact research would ... supply us with a true theory of the world. But that science cannot do. It is becoming more and more evident that it does not realize such an all-comprehensive world-view that will satisfy both feeling and imagination.... Such disappointment is widespread [Paulsen 1906, 66–67].

The body of knowledge was becoming fragmented; a university was impoverished if it could produce only "endless facts and no ideas," argued Karl Jaspers. Eduard Spranger regretted that "the metaphysical totality of learning" was collapsing into "a sum of specialized disciplines" (cited in Ringer 1969, 257; cf. 254–55). There was widespread unease at the scholarly trend away from comprehensive knowledge and judgment toward "narrow expertise" and "conscientious but pedestrian" work (*gewissenhafte Kleinkrämerei*) (Tompert 1969, 27; cf. 66). It was not merely that members of different disciplines could no longer understand one another's work; that had been largely accepted as an inescapable consequence of the growth and differentiation of scholarship as a whole. Rather the critics feared that even *within* each discipline there was little appreciation of the overarching ideas which could lend coherence. If intellectuals had less influence in society than a century earlier, Alfred Weber warned, it was because they were undertaking *partial* analyses of technical problems without being able to provide integral solutions to the important issues of life (Ringer 1969, 265). They were producing knowledge devoid of wisdom.

In a desperate attempt to stem the decline, Ringer has argued, academics from all disciplines called for a reorientation of scholarship. While the recipes for change varied in different sectors of the professoriate, all were agreed on the need for synthesis. Somehow the fragments of knowledge had to be put back together. Some called for a synthesis of reason and faith; others sought a counterweight to the growth of positivism in the humanities. Many believed the problem would be solved only by a philosophically informed striving for the totality behind the particulars in each field. "Wholeness" and "synthesis" became the slogans of this campaign for the rejuvenation of scholarship (Thomas 1985; Tompert 1969, 67–71; Ringer 1969).

Inevitably the First World War simply intensified academics' sense of crisis. Quite apart from material and psychological devastation, the prospects for a unification of German-speaking peoples had worsened, and the replacement of the monarchy by a democratic republic seemed to undermine the last vestiges of social order. At war's end Spengler's *Decline of the West* reiterated many of the themes of late-nineteenth-century cultural pessimism and found considerable resonance among educated circles, prompting hundreds of articles and reviews and selling one hundred thousand copies by 1926 (Hughes 1952, 90ff.). While almost all academics were critical of particular aspects of Spengler's thesis, their own diagnoses of the Weimar Republic's ills were little different. At the height of the inflation the University of Hamburg's rector saw "intellectual and spiritual indiscipline" (*Haltlosigkeit*) everywhere (Töpner 1970, 26, n. 64). With the arrival of financial cutbacks a few years later, the senate of the University of Göttingen complained that "the German universities are the recognized guardians of the nation's cultural heritage [*geistigen Besitzes*]."[11] Once again academics attributed social disorder to cultural crisis. "The emergencies of our time," staff at the University of Breslau declared, "are not simply nor even primarily economic. It is becoming increasingly clear that we find ourselves in a crisis of *Weltanschauung*."[12] The professor of theology Reinhold Seeberg agreed (Seeberg 1930, 178). The increasingly academic character of religious instruction plus the decline of *Bildung* in secondary schools failed to satisfy that basic need for spiritual certitude so eagerly sought by the youth movement. A worldview was the sine qua non of a scholar; without it no meaning could be assigned to empirical objects or events. It was, therefore, the university's obligation to encourage the individual's search for a worldview.[13] Specialized science was of little use for this purpose, but a greater emphasis upon philosophy could both help students to find their spiritual bearings and counteract the tendency to one-sided specialization in scholarship. In this way the uni-

---

11. Proclamation by the rector and senate of the University of Göttingen, dated 15.12.31 (Math.-Naturw. Fakultät Nr. 41, Universitätsarchiv Göttingen).

12. Statement from rector and senate of the University of Breslau to all German universities, dated 5.11.31, Math.-Naturw. Fakultät Nr. 41, Universitätsarchiv Göttingen.

13. Scientists were not exempt. The plant physiologist Erwin Bünning (1906–?) recollected that "every well-educated man was expected to have and to passionately defend his own *Weltanschauung*" (Bünning 1977, 4).

versities could set an example: It was the task of educated youth to overcome the self-destructive inclinations of the older generation and create a "unity of national will." Eduard Spranger's assessment was similar: "Research and teaching are falling apart through a process of specialization which is indifferent to philosophy" (Spranger 1930, 12). And again academics stressed the need to tackle large themes in overviews and syntheses (e.g., Wach 1930, 204).

8.2 Modernization Diversifies the Professoriate

It should by now be clear that the comprehensive style of thought within the genetics community was simply one variant of the ideology of *Bildung* which characterized the class to which most academics belonged. But what, then, is the significance of the pragmatic style? Was it an aberration peculiar to the genetics community? Evidently not. By the turn of the century the concept of *Bildung* was in dispute, and critiques of the traditional meaning of that term displayed several elements of the pragmatic style. This debate over *Bildung* was at the same time a debate over the politics of modernization, a subject on which the professoriate could not agree. And from this perspective, the pragmatic style of thought was simply an indicator within the genetics community that the social cohesion of the educated middle class was breaking down.

The "hunger for wholeness," as many historians have noted, was found right across the political spectrum among the Weimar Republic's artists and intellectuals (Gay 1974; Schorske 1981; Lebovics 1969; Laqueur 1974). It would be wrong, however, to conclude that *Bildung* meant the same thing to all German academics during this period. As Ringer has observed, although the majority of professors endorsed the virtues of *Kultur*, there was always a minority "modernist" faction who, although embracing their traditional role as culture-bearers, were nonetheless more sympathetic to the values of *Zivilisation*.[14] The Prussian Minister of Education, C. H. Becker, nicely illustrates many features of this modernist outlook. Becker agreed that *Bildung* was the key to Germany's regeneration, but his was not the traditional conception. The conventional view endorsed in the universities, he argued, was one-sidedly intellectual; the harmony of the whole person, however, demanded a broader form of cultivation which embraced character and creativity. To achieve this, a reform of higher education was necessary

14. Ringer 1969; Spranger seems to have recognized this split within the professoriate at that time, manifested as different conceptions of scholarship and *Bildung* (1930, 33–34).

which would overcome the existing isolation of academic and vocational institutions and democratically extend *Bildung* to a wider sector of the population. At the secondary level, comprehensive schools would help to create a solidaristic society (*Volksgemeinschaft*). This modern conception of *Bildung*, Becker felt, offered a solution to the cultural crisis and a way to reduce the social tension between elite and masses.[15]

Some scientists, too, proffered modern syntheses. Felix Klein (professor of mathematics at the University of Göttingen from the late nineteenth century) argued that pure mathematics could create a unified picture of nature by linking diverse areas of the sciences. It thus deserved to replace classical learning as the core of the secondary school curriculum. With this proposal Klein sought to defuse more radical and utilitarian calls for reform of the curriculum.[16] But others were less accommodating. Alois Riedler (rector of the *Technische Hochschule* at Berlin), for example, insisted on stretching the concept of *Bildung* yet further. The domination of educational institutions by humanistic *Bildung*, he objected, served the interests of "certain circles" but not that of the state. While conventional *Bildung* placed language, history, and speculative subjects at the heart of culture, "technical *Bildung*" perceived the cultural development of mankind in the exploration and control of nature. Technical *Bildung* was actually far broader than humanistic *Bildung*. Top engineers had to master *all* aspects of the problems which they tackled; not just the technical or theoretical issues, but economic and social ones as well. There was a gap between the 'merely intellectual' and the 'productive' (*schaffenden*) worlds, Riedler complained, and the importance of the former was enormously exaggerated.[17] Other modern-minded defenders of "a scientific worldview" shared Riedler's distrust of traditional *Bildung* and were in no doubt as to its political associations. The Vienna Circle's manifesto declared:

The increase of metaphysical and theologizing leanings which shows itself today in many associations and sects, in books and journals, in talks and university lectures, seems to be based on the fierce social and economic struggles of the present: one group of combatants, holding fast to traditional social forms, cul-

---

15. Becker 1919, 1930. Friedrich Paulsen held similar views on education (Paulsen 1921, 682–92). For other Weimar modernists' perceptions of crisis see Döring 1975, 205–6.

16. Pyenson 1983, 53–56. On Klein's attempts to bring universities and industry closer together, see Manegold 1970.

17. Hunecke 1979. The applied entomologist Karl Escherich voiced similar irritation at the "narrow and unworldly" work and interests of some scholars (Escherich 1944, 25–27).

tivates traditional attitudes of metaphysics... whose content has long since been superseded; while the other group, especially in central Europe, faces modern times, rejects these views and takes its stand on the ground of empirical science. This development is connected with that of the modern process of production, which... leaves ever less room for metaphysical ideas. It is also connected with the disappointment of broad masses of people with the attitude of those who preach traditional metaphysical... doctrines [Hahn, Neurath, and Carnap 1973, 19].

European intellectuals of the left during the interwar period were in no doubt that the educated middle class was increasingly divided. In his reflections on the modernization of Italy from the late nineteenth century, Antonio Gramsci identified two sectors within the intelligentsia. Although the older "traditional intellectuals" (including doctors, lawyers, priests, and teachers) were in fact historically associated with the landed aristocracy, they appeared to be a classless group. Their idealist philosophy resonated with the fact that they thought of themselves as an autonomous social stratum whose historical continuity survived the vagaries of political and social change. Concentrated in rural and precapitalist parts of the country, traditional intellectuals served to link the masses with the state. Industrialization, however, generated a newer stratum of "organic intellectuals" whose role was to link workers and management. This stratum—of "technicians," as Gramsci sometimes calls them—identified far more closely with the emergent bourgeoisie which sustained it than with the traditional intelligentsia which preceded it (Gramsci 1971, 5–23 and passim). Karl Mannheim's analysis of the German intelligentsia was similar. As a Hungarian emigré and a social democrat to boot, he was a marginal figure among Weimar Germany's professors. But marginality has its compensations, and as an outsider Mannheim may have been better able to understand the origins of his colleagues' preoccupations with synthesis:

The... fragmented outlook of the contemporary intellectual [is] not the upshot of a growing scepticism, a declining faith or the lack of ability to create an integrated *Weltanschauung,* as some writers regretfully maintain. Quite the contrary, secularization and the multipolarity of views are the consequence of the fact that the group of the learned has lost its caste organization... [Mannheim 1956, 117].

How did that loss occur? Following the early-nineteenth-century educational reforms, the classical grammar school (*Humanistisches Gymnasium*) became the normal destination for sons of the educated middle class. It was there that *Bildung* was formally transmitted, and entrance to university was restricted to graduates of this type of school. In the course of the century, however, the classical school's dominance was increasingly challenged as two kinds of modern or "realistic" school emerged: the *Realgymnasium* (in which modern languages displaced Greek from the curriculum) and the *Oberrealschule* (in which both Latin and Greek gave way to modern languages and natural sciences) (Ringer 1979). Defenders of the modern schools—among newer urban and industrial strata [18]—complained of the classical school's irrelevance for the modern world and called upon the government to open the universities to graduates of modern schools. The professoriate reacted sharply, insisting on the necessity of *Bildung* for university preparation:

> As praiseworthy and exemplary as many North American virtues are... so may the good Lord in all His mercy preserve us from sinking down into Americanism, toward which the government equalization of modern and humanist teaching or the equation of the principles of usefulness and ideality would be a first step.[19]

Nevertheless, the government gradually responded to demands for reform, and the classical grammar school began to lose its monopoly as the entry route to the university. From 1870 *Realgymnasium* graduates were allowed to enroll at the university in certain disciplines (including the sciences), and in 1900 the graduates of both types of modern school were formally granted essentially the same rights of access to the universities which classical school graduates had enjoyed. The reforms of 1900 made the modern schools far more attractive to some parents; over the next decade the secondary school population increased rapidly, with the modern school growing much faster than the classical school.[20] That is not to say that the prestige of the classical schools disappeared overnight; two-thirds of the educated middle class, for example, continued to favor the classical schools (Lundgreen 1981, tables 1b and 2). As the geneticist Gerda von Ubisch recalled, "'better' circles around the turn of the century always sent their sons to a classical grammar school, and it was practically a sign of lesser ability if a boy from the upper

---

18. Jarausch 1982, 105. On criticisms of the traditional mathematics curriculum, see Pyenson 1983.

19. Professor Zacher, on behalf of the philosophical faculty of the University of Halle in 1878, quoted in Jarausch 1982, 102.

20. Between 1899 and 1909—against the backdrop of a 15 percent increase in population—the number of pupils in the classical schools increased 14 percent while the numbers in the *Realgymnasien* and *Oberrealschulen* increased by 104 percent and 61 percent, respectively (Pyenson 1983, 78).

middle class [*gehobenen Mittelstand*] went to a modern school" (von Ubisch 1956, 415). But attitudes gradually began to shift. Toward the end of the century the modern schools began to be taken seriously as a second choice by newer strata: engineers, industrialists, and those in commerce. Thereafter these schools steadily increased their popularity at the expense of the classical grammar schools. The trend was evident in all social strata, but particularly marked among commercial and industrial sectors. By 1932, for example, 60 percent of entrepreneurs and managers were sending their children to modern schools.

The effects of secondary school reform were soon evident within the universities. Between 1905 and 1911 the proportion of science students who had graduated from a classical school decreased from 54 percent to 48 percent, while the proportion from *Oberrealschulen* increased from 18 percent to 28 percent, and by 1929 the latter had achieved parity with the former (Titze 1987, tables 104 and 105; cf. Jarausch 1984, 76). Given the persistence of differential class preferences, the shift toward modern schools meant that over the course of the nineteenth century and into the twentieth, students were increasingly drawn from newer commercial strata and less so from the educated middle class. Among science students at Prussian universities, for example, sons of higher civil servants declined between 1895 and 1927 from sixteen percent to 10 percent of the student body, while sons of lower- and middle-level company employees increased from 0.2 percent to 10 percent.[21] The changing social profile of the student body was soon evident among teaching staff at the universities. Between 1860 and 1933 the proportion of university teachers from industrial and commercial backgrounds increased from 28 percent to 36 percent, while those from the educated middle class declined from 63 percent to 48 percent.[22] Had academics been indifferent to questions of social status, these shifts might have had little effect upon the social cohesive-

---

21. Titze 1987, table 125 (cf. table 117). See also Lundgreen 1985, table 10. For comparable changes in the class background of students in all fields at the universities of Freiburg and Heidelberg between 1869 and 1910, see Riese 1977, 352–53. That this trend continued through the 1920s is evident from the figures in Jarausch 1984, 133–36.

22. Von Ferber 1956, table 26. See also pp. 171–72 and table 27 for the same shift among science staff. By 1907 the same trend was evident among staff at the University of Freiburg (Nauck 1956, 47). The trend is sometimes sharper once the data are disaggregated. Between 1859 and 1923, for example, the proportion of university staff who were the sons of clergymen declined from 15 percent to 6 percent, while the proportion who were sons of factory owners and those in higher positions in commerce increased from 4 percent to 19 percent (Kaelble 1973, 56–57).

ness of the professoriate. In fact, however, academics seem to have been highly sensitive to differences of social origin within their ranks. At Heidelberg before 1914, social acceptability in university circles was a delicate matter. Bankers' sons among the staff were at a disadvantage compared to academics' sons. Inbreeding among educated middle-class families was high, and for a young *Dozent* to marry the daughter of a primary school teacher raised the occasional eyebrow (Tompert 1969, 35). Since primary school teachers were trained in nonuniversity institutions following secondary education in a modern school, they were regarded condescendingly in academic circles as "half-educated." Few sons of primary school teachers were able to break this pattern of social exclusion, since the cost of a university education around the turn of the century was nearly equal to their father's salary.[23]

Catholics were another group which is said to have hovered on the margins of the Wilhelmian educated middle class, ambivalent about the traditional values of *Bildung* (Vondung 1976; cf. Conze and Kocka 1985, 26). And the educated middle class's hostility toward industrialization and its insistence that cultivation and utility were incompatible effectively damned one substantial sector within the academic community: engineers and applied scientists. Prior to 1900 many modern school graduates in search of higher education had turned to the colleges of engineering, agriculture, or commerce for training in applied science or vocational subjects. Thus the social segmentation which we have already seen at secondary level was perpetuated in higher education: Engineers and applied scientists were more likely to come from industrial and commercial strata than were university graduates in "pure" subjects (Ringer 1979). Although the engineers were highly educated—engineering colleges could grant doctorates from 1900—and felt themselves a part of the educated middle class, their entitlement was doubted by many classically educated members of that class. The specialized character of applied knowledge could not confer true *Bildung*. In response, many engineering colleges "academicized" their curricula, and spokesmen for the profession sought to show that engineering bridged the gap between *Zivilisation* and *Kultur:* The engineer was more an artist than an agent of capital (Manegold 1970; Herf 1984).

This pattern of declining social and ideological cohesion within the

---

23. Ringer 1979, 33. When the new Weimar government's plans for university reform proposed, among other things, to train such teachers at universities, the professoriate was outraged (Ringer 1969, 244, 256–58, 286, 288).

Table 8.1 Social Class Background and Secondary Schooling of Pragmatics and Comprehensives

	Born	Class	School	Source
Pragmatics				
Bauer	1904	ILMC[a]	modern[b]	Wagenitz 1988
Grüneberg	1907	EMC[c]	mod-clas	Lewis/Hunt 1984
Kappert	1890	ILMC	classical[d]	Kappert 1978
Schwemmle	1894	ILMC	classical	Ziegler 1980
Stern	1902	ILMC	modern	Stern papers
Comprehensives				
Belar	1895	EMC	classical	Caspari, DSB
Brieger	1900	EMC	classical	Linskens 1986
Correns	1864	EMC	classical	Correns papers
Goldschmidt	1878	ILMC	classical	Goldschmidt 1960
Haecker	1864	EMC	classical	R. Haecker 1965
Jollos	1887	EMC	classical	Goldschmidt[e]
Oehlkers	1890	EMC	classical	Marquardt 1974
Renner	1883	ILMC	classical	Cleland 1966
Timofeeff-Ressovsky	1900	EMC	classical	Timoffeeff 1959, Medvedev 1982
Winkler	1877	EMC	classical	Brabec 1955
Woltereck	1877	ILMC	classical	Zirnstein 1987
Unclassifiable				
Hämmerling	1901	EMC	classical	Harris 1982
Lehmann	1880	ILMC	classical	*Wer Ist's* 1935

[a] ILMC = industrial, commercial, or lower middle class.
[b] modern = *Realgymnasium* or *Oberrealschule*.
[c] EMC = educated middle class.
[d] classical = *Humanistisches Gymnasium*.
[e] For Goldschmidt's biography of Jollos, see chap. 3.

professoriate is clearly reflected within the interwar genetics community. On the whole, older geneticists were more likely to be of educated middle-class background and to display a comprehensive style of thought, while younger ones were more likely to come from lower- or industrial-middle-class strata and to be identifiable as pragmatics (tables 8.1 and 8.2). For example, among the older generation of geneticists born in 1900 or earlier (for whom social class background is known), 65 percent were from the educated middle class. Of those born after 1900, however, this proportion had dropped to 50 percent. Simi-

Table 8.2 Social Class Background and Secondary Schooling of Members of the Schools of Erwin Baur, Alfred Kühn, and F. von Wettstein

	Born	Class	School	Source
The Baur School				
Baur	1876	ILMC[a]	classical[b]	von Rauch
Stein	1879	ILMC	(neither)	Schiemann 1955
Schiemann	1881	EMC[c]	modern[d]	Schiemann 1964
von Ubisch	1883	EMC	classical	von Ubisch 1956
Hertwig	1889	EMC	modern	Nachtsheim 1959
Nachtsheim	1890	~EMC[e]	classical	Nachtsheim 1966
von Sengbusch	1898	~EMC	modern	Sengbusch papers
Stubbe	1902	EMC	modern	H. Stubbe 1959
Kuckuck	1903	~EMC	classical	Schiemann 1963
Schick	1905	~EMC	modern	H. Stubbe 1985
Michaelis	1900	EMC	modern	W. Stubbe 1987
von Rauch	1905	EMC	modern	von Rauch 1935
Husfeld	1900	?	?	
Gruber	1904	?	?	
von Rosenstiel	1905	?	?	
Hackbarth	1906	?	?	
The Kühn School				
Kühn	1885	EMC	classical	Kühn 1959
Henke	1895	ILMC	modern	Wagenitz 1988
Caspari	1909	EMC	classical	Eicher 1987, interview
Schwartz	1907	ILMC	modern	Wagenitz 1988
Piepho	1909	ILMC	modern	pers. commun.
Plagge	1911	ILMC	modern	Wagenitz 1988
E. Becker	1913?	?	modern?	pers. comm. from Schwartz, Piepho
The von Wettstein School				
von Wettstein	1895	EMC	classical	H. Stubbe 1951
Pirschle	1900	EMC	classical	H. Stubbe 1955, BDC[f]
E. Kuhn	1904	EMC	modern	Knapp 1943
G. Becker	1905	ILMC	classical	Wagenitz 1988
Melchers	1906	ILMC	classical	Wagenitz 1988
Knapp	1906	ILMC	modern	Butterfass 1980
Schlösser	1906	EMC	?	
Barthelmess	1910	?	?	

[a] ILMC = industrial, commercial, or lower middle class.
[b] classical = *Humanistisches Gymnasium*.
[c] EMC = educated middle class.
[d] modern = *Realgymnasium* or *Oberrealschule*.
[e] ~EMC = groups or occupations marginal to educated middle class.
[f] BDC = Berlin Document Center.

larly, the average birth date of the comprehensives listed in table 8.1 is 1883, while that of the pragmatics is 1899. Are social class and style of thought, therefore, correlated? Yes: Among geneticists born into the educated middle class, 62 percent are identifiable as comprehensives and 19 percent as pragmatics.[24] Conversely, among geneticists from lower- and industrial-middle-class strata or from groups marginal to the educated middle class, 38 percent are comprehensives and 57 percent are pragmatics.[25] Furthermore, the age difference between comprehensives and pragmatics suggests that the proportion of pragmatics within the genetics community might have been growing, a guess borne out by the fact that pragmatics make up only 31 percent of the older generation (i.e., those born in 1900 or earlier), but 50 percent of the younger one.[26] Thus geneticists's age and class backgrounds sustain my theory that the two factions within the genetics community mirror the progressive fragmentation of the professoriate.

Although I am claiming that the comprehensive and pragmatic factions were divided over the merits of modernization, however, I have so far shown only that they differed in class background. And one might reasonably object that social class is not a particularly good indicator of attitudes toward modernization; as we saw above, a minority of the educated middle class sent their sons to modern schools, while many industrial-middle-class families preferred a classical education for their sons (at least before the First World War). Arguably, therefore, school choice might provide a better indicator of attitudes toward modernization than does social class. In fact, the results are much the same. Among comprehensives 70 percent had been to a classical school and

24. Of twenty-one persons in the sample, thirteen are identifiable as comprehensives, four are pragmatics (Grüneberg, Stubbe, von Ubisch, and Hertwig), and four are ambiguous or unknown (Hämmerling, Michaelis, von Rauch, and E. Kuhn). For reasons to be discussed shortly, those whose class membership is categorized as "+ or − EMC" have been omitted from the sample.

25. In this sample of twenty-one there are eight comprehensives (three from table 8.1, G. Becker and four members of the Kühn school), twelve pragmatics (four from table 8.1, six from the Baur school, plus Knapp and Melchers), and one who is unclassifiable (Lehmann).

26. That is, of twenty-six geneticists in the older group, eight are identifiable as pragmatics: Baur, von Sengbusch, Nachtsheim, Kappert, and Schwemmle are clear-cut cases, and although I lack sufficient information on von Ubisch, Hertwig, and Stein, they can be treated as pragmatics here in order to err on the conservative side. Of the sixteen born after 1900, eight (Knapp, Melchers, Stubbe, Schick, Kuckuck, Bauer, Grüneberg, and Stern) were pragmatics. The remainder, of course, cannot simply be classed as comprehensives since some do not fit neatly in either category, and I lack sufficient information on others (cf. chap. 7).

292   Styles of Thought

30 percent to a modern one.²⁷ Pragmatic geneticists had attended modern and classical schools in about equal proportions.²⁸ Looking at the issue the other way around, the type of school which a geneticist attended is quite a good predictor of his or her style of thought. Of those from classical schools, 64 percent became comprehensives and 28 percent pragmatics;²⁹ of those from modern schools, 29 percent became comprehensives and 47 percent pragmatics.³⁰

But let us now shift away from statistics on the genetics community as a whole to look in detail at the Baur school (table 8.2). A high proportion of its members came from groups marginal to the educated middle class or outside it altogether. In view of the agricultural colleges' practical curriculum, the two-year apprenticeship required for entry, and their willingness to accept entrants who lacked any secondary school diploma, this is not surprising.³¹ Baur himself was the son of a Catholic pharmacist who had only one year of higher education at a polytechnic, and his mother was the daughter of an innkeeper/butcher. Coming from such a modest background, he had worried in 1910 that he lacked the family connections that might be necessary in order to get a chair.³² Schick's and Kuckuck's fathers were both college-educated

---

27. Of twenty-three in the sample, sixteen had been to classical schools (eleven in table 8.1 plus Kühn, Caspari, von Wettstein, Pirschle, and G. Becker) and seven to modern ones (Schiemann, Michaelis, Henke, E. Kuhn, Schwartz, Piepho, and Plagge).

28. Of fifteen in the sample, eight attended modern schools (Hertwig, von Sengbusch, Stubbe, Schick, von Rauch, Knapp, Stern, and Bauer), while seven attended classical ones (Baur, von Ubisch, Nachtsheim, Kuckuck, Melchers, Kappert, and Schwemmle). Do these seven "anomalies" undermine my hypothesis? Not necessarily; five of them came from lower- or industrial-middle-class backgrounds, and there were several reasons (other than ideological conviction) why families in those strata might have chosen a classical school: lack of an alternative—especially in small towns before 1914—or the search for social advantages.

29. Of twenty-five persons in the sample, sixteen were comprehensives (eleven from table 8.1 and five from the Kühn/von Wettstein schools), and seven were pragmatics (Kappert, Schwemmle, Baur, Kuckuck, Nachtsheim, von Ubisch, and Melchers).

30. This sample consists of seventeen individuals: five comprehensives (four from Kühn's school plus Schiemann), eight pragmatics (three from table 8.1, four from the Baur school plus Knapp), and four unclassifiables (Michaelis, von Rauch, Stein, and E. Kuhn).

31. In 1926–27 only 19 percent of the students at German agricultural colleges had graduated from a classical school, while 56 percent lacked any secondary school qualification. The corresponding proportions for university students were 47 percent and 7 percent (Graven 1930, 330). The figures for Prussia are similar (Titze 1987, tables 105 and 108).

32. Baur to Correns, 21.11.10, quoted by von Rauch 1944, 92 (cf. 25). On his family background see von Rauch 1944, 4–21. According to Kuhn's 1963 *Soziologie der*

Mandarins Confront Modernization   293

(the latter with a doctorate), but in engineering.[33] Strictly speaking, von Sengbusch's father (a physician) would be categorized as educated middle class, but the five generations before him had all been businessmen: The family import-export firm was founded in 1769. The extent of his father's allegiance to the ideology of *Bildung* may be seen from the fact that the father opened a workshop for making prosthetic limbs and chose to send his son to a modern school.[34] Stubbe, Michaelis, and von Rauch are similar cases: Their fathers were educated middle class by profession, but sent their sons to modern schools.[35] As the son of a higher civil servant and graduate of a classical school, on the other hand, Nachtsheim looks like a more serious anomaly. The fact that he was Catholic perhaps explains in part his apparent indifference to *Bildung*.[36]

The women in the school constitute a deviant minority in several respects. Apart from being twenty years older than most of the male assistants, they came from elite family backgrounds.[37] Schiemann and Hertwig were the daughters of well-known professors at the University of Berlin, while von Ubisch's father was an art historian and director of a major Berlin museum. Stein was financially independent, thanks to her father's industrial wealth. Such social advantages did not, however, spare them the difficulties of gaining entrance to a university or obtaining an academic post. For these geneticists were among the first women to receive doctorates at German universities. Since none of them was

---

Apotheker, pharmacy began to be taught at some German universities in the nineteenth century, but such training was not compulsory until 1875, and even then only four semesters of study were required. Pharmacy was not highly regarded in academic circles, since entrance to its courses did not require a secondary school diploma until 1921. (For the school backgrounds of pharmacy students at Prussian universities in the late nineteenth century, see Titze 1987, table 100.) As Paulsen put it, the apothecary's job was one of those practical occupations—like the chemist's—which bridged the gap between scholar and businessman (Paulsen 1921, 698).

33. Schiemann 1963; H. Stubbe 1985. Georg Melchers, a pragmatic from von Wettstein's group, had originally intended to study engineering (Melchers 1987a).

34. Ordner 1a and 1b, von Sengbusch papers.

35. H. Stubbe 1959b; on Michaelis see W. Stubbe 1987; for von Rauch's curriculum vitae see his doctoral dissertation (1935, p. 62).

36. Cf. Nachtsheim's *Lebenslauf*, Sign 20281, Rep 87B, ZStA-Merseburg. Hans Kappert, Baur's successor in the chair of genetics at the Berlin Agricultural College, was also Catholic (cf. questionnaire dated 31.3.48, box 1, Kappert papers).

37. This gender difference in social background was common among Prussian science students before the 1920s: Female students were more likely than males to come from the educated middle class and less likely to come from industrial and commercial strata (Titze 1987, figs. 115 and 121).

trained in agricultural science and only Schiemann had been a student of Baur's, there was nothing "natural" about their taking up assistantships with him at the Berlin Agricultural College. Posts open to women were few and far between, and the wealthier among them were probably prepared to accept unpaid assistantships or posts in less prestigious institutions until something better came along.[38] It is thus not surprising that in terms of class background Schiemann et al. were "misfits" in Baur's institute.[39]

Apart from its female members, therefore, the Baur school shows a social profile which neatly illustrates how the pragmatic style of thought was rooted in promodern social strata. Unfortunately, however, neither the Kühn nor the von Wettstein school provides a clear-cut contrast, since neither is as socially homogeneous as the Baur school. Although Kühn and von Wettstein themselves display the expected association between social class, schooling, and style of thought, very few of their assistants do. Moreover, among Kühn's assistants industrial-middle-class backgrounds are even more prevalent than in the Baur school.[40] It is noticeable that Kühn's and von Wettstein's assistants are among the youngest members of my sample, most having attended secondary school during the more socially progressive atmosphere of the Weimar Republic. If an increasing proportion of the science students within this age cohort came from more modest class backgrounds or modern schools, then a "generational gap"—manifest in the social backgrounds of directors versus assistants—might be more marked in these two schools than in Baur's. Whether one should conclude that the correlation between social origins and style of thought had substantially weakened by the 1930s is not clear; as is evident from table 8.2, there is so little information available on most of Kühn's and von Wettstein's assistants that assigning them to comprehensive or pragmatic categories is pointless.[41]

38. On Stein and von Graevenitz, see chapter 6. Paula Hertwig worked as an unpaid assistant in her father's Anatomical Institute from 1916 until Baur was able to offer her a salaried post in 1921 (Kosswig 1959).

39. On Schiemann see Kuckuck 1980 and Schiemann 1960. On Hertwig see Kosswig 1959; Nachtsheim 1959; and Schiemann, Veröffentlichung Nr. 117 (letter to Paula Hertwig on her seventy-fifth birthday, 1964), Schiemann papers. On Emmy Stein see Schiemann 1955 and the Stein papers. On von Ubisch see her 1956.

40. Since Erich Becker took his doctorate at a *Technische Hochschule* (Hans Piepho to me, 20.10.87; V. Schwartz to me, 1.10.87), it is quite likely that he, too, graduated from a modern secondary school.

41. There is no reason to assume that assistants will necessarily share the institute director's style of thought. One can, after all, work on problems in developmental and/or

8.3 The Politics of the Professoriate

So far I have argued that comprehensive and pragmatic geneticists' social backgrounds and attitudes (toward science and culture) make sense in the context of the declining social and ideological cohesion of the professoriate from the late nineteenth century. What remains to be explained is the pattern of political differences between these factions. The theory which I have been developing in this chapter might lead one to expect that pragmatics would simply be politically to the left of comprehensives. But as we saw in the previous chapter, the actual pattern is more complex; the two factions differ in terms of party-political *activity* rather than along a left-right spectrum. The questions which I will address in this section are (a) how does the pattern of political differences observed within the genetics community compare with that found within other sectors of the academic community during this period? and (b) how might my theory of modernization account for such differences? To develop this argument systematically and convincingly would require a book on its own; my aim here is merely to show that the existing evidence makes such a hypothesis plausible.

To begin with, it is clear that substantial differences of political outlook separated university staff from artists and writers. The predominantly conservative and nationalist sentiments within the professoriate during the Wilhelmian and Weimar periods are well documented. Support for the Social Democratic Party among academics was extremely

---

evolutionary genetics without subscribing to the comprehensives' metaphysical premises; Baur is a case in point. Georg Melchers is probably another. Although he devoted his first ten years in von Wettstein's laboratory to evolutionary and physiological genetics, Melchers was in many respects an archetypal pragmatic. From his autobiography one learns that the work on tobacco mosaic virus which he initiated in 1937 was frequently criticized by zoologists and botanists as too specialized. And Melchers had little time for holism; impatient with the philosophical discussion of biological problems, he has recently described his views—with characteristic irony—as "hopelessly reductionist" (Melchers, "Joseph Straub: Lehrer, Forscher, Wissenschaftspolitiker," typescript, 1981). From the autobiography and other essays it appears that his sole interest outside science was politics (Melchers 1987a, esp. 376, 384–85, and 387–88).

Once more is known about Kühn's younger assistants, similarly, it is quite possible that they will turn out not to have been comprehensives in their master's mold. If so, perhaps Kühn had found by the mid-1930s that if he were going to pursue his research program through *teamwork,* he could just as well do so using students from more modern strata. Despite their lack of traditional *Bildung,* such young pragmatics might have been not only just as clever as their comprehensive peers, but also more willing to take their specialized places in a division of research labor.

rare, not least because it could seriously jeopardize one's career.[42] And even those in the middle-of-the-road Catholic Center Party or German Democratic Party (founded in 1918) were in a minority. But the distribution of political sympathies among intellectuals and artists outside the universities—at least during the Weimar Republic—was skewed much further to the left (e.g., Laqueur 1974, 43ff.). As the case of the magazine *Die Weltbühne* illustrates, such intellectuals were not necessarily members of left-wing parties, but some did become involved in street-level politics.[43] And numerous memoirs testify to the leftish enthusiasm—at least for a while—of many others.[44]

But to what extent was the professoriate itself split along political lines? The best starting point is Fritz Ringer's analysis in *The Decline of the German Mandarins* (Ringer 1969). Unlike most studies of the structure of the German professoriate during this period, Ringer's does not focus narrowly upon either their writings or their political inclinations.[45] Instead he integrates both of these spheres in a historical sociology of knowledge. Ringer's thesis is that the prominence of holistic thinking in the humanities and social sciences, as well as the recurrent concern for scholarly synthesis and cultural renewal in the service of national integration, are explicable in terms of the professoriate's sociohistorical situation. Nineteenth-century "mandarins" saw themselves as a cultivated elite whose *Bildung* entitled them to play an important role in the leadership of German society. As industrialization and democratization (from 1918) eroded their influence, they reasserted their traditional role more forcefully. But Ringer is fully aware that a minority of mandarins deviated from this pattern: He designates members of this minority as "modernists," in contrast with the "orthodox" majority. Although modernists also endorsed the need for synthesis and

---

42. On the difficulties encountered by Leo Arons, Robert Michels, and Emil Gumbel, see Busch 1959, 114–15; Gerth and Mills 1970, 19; and Benz 1983, respectively.

43. The central role of writers in the short-lived revolutionary government of 1918–1919 in Munich is a case in point (e.g., Phelan 1985). On the *Weltbühne* see Deak 1968; Poor 1968.

44. One thinks, among others, of Georg Grosz, Arthur Koestler, Axel Eggebrecht, and Manes Sperber. Of course the literati were no more politically homogeneous than any other occupational group; right-wing writers were also in evidence—and sometimes widely read, as with Ernst Jünger—but they seem to have felt themselves outnumbered (Laqueur 1974, 79).

45. Among the other studies of the Wilhelmian and Weimar professoriate, vom Bruch 1980 is primarily a contribution to the history of ideas (for his antipathy to social history as a "fashion," see pp. 28, 32, and 424), while Schwabe 1969, Töpner 1970, and Döring 1975 focus upon the professors' political activities.

often developed holistic approaches in their scholarly work, they were nonetheless distinctive in two respects. On the one hand, in their scholarly writings as well as in their commentaries on the crises of their time, they were wary of the heavy emphasis placed by most of their colleagues upon speculative, metaphysical, romanticist, or quasi-mystical doctrines as guides to the moral and cultural renewal of Weimar society. Placing more reliance upon the virtues of "cold reason," modernists were more favorably inclined to the values of *Zivilisation* than were their orthodox brethren (Ringer 1969, 207, 213–14). Although *Bildung* was still to play a key role in societal renewal, modernists broadened its meaning so as to encompass not only classical but modern education (cf. section 1). Furthermore, to form the "whole pupil," one-sidedly intellectual curricular aims would have to be balanced by vocational ones. In this way the social distance between cultivated elite and masses would be bridged, and social harmony would be achieved (Ringer 1969, 273, 277–79). The other difference between orthodox and modernist mandarins was that the latter were less hostile to modernization. In terms of party affiliation, most modernists were close to the middle-class, left-liberal German Democratic Party, with its sympathy for the Weimar Republic, while orthodox mandarins' hostility to democracy placed them close to the German Nationalist Peoples' Party (*Deutschnationale Volkspartei*) along with large landowners, the aristocracy, and the top echelons of the civil service and the military (Ringer 1969, 200–201).

This division of the academic community into two mandarin factions is borne out by the work of several other historians of the German professoriate (summarized in figure 8.1). Klaus Schwabe's analysis of the professoriate's political views during the First World War, for example, reveals a split very similar to that between orthodox and modernist mandarins (Schwabe 1969). When war broke out in August 1914, it seemed for a while as if the divisions within Wilhelmian society—in particular the widespread fears in middle and upper strata of the growing strength of the Social Democratic Party—had been healed. The Party's internationalism was shelved, and workers enlisted as a wave of patriotic sentiment swept the country. The professoriate was jubilant: At last the nation was truly one, and German *Kultur* would display its superiority to western *Zivilisation*. Prewar tensions within the academic community were put aside, and professors of all political colorations brandished their pens in a flourish of activity aimed at refuting foreign propaganda and maintaining morale at home. As the war dragged on, and a German victory seemed increasingly remote, professors began to voice dissatisfaction with the apparent lack of coor-

Styles of Thought

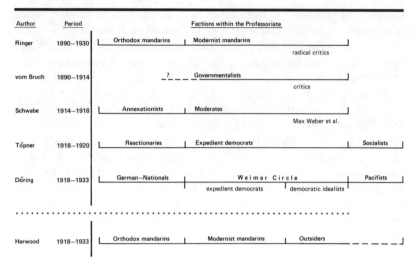

Figure 8.1 Structural Analyses of the German Professoriate

dination between government and military, and the beginnings of revolution in Russia renewed doubts about the loyalty of the German working class. In recommending changes of government policy, academics' hitherto suppressed prewar differences resurfaced. The majority of them (whom Schwabe calls "annexationists") placed primary emphasis upon territorial expansion, which they thought would simultaneously consolidate state power and unify the nation (since expansion was thought to be popular with the working class). The minority "moderate" faction also sought an acceptable conclusion to the war and national unity, but expansionism, they believed, would instead undermine unity. The only realistic way of achieving this was through political concessions to the working class. Reason, not emotion, should dictate policy.

The period 1918 to 1920, like the latter half of the war, is of particular historiographical interest, since the abortive government of workers' and soldiers' councils, the new Weimar constitution, and the first Social Democratic government provoked position statements from many academics. In categorizing professors' stances on these issues, Töpner (1970) found that they fell into three categories, corresponding closely to existing party politics. The largest faction, close to the reactionary German Nationalist People's Party, was unremittingly hostile, while a minority of "expedient democrats" (*Vernunftrepublikaner*), who defended the Republic on pragmatic grounds rather than by conviction, were close to either the Catholic Center Party or the German Democratic Party occupying the political middle ground. A tiny handful of "socialists" of various complexions were not openly hostile to the Weimar Republic but hoped for the arrival of "real" socialism. Wherever Töpner and Schwabe look at the same persons, they invariably agree: Töpner's expedient democrats had all been wartime moderates, while those hostile to the Republic and close to the German Nationalist People's Party had been wartime annexationists. For the Weimar period, Herbert Döring's study of those professors loyal to the new republic—a category which he regards as equivalent to Ringer's modernists—explicitly addresses this question of continuity (Döring 1975). By identifying all of those individuals who signed political proclamations between 1914 and 1933, Döring can demonstrate that the wartime split between moderates and annexationists persisted largely unchanged throughout the Weimar Republic. Thus despite the different historical periods which they have examined and the differing criteria which they have employed in identifying factions within the professoriate, the analyses by Schwabe, Töpner, and Döring all confirm that the professoriate during the Wilhelmian and Weimar periods was split politically along much the same line that divided modernist from orthodox mandarins.[46]

It is clear that Alfred Kühn, along with the other geneticists to whom I have imputed a comprehensive style of thought, corresponds closely to Ringer's characterization of the mandarin. Baur and the other pragmatic thinkers, however, pose a problem, since they bear very little resemblance to the mandarins. Obviously a number of solutions are possible. One is that the genetics community—and perhaps German scientists more generally—was simply more diverse in social background and ideology than those in the humanities and social sciences. Indeed some evidence suggests that scientists may have been more liberal than their colleagues in the humanities.[47] For example, among professors who were politically active between 1914 and 1933, Döring

---

46. Ringer emphasizes this continuity (1969, 150, 190–92, 196, 201–2), as does vom Bruch (1980, 201, 322).

47. On *a priori* grounds Ringer and others have suggested that this might prove to be the case (Ringer 1969, 6). Barkin (1971, 282) suspects that orthodox mandarins were less common outside the philosophical faculty, and Ringer's evidence on social scientists certainly bears this out. In contrast, Burchardt claims that the vast majority of Wilhelmian scientists accepted the "sociopolitical *status quo*," but presents no relevant evidence (Burchardt 1988, 204, 211).

found that over two-thirds had signed reactionary manifestos, the remainder signing "moderate" and prodemocratic ones (Döring 1975, appendix, tables 1 and 2). If one focuses on that subset of Döring's signatories who were scientists, however, the proportions are roughly reversed.[48] In my sample of geneticists, similarly, more than half of those who actively supported a political party were social democrats or liberals.[49] This result is at least consistent with the fact that as early as the mid-nineteenth century, science students were much more likely to come from industrial and commercial backgrounds than were students of medicine, theology, law, or other nonscience subjects (Borscheid 1976, 69). Nevertheless, until more analyses of the politics of German scientists are available, we cannot be certain how closely the distribution of their ideological predilections may have corresponded to that of their colleagues in the humanities.

Another possibility, which I will explore in detail here, is that Ringer has overlooked the significance of a substantial number of nonmandarins, even among humanists and social scientists. Clearly, his principal concern was to illuminate the *differences* between orthodox and modernist mandarins. In consequence, however, he tends to overlook the amount of common ground between these camps, a point made by several reviewers of his book (Barkin 1971, 278; vom Bruch 1980, 322–23; Willey 1977, 191; Habermas 1971, 426). To be fair, Ringer acknowledges that some of the camps' concerns were shared: Modernists also saw themselves as an unpolitical repository of wisdom whose advice would be crucial in bringing about the ethical renewal and unification of German society, and they called accordingly for scholarly syntheses to stem the tide of academic specialization. Modernists, Ringer notes, were merely more "enlightened conservatives," and their support for democracy and modernization was expedient rather than principled (Ringer 1969, 106, 130–40, 228, 241–43, 260–68, 281–82, 342–51). Had Ringer focused upon the similarities between these two factions, however, he might have paid more attention to those individuals within the mandarin category whom he called "radical critics." Since geneticists of pragmatic persuasion were so *un*mandarin in out-

---

48. Of the academics in tables 1 and 2 of Döring's appendix who are known to me as biological or physical scientists, five were annexationists and ten were moderates.

49. This evidence can only be suggestive; various methodological problems with my sample (e.g., its small size, the high proportion of junior staff in it) mean that it may not be representative of senior geneticists as a whole, to say nothing of science professors in general.

look, it is worth looking closely at what Ringer has to say about these radical critics.

A few modernists, he concedes, were genuine democrats by conviction. And social scientists like Ferdinand Tönnies, Leopold von Wiese, and the socialists Emil Lederer, Ernst von Aster, Franz Oppenheimer, and Karl Mannheim criticized the mandarins' obscurantist appeals to *Gemeinschaft* and their attacks upon "selfish individualism" (Ringer 1969, 202–3, 230ff, 356–57, 391–92). But Ringer makes surprisingly little of the fact that another radical critic (Max Weber) rejected the mandarins' claim to be prophets above politics, backed the industrialization of Germany, and insisted upon the inexorability of conflict within a modern society. Furthermore, it is surely worth noting that Weber was both proud of his bourgeois origins and contemptuous of the tendency for intellectuals and the bourgeoisie to ape aristocratic lifestyles.[50] Nevertheless, in his two-page discussion of these radical critics, Ringer merely observes that some of them had strong ties to the intelligentsia outside the universities, and others were Jewish and thus sympathetic to modernization. "Typical outsiders," he concludes, without pursuing the matter further.[51]

What is striking, however, is that such "outsiders" can be found in the liberal wing of the professoriate throughout the period under discussion. In his analysis of Wilhelmian academics' forays into the political arena between 1890 and 1914, Rüdiger vom Bruch has discussed a major schism which divided the Social Policy Association (vom Bruch 1980). The politically more conservative "governmentalists" (*gouvernementale Intelligenz*), such as Gustav Schmoller, Hans Delbrück, Adolf Harnack, and Hermann Oncken (all of whom Ringer treats as "modernists"), aimed their scholarly analyses of "the social question" at top civil servants whom they took to be—like themselves—politically neutral. Sensible social reforms, they hoped, could integrate the labor movement without radical political change (vom Bruch 1980, 359–63, 414–15). Their left-liberal critics, however—including Lujo Brentano,

50. Mommsen 1974, 97–107, 109, 113, 123, 474. At a memorial service for Weber at the University of Heidelberg in 1920, the philosopher Karl Jaspers described Weber as a fragmentary figure. In his scholarship as in his political activity, he created no synthesis, believing it impossible to grasp the whole (Green 1974, 287).

51. Ringer 1969, 238–39. In view of the fact that the concluding chapter of Ringer's book emphasizes the continuity between the mandarin ideology and German academics' accommodation to fascist rule after 1933, it is yet more puzzling that he spends so little time on these "radical critics." Surely we need to understand the conditions which fostered the emergence and survival of such a minority prior to 1933.

302   Styles of Thought

Max and Alfred Weber, and Ferdinand Tönnies—addressed their scholarly work to the Social Democratic Party on the grounds that only a strengthened labor movement could force the necessary social and political concessions from a state which was controlled by reactionary strata.[52] Much the same split opened up in the final year of the war. As Schwabe has shown, Max Weber's call in 1917 for Prussian electoral reform (which would have given the working class more representation within the *Landtag*) enjoyed general support among "moderates" (read "modernists"), but his insistence that the *Reichstag* should be empowered to shape government policy (along the lines of parliamentary democracies) did not. This proposal for a powerful, Westernized *Reichstag* was endorsed by Alfred Weber, Lujo Brentano, and the pacificist F. W. Foerster, but was too extreme for most other "moderates" (e.g., Meinecke, Delbrück, Troeltsch, and Harnack), and a series of subsequent issues from 1917 until war's end continued to divide the moderates along the same lines (Schwabe 1969, 27–33, 158ff, 198).

The most extensive and systematic analysis of this split within the modernist/governmentalist/moderate camp after 1918 is to be found in Herbert Döring's study of the Weimar Circle, a group of professors close to the German Democratic Party who sought to cultivate support for the Weimar Republic from the entire range of political opinion within the universities. The Weimar Circle was dominated by "expedient democrats" like Meinecke and Delbrück, whose modernist-mandarin views are already familiar to us through the works of Schwabe and Ringer. A minority faction within the Weimar Circle, however, was fully committed to the democratization of German society and critical of key mandarin assumptions. Among these "democratic idealists," as Döring calls them, were the Weber brothers, Tönnies, G. Anschütz,

---

52. vom Bruch 1980, 200–205, 278–93, 316–17. Attempting to compare vom Bruch's account of the German professoriate with Ringer's is particularly difficult because the former is so reluctant to identify categories of academics. Indeed, while vom Bruch is dissatisfied with various classifications of the Wilhelmian professoriate—including Ringer's—he admits that he is unable to offer an alternative (e.g., 422). Instead of attempting to clarify the relationship between his own loosely defined category of "governmentalists" and Ringer's mandarins, he notes rather limply that governmentalists and orthodox mandarins were similar "in many respects," as were the governmentalists' critics and modernist mandarins (322). On the other hand, many of the figures whom vom Bruch actually places in his governmentalist category (e.g., Schmoller, Harnack, Delbrück, Oncken) are regarded by Ringer as modernist mandarins and by Schwabe as moderates. Thus the boundaries of the governmentalist category are not clear, but they appear to include both orthodox and modernist mandarins, a fact which I have sought to convey in figure 8.1.

Brentano, O. Baumgarten, and Hugo Preuss. (There appears to be good agreement on membership of this nonmandarin category; Ringer identifies the first three of these figures as radical critics of the mandarin perspective, and Schwabe notes that the last four joined Max Weber in pressing for the democratization of the *Landtag* in 1917–1918.)[53] Unlike the expedient-democratic mainstream of the Weimar Circle, these "democratic idealists" did not accept that the state was—or ever had been—politically neutral; future governments had to be accountable to a Western-style parliament. Furthermore, they were open to joint action with pacifists and socialists. And inspired by Max Weber's "Science as a Vocation,"[54] they saw no contradiction between their academic role and party-political activity. Although both sectors of the Weimar Circle envisaged an important role for the professoriate as counselors to government, at least some democratic idealists joined Weber in seeing the academic's role as necessarily that of analyst or expert rather than prophet. Moreover, the sectors differed as to the direction in which German society ought to develop. Hoping that the process of modernization could be slowed, expedient democrats regarded the traditional ruling stratum (landowners, the military, top civil servants) as best equipped to preside over the young German democracy and temper its excesses. Democratic idealists, on the other hand, welcomed modernization wholeheartedly and insisted upon government by an alliance of the newer working class and industrial middle class.

Clearly it would be useful to have a term which characterizes this rather motley assemblage, consisting of Ringer's radical critics, the left wing of the Social Policy Association, the left wing of Schwabe's moderates, Töpner's socialists, and Döring's democratic idealists, along with the handful of pacifists whom Döring places outside the Weimar Circle. I have chosen the term *outsiders* to describe this small minority sympathetic to modernization, since it conveys their ideological distance from the dominant mandarin tradition *without* implying any party-political uniformity.[55] Once Ringer's dichotomous schema is modified through the addition of this "outsider" category, it better reflects the structure not only of the professoriate as a whole, but also of the genetics

---

53. Töpner (1970, 207, 211–13) notes that among middle-of-the-road supporters of the first Weimar government, Preuss was one of the few who was a democrat by conviction rather than expediency.

54. Weber, in Gerth and Mills 1970, 129–56.

55. In terms of size, Döring estimates that among approximately 6,000 university staff during the 1920s, about 300 were members or sympathizers of the German Democratic Party and 50 of the Social Democratic Party (Döring 1975, 74–75).

community. Comprehensive geneticists are now best understood as a broad group, including both orthodox mandarins (e.g., von Wettstein, Renner, and Goldschmidt) and modernist ones (e.g., Kühn and Oehlkers). The pragmatic faction's indifference to *Bildung* and high level of party-political activity—whether as liberals (e.g., Hertwig and von Ubisch) or social democrats (e.g., Stubbe, Kuckuck, and Schick)—identify them as outsiders.

As it stands, however, this outsiders category is not yet quite broad enough to incorporate all of the pragmatic geneticists, since quite a few of the latter were attracted to the radical right before 1933. Where to situate those on the radical right was a problem for Ringer, too. Though politically far to the right of his modernist mandarins, the radical right could hardly be grouped with his orthodox mandarins, since the radicals rejected the ideals of classical *Bildung* and advocated a wider distribution of power among university staff.[56] As a variety of historians have noted, the ideology of the radical right was quite different from traditional conservatism of the mandarins' variety (Stern 1961; Struve 1973; Hughes 1952; Lebovics 1969). As I see it, the reason why Ringer's scheme has difficulty accommodating the radical right is that he distinguishes his two categories in terms of *both* party-political sympathies *and* attitudes toward modernization. By conflating the two, he misses the fact that the most vocal advocates of modernization (outsiders) were extremely diverse in their party affiliations. My aim in creating the category of outsiders is to unpack these two dimensions, enabling us to make room for diverse attitudes toward modernization ("mandarins" versus "outsiders") at any point along the conventional left-right political axis (for a graphic representation of these relationships, see figure 8.2).

But where would academics of the radical right belong in my classification? To place them alongside left-liberals, pacifists, and social democrats might at first seem absurd, but a reasonable case can be made for stretching the outsiders category so as to accommodate the radical right. While the earlier claims by David Schoenbaum and others that National Socialism effected revolutionary changes in Germany are now disputed by historians, there is widespread agreement that the Party leadership at least *sought* to restructure the society, replacing traditional elites with Party functionaries. Accordingly, Nazi ideology contained a

---

56. Ringer 1969, 290. For a sketch of Ernst Krieck's idea of the National Socialist university, see Seier 1988, 257–58, 261.

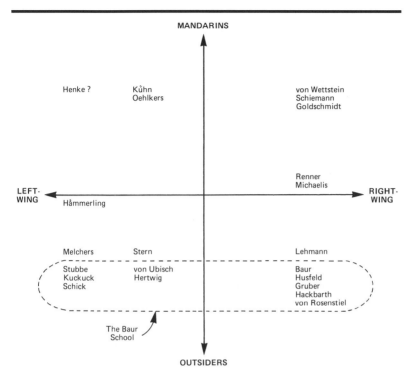

Figure 8.2 The Relationship between Party Politics and Attitudes toward Modernization among German Academics

number of modernizing elements which were important in recruiting members and voters before 1933.[57] Within the academic community, similarly, Nazism appealed to outsiders. At the University of Frankfurt between 1933 and 1939, for example, the highest rates of academic promotion were among staff from working-class backgrounds, due in part to their markedly higher level of Party activism.[58] Furthermore, the Party was particularly successful in attracting support from applied disciplines whose members resented their low status within the academic

---

57. For a review of the debate over National Socialism as a modernizing force, see Kershaw 1985, chap. 7.
58. Zneimer 1978. Figures on the class background of university staff before and after 1933 are consistent with this picture (Seier 1988, 255, cites the data in von Ferber 1956, 177).

community and the educated middle class more generally.[59] There are also signs that the Party's egalitarian rhetoric and its insistence that scholarship be accountable to the lay public were particularly attractive for academics of lower-class origins. Within the genetics community the right-radical inclinations of Erwin Baur and Ernst Lehmann before 1933 are consistent with this pattern, as is the fact that after 1933 National Socialism appealed to several pragmatic geneticists who were the sons of primary school teachers.[60]

Despite the great diversity of their party-political affiliations, therefore, outsiders were a new breed of academic, distinct from the mandarins in both social background and ideology. Born into more modest and/or newer strata, they were inclined to look more favorably upon the political changes wrought by modernization. And in keeping with those social origins, their self-understanding as academics was far removed from that of the "culture-bearers." Instead of striving for intellectual synthesis, allegedly necessary for moral renewal and social harmony, outsiders settled for the more modest role of "expert." Scholarship, they agreed with Weber, can guide us in the choice of means, but not ends.

8.4 Integrating Institutional and Societal Explanations

Having argued that the comprehensive and pragmatic approaches to genetics are the products not only of institutional structure but also of social class and culture, I now turn to consider how these two forms of explanation relate to each other. The general issue of how different levels of social structure (or arenas of social experience, if one prefers) interact to mold cognition is a major theoretical problem in sociology of knowledge. The contemporary scientist, for example, is a member of many collectivities, both inside and outside the scientific community:

59. On engineers see Ludwig 1979; Herf 1984, chap. 7. On agricultural scientists see H. Becker 1987.

60. Joachim Schwemmle's decision to join the SA Reserve in 1933, for example, was apparently motivated by the hope that the SA might reduce what he regarded as the unfortunate social separation in Erlangen between "town and gown" (Schwemmle to Curt Stern, 4.4.47, Stern papers). Edgar Knapp is perhaps a similar case. His public support for National Socialism is evident from various gratuitous comments in his 1943 (p. 346) and from his difficulties in continuing his academic career after the war (A. Lang to Klaus Pätau, 25.9.45; Georg Melchers to Pätau, 16.11.45; Pätau to Melchers, 4.12.45; Melchers to Pätau, 14.3.45, all in the Pätau correspondence). On Knapp's postwar employment in commercial plant breeding until 1952, see his obituaries by Abel (1980) and Butterfass (1980).

laboratory, specialty, discipline, family, social class, nation, etc. At present we have no means of specifying which of these memberships will be most influential in shaping knowledge production under a given set of circumstances. As a result, most students of knowledge construction in science get around this problem by focusing exclusively upon the participants' social location(s) *within* the scientific community (cf. chap. 1). Given the complexity of the problem, one can certainly sympathize with this choice, but the fact remains that it is generally an arbitrary one, lacking any theoretical justification.

How might one integrate institutional and social explanations to account for German geneticists' styles of thought? To say that a style is "overdetermined" by both institutional and societal factors is of little use. It is more fruitful to treat institutional structure and societal context as interacting forces which shape each other. For example, when a university is established, its founders must presumably formulate their proposals in a manner which will resonate with potential patrons' taken-for-granted assumptions about the role of knowledge in the society in question. If university reforms in early-nineteenth-century Germany were justified in terms of *Bildung,* the organization of the new privately endowed American universities of the late nineteenth century had to be client-centered and flexible. And once these universities were formed, their structure, as we saw in chapter 4, would continue to shape the thought and behavior of the staff and students who passed through them. In this way an institution reproduces the values which spawned it. At least to some extent; by the end of the nineteenth century, of course, German society had become more complex. Increasingly, university students were coming from new strata where the traditional ideal of *Bildung* held little cachet. What are the consequences for the formation of a style of thought in such a situation, when class background and university structure foster conflicting tendencies? In this section I sketch a simple integrative theory which treats early socialization in the family and school as predisposing young scholars to favor comprehensive or pragmatic styles of thought, while the structure of the institutions where they are subsequently trained or employed either reinforces or weakens that predisposition. In short: Class proposes, institution disposes. Although the data available to me are inadequate for testing the theory, it is nevertheless plausible enough to justify outlining here. What does the theory claim?

Prospective young geneticists will have acquired a sense of the importance (or otherwise) of *Bildung* from their family upbringing, sustained in most cases by the secondary school to which their parents sent them. At the university, however, the young student was removed from these powerful socializing forces and may have encountered a range of disciplines and viewpoints, in some of which the emphasis on *Bildung* was greater than in others. Having settled on a discipline, the student then began to look around for a *Doktorvater* and perhaps to think about careers. Should an academic career be envisaged, the candidate needed to find a topic which would not compromise his or her chances of obtaining an assistantship. When choosing a topic for the *Habilitation* a few years later, the young assistant had to consider which ones would entitle him or her to teach in areas which would attract reasonable numbers of students, be acceptable to his or her professor, and equip him or her to compete for a tenured post. As the young geneticist progressed through successive stages of an academic career, therefore, he or she would be repeatedly confronted with institutional constraints which would sustain or erode the cognitive predispositions conferred by family and schooling.[61]

Consider a few examples. In chapter 4 we saw how the structure of the university tended to induce (or reinforce) cognitive breadth among would-be academics. Elsewhere, the Berlin Agricultural College and the KWI for Breeding Research provided homes for "outsiders" whose social backgrounds, outside or peripheral to the educated middle class, predisposed them to a pragmatic style of thought. The other niche in which pragmatics could thrive was the KWI for Biology. *Drosophila* genetics epitomized the pragmatic approach which was developing so rapidly in the United States during the 1920s and which many German biologists found unacceptably narrow. Nevertheless, partly due to the freedom which assistants enjoyed at the KWI for Biology, Curt Stern was able to work exclusively on *Drosophila* from 1924 and to supervise doctoral candidates in that area. Some might have regarded Stern's choice of research topic as too risky: There were, after all, exceedingly few tenured posts earmarked for geneticists. That Stern persisted with his pragmatic research strategy made sense in terms of the career options which were open to him. After spending two postdoctoral years in T. H. Morgan's laboratory at Columbia, he was offered a job in Morgan's new department at the California Institute of Technology in 1928.[62] Despite this opportunity, many Germans might have hesitated before opting for a career in the United States—one thinks of Karl

61. It bears emphasizing that this is a probabilistic, not a deterministic, model. None of these way stations along a career path can be expected to fix an individual's style of thought, merely to channel it, other things being equal.
62. Calvin Bridges to Curt Stern, 24.5.28, Stern papers.

Figure 8.3 Curt Stern at the University of Rochester in 1938 (photo by Howard S. Brasted, courtesy: American Philosophical Society).

Belar or Hans Gaffron (cf. chap. 5)—but Stern seems not to have shared their cultural and ideological reservations. As one American geneticist put it, while Richard Goldschmidt never fit into "the spirit of American science," Stern fit "perfectly from the beginning."[63] In 1931 he married an American woman, returning the following year to Morgan's laboratory.

Hans Bauer was another pragmatic geneticist at the KWI for Biology. On receiving his doctorate for a dissertation in cytogenetics in 1931, Bauer found himself unemployed. Had his financial situation obliged him to take on an assistantship in an established biological discipline, he might have had to abandon cytogenetics. As it happened, he went to the Institute of Tropical Medicine in Hamburg for a year and

63. Interview with Bentley Glass. Glass knew both men personally, having spent several months in Goldschmidt's department at the KWI for Biology in 1933. Ernst Caspari, similarly, described Stern as "half-American" (interview).

a half, apparently as a self-financed visiting fellow. Blessed with such freedom, he was able to remain in cytogenetics, and there he began the work on giant salivary chromosomes which was later to make his reputation. In 1932 he took up a small fellowship at the KWI for Biology, and within a year the forced emigration of Jewish staff left vacant an assistant's post in Max Hartmann's department which Bauer held until becoming a tenured department head in 1942.[64] Like Curt Stern, therefore, Bauer was able to resist the institutional pressures for cognitive breadth through a combination of good luck and a job at the KWI for Biology.

How might the model outlined above be tested? Table 8.3 presents a series of "career trajectories" which were taken by comprehensive and pragmatic geneticists, listing those way stations which the model suggests should be formative: class, schooling, higher education, and employment. By comparing the styles of thought of individuals who followed slightly different paths, one can infer the significance of each of these way stations (or combinations of them) in molding thought. To begin with, we can ask whether class background was important. By comparing path 1 with path 4 (or 7 with 8, or 10 with 11), we are looking at pairs of geneticists whose career trajectories were identical except for class background. Since both members of each pair displayed the same style of thought, it would seem that class by itself did not influence cognitive style, at least for geneticists employed in universities or agricultural institutes. This does not mean, however, that class had no influence upon cognitive style; the model predicts that social origins might become manifest where a geneticist's place of work did not curb its members' predispositions. The KWI for Biology was a "permissive" institution of this kind. When we compare path 2 with path 6, accordingly, we find that cognitive style correlates with class origin.

Does schooling make a difference? A comparison of paths 3 and 8 (or 4 and 5, or 9 and 11) yields contradictory results. Moving on to the site of higher education, comparison of paths 3 and 10 suggests that this variable might be important, but the data are weak. Finally, we can ask whether the kind of institution in which geneticists were employed was as important as the model would predict. Comparing paths 1, 2, and 8 (or 5 and 6, or 4 and 7) suggests that it is.

Another way in which the model should be testable is to look for *shifts* in problem choice between different phases of certain individuals'

---

64. Bauer 1966; see also Emmy Stein's draft manuscript on the history of the KWI for Biology, Correns papers.

Table 8.3 Career Trajectories

Path	Class	School	Higher Educ.	Job[a]	Examples
		Comprehensives			
1	EMC[b]	classical[c]	univ[d]	univ	von Wettstein, Kühn
2	EMC	classical	univ	KWIB[e]	Correns, Jollos, Belar
3	EMC	modern[f]	univ	agr[g]	Schiemann, Michaelis
4	ILMC[h]	classical	univ	univ	Renner
5	ILMC	modern	univ	univ	Henke
		Pragmatics			
6	ILMC	modern	univ	KWIB	Stern, Bauer, Knapp
7	ILMC	classical	univ	agr	Baur, Kappert
8	EMC	classical	univ	agr	Nachtsheim
9	~EMC[i]	classical	agr	agr	Kuckuck
10	EMC	modern	agr	agr	Stubbe
11	~EMC	modern	agr	agr	von Sengbusch, Schick

[a] job = principal employer, 1925–1935.
[b] EMC = educated middle class.
[c] classical = *Humanistisches Gymnasium*.
[d] univ = university.
[e] KWIB = Kaiser-Wilhelm Institute for Biology.
[f] modern = *Realgymnasium* or *Oberrealschule*.
[g] agr = agricultural college or KWI for Breeding Research.
[h] ILMC = industrial, commercial, or lower middle class.
[i] ~EMC = occupations marginal to the educated middle class.

careers. In the case of the geneticists associated with paths 3, 6, 7, and 8, for example, it would be of interest to know whether their move from a university education to employment in a different kind of institution was accompanied by a narrowing of cognitive style.

Lastly, the model could be explored by focusing upon the careers of "misfits": geneticists whose social background and/or training were atypical for their place of employment. Misfits should have experienced a tension between their own interests—scientific or otherwise—and those of their milieu. One such might be Peter Michaelis, who was the only department head at the KWI for Breeding Research during the mid-1930s whose doctorate had been awarded by a university rather than an agricultural college. Interestingly, he was also the only one to work on evolutionary and developmental problems. Or consider Joachim Schwemmle's early years as a young professor at Erlangen. As the son of a primary school teacher, as a scientist who was averse to generalizations or to "syntheses" (cf. chaps. 7 and 9), and as a professor who regretted the ivory-tower mentality of his colleagues, Schwemmle may have found life in Erlangen's small philosophical faculty somewhat awkward. Finally, Baur's willingness to employ able people wherever he could find them provides us with a most interesting category of misfits: his female assistants (cf. chap. 6). Not only their gender but also their class origins and university education made these women anomalous within Baur's school, and it may be no coincidence that all of them were left behind when he moved to the KWI for Breeding Research in 1928. One might guess, therefore, that they were closer to the comprehensive than to the pragmatic style.[65]

Although none of the data discussed in this section can be said to have proved that "class proposes and institution disposes," I hope to have persuaded others that the model is worth exploring systematically.

8.5 Conclusion

Clearly the implications of my argument in this chapter go well beyond the German genetics community. Once Ringer's dichotomous scheme of the German professoriate is modified through the addition of the outsider category, it more fully reflects the ideological divisions among the humanists and social scientists with whom he, Schwabe, Töpner,

65. Although Elisabeth Schiemann worked primarily in transmission genetics in Baur's institute, her studies of the evolution of crop plants demanded not only broad knowledge in biology (e.g., in systematics and evolution), but also an understanding of relevant issues in history, archaeology, linguistics, and ethnology. Since all of her biographers, furthermore, attest to the depth of her interest in the humanities (Kuckuck 1980, Stubbe 1972, Schiemann 1960, Hertwig 1956), she clearly belongs closer to the comprehensive pole. Apart from Paula Hertwig, who would appear to have endorsed more "pragmatic" views (cf. chaps. 6 and 7), lack of information makes it impossible to identify the styles of thought of Baur's other female assistants (cf. table 7.2).

vom Bruch, and Döring were primarily concerned. In addition, the outsider concept enables us to explain why that particular constellation of attitudes toward science, culture, and politics which I have called the pragmatic style of thought should have been associated with geneticists from a particular range of social backgrounds. (In my usage, therefore, "mandarin" and "outsider" are more inclusive concepts, denoting both an academic's style of thought *and* his or her social origins and schooling.) More generally, the existence of a recurring pattern—be it comprehensive or pragmatic—linking geneticists' philosophical and political views, on the one hand, with their choice of research problem and conception of their discipline, on the other, serves as a reminder that we ought not to treat scientific thought as if its development were necessarily isolated from that of general culture. Following Mannheim, who argued that styles of political thought have been formulated by particular social strata faced with historically specific problems (Mannheim 1953), I have shown how comprehensive and pragmatic styles of thought reflected the differing class backgrounds and experience of two sectors of the German academic community during a period of rapid social change.

If the argument outlined above is correct, it should be possible to trace similar changes in problem choice and role conception in other areas of German scholarship from the late nineteenth century. This will, of course, require historians to look rather more closely for diversity *within* disciplines than they have so far. As one recent collection of essays on Wilhelmian academics demonstrates, Ringer is by no means the only historian who tends to downplay the significance of outsiders within the professoriate.[66] Although almost all of the contributors draw attention to the fact that the professoriate was heterogeneous in one or another respect, none of them systematically pursues the corollaries of that diversity. For example, some authors are aware of the professoriate's diverse class backgrounds but do not consider the possible intellectual or political consequences thereof. Others note the political differences among Wilhelmian academics but make no attempt to relate these to disciplinary membership, scholarly debates, or class backgrounds. And although one contributor maps various divisions within the professoriate, acknowledging the connection between scholarly debates and party-political allegiances, he fails to explore the origin of

---

66. See the essays in Schwabe 1988 by Fritz Ringer, Lothar Burchardt, Kurt Sontheimer, Bernd Faulenbach, Rüdiger vom Bruch, Charles McClelland, and Bernhard vom Brocke.

such divisions (e.g., whether they might be related to the protagonists' class backgrounds). By urging historians to take note of variation, not only among German academics' scholarly views and political inclinations but also among their social backgrounds, I certainly do not mean to suggest that class or educational experience *determines* political views, philosophical assumptions, or academic role conceptions. That is too crude and also demonstrably wrong. Studies of the social stratification of German secondary education during this period, however, indicate that social origins affected educational destinations, and the evidence presented in this chapter makes it appear very likely that both of these forms of experience shaped academics' outlook, to some extent, on a wide variety of issues.

Finally, it would be foolish to assume that genetics was typical of all German disciplines at that time. On the contrary; my argument implies that the distribution of outsiders and mandarins within higher education between the world wars was probably quite uneven.[67] One would expect to find fewer outsider academics in the universities, for example, than in technical institutions such as colleges of engineering, agriculture, and commerce or in applied research institutes of various kinds. And in the universities, too, outsiders are more likely to have been concentrated in those disciplines which, because of their vocational character (e.g., chemistry and agricultural sciences) or their relative youth and low status (social sciences), tended to recruit more of their students from commercial and lower-middle-class strata.

67. Some evidence, for example, suggests that Wilhelmian historians were overwhelmingly mandarin in outlook (vom Bruch 1988).

9 The Politics of Nuclear-Cytoplasmic Relations

My aim so far has been to show that *problem* choice in genetics was shaped by geneticists' social and institutional context. By situating genetics within a broader historical context, however, we also gain insight into conceptual change. In this chapter I return to intellectual history in order to show that the explanatory framework of part 2 can also illuminate the process of *theory* choice.

Ringer's analysis is particularly interesting in this regard, since he shows how widespread were holistic assumptions in the work of both orthodox and modernist mandarins. The evidence presented in chapter 7 is consistent with Ringer in that holistic inclinations are evident among at least some comprehensive thinkers, but not among pragmatics. This raises the possibility that, confronted with a choice between holistic and atomistic concepts or theories with which to interpret a given set of data, comprehensives and pragmatics might choose differently. In order to explore this possibility, one needs to identify a research problem which attracted the attention of both pragmatics and comprehensives and in which both holistic and atomistic solutions were tenable. Such problem areas are relatively rare, since, as we have seen, pragmatics and comprehensives tended to work on different kinds of genetic problems. Fortunately, however, the debate over the plasmon theory of cytoplasmic inheritance is ideal for this purpose. It attracted members of both groups, since its theme—the sufficiency of Mendelian genes in explaining patterns of inheritance—was of general interest among geneticists. During the 1930s, furthermore, the advocates of the theory disagreed among themselves over the structure of the plasmon: Some suggested a unitary structure and others a particulate one. In this chapter, therefore, I will analyze the debate over the plasmon theory and advance the following argument:

(i) Three basic groups can be identified in the debate: advocates of the orthodox plasmon theory, "revisionist" defenders of a modified plasmon theory, and the critics of the plasmon theory who defended *Kernmonopol* (table 9.1). None of these positions could be justified solely in terms of the available empirical evidence. Each of the three groups, therefore, made simplifying assumptions, albeit different ones.

(ii) Against the backdrop of the previous chapter, moreover, it is

316 Styles of Thought

Table 9.1 Positions Adopted in the Debate over the Plasmon Theory

Participants	View of Plasmon theory	Academic role
Correns		mandarin
von Wettstein		mandarin
Oehlkers	orthodox	mandarin
(Kühn)[a]		mandarin
(Woltereck)		mandarin
Renner		marginal[b]
Michaelis		marginal
Schwemmle	revisionist	outsider
(Knapp)		outsider
(Baur)		outsider
Stern		outsider
Goldschmidt		mandarin
Hertwig		outsider
Hämmerling	critical	?
Lehmann		outsider
Jollos		mandarin
Nachtsheim		outsider

[a] Parentheses around a name indicate that, while not a major participant in the debate, the individual was sympathetic to a particular view of plasmon theory.

[b] marginal = marginal to the mainstream mandarin tradition.

evident that each of these groups tended to subscribe to different conceptions of the academic's role. Orthodox plasmon theorists subscribed to the mandarin role; critics were largely outsiders, and revisionist plasmon theorists include some of each.

(iii) The metaphors with which nuclear-cytoplasmic relations were described reflect each group's view of the proper relationship between politicians and the professoriate in the maintenance of social order. Perceiving the cell as a microcosm of German society, each group ascribed to the plasmon particular properties analogous to those political rights and responsibilities which it granted to the professoriate. In this debate, therefore, theory choice was channeled by geneticists' self-understanding as mandarins or as outsiders.

9.1 Critics of the Plasmon Theory

As we saw in chapter 2, the advocates of cytoplasmic inheritance came under attack from the defenders of *Kernmonopol* ("nuclear monopoly"). One of these criticisms was that cytoplasmic inheritance was a relatively rare phenomenon and thus of limited biological significance. Kernmonopolists emphasized that interspecies hybrids were more often reciprocal than nonreciprocal. This could not, of course, be a conclusive argument, because, as plasmon theorists observed, such reciprocity might simply mean that the cytoplasmic heredity of the two species happened to be identical. After all, chromosomal genes, too, could only be detected via crosses when they *differed* between the two parents. Furthermore, the very few instances of cytoplasmic inheritance in animals did not prove that the phenomenon was limited in scope; it might simply be a reflection of the fact that interspecies hybrids in animals were less viable than those in plants.

Perhaps because they realized such criticisms could not be conclusive, the Kernmonopolists went further, challenging the existence of cytoplasmic inheritance even in the small number of well-analyzed cases. The interesting thing about many of these criticisms is their *a priori* character. A frequent objection was that one could not rule out some kind of predetermination in which it took many generations before cytoplasmic substances of nuclear origin were finally diluted out. Thus what appeared to be a stable cytoplasmic effect would eventually prove to be no more than a dauermodification. But often the critics offered no evidence in favor of predetermination, acknowledging that the issue was still open.[1] Joachim Hämmerling, for example, insisted that his now famous experiments with *Acetabularia* provided no evidence for the plasmon theory, while admitting that they were nevertheless compatible with it.[2]

Confronted with the evidence for cytoplasmic inheritance, Goldschmidt offered no specific criticisms, but dismissed von Wettstein's work as not only "irrelevant" (*überflüssig*) but dangerous, asserting that it was as yet unproven and "extremely unlikely" (Goldschmidt 1933a, 4). It was "absolutely certain that the decisive role in inheritance is played by chromosomal genes which set in motion the processes which

---

1. For example, W. Schleip 1927, 3, 75; Hertwig 1934, 427; Jollos 1939, 83. *A priori* objections were not peculiar to German skeptics. T. H. Morgan also asserted that the cytoplasm was under the chromosomes' influence and thus played "a minor role in inheritance" (Morgan 1923a). The single experiment which he cited in evidence (cf. p. 627), however, was utterly irrelevant to his conclusion.

2. Hämmerling 1934a, 835–36; 1934b, 262–64; 1935, 455–56, 459–60. While sharing what he called Hämmerling's "opinion" that development in *Acetabularia* was directed by nuclear genes, Curt Stern warned that the data did not rule out the possibility that unspecific nuclear stimuli were triggering specific morphogenetic processes in the cytoplasm, perhaps laid down by a plasmon (Stern to Hämmerling, 13.5.39, Stern papers).

are typical of development." (1933a, 5). The following year he was slightly more charitable, conceding that the problem deserved experimental study and left a good deal of room for interpretation, not hesitating to offer his own: "I can conceive of the facts which have been reported only in one way: I *assume* that genes act in controlling the differentiation of hereditary traits... through the control of chains of reaction of definite velocity" (Goldschmidt 1934, 14 [emphasis supplied]; see also 1938, 263–78). Faced with the stable cytoplasmic effects observed by von Wettstein and Michaelis, Goldschmidt could only hopefully suggest that "it *might* be quite possible that longer continued experiments would show" that the effects decline under the influence of foreign nuclei (1934, 16; emphasis supplied). Nearly thirty years later Goldschmidt was still relying on wishful thinking; dismissing the evidence for a plasmon, he assured his readers that "it is extremely improbable that such a case will ever be found" (1961, 217).

These critics were stuck. Since there was no agreed criterion as to how many generations of backcrossing were necessary to be sure that cytoplasmic effects were *truly* stable and thus heritable—indeed, no such criterion could have been decisive, just as we cannot rule out the possibility that the sun will fail to rise tomorrow—the critics were forced to claim that future results would somehow vindicate their skepticism. Not all criticisms of cytoplasmic inheritance, however, were *a priori*. Ernst Lehmann and his students at Tübingen isolated strains of *Epilobium* which, when crossed with a marker strain, varied in the extent to which their cytoplasm inhibited the growth of species hybrids. Since such strain differences were inherited along Mendelian lines, and his students could show that the cytoplasm of those strains which displayed the most inhibition had lower levels of growth hormone, Lehmann argued that the apparent evidence of a plasmon's inhibitory effects was in fact due to the presence of growth inhibitors in the cytoplasm whose level was controlled by nuclear genes (Lehmann 1928; 1932; 1936a; 1938–39). This attempt to retain *Kernmonopol* was, however, not without its own difficulties. A serious defect with his theory, as Lehmann conceded, was that it could not explain how cytoplasmic effects of nuclear origin could persist in species hybrids over many backcross generations, since cytoplasmic growth inhibitors were not known to replicate and should, therefore, have been quickly diluted out.

Had Lehmann succeeded in building a school of disciples during the 1920s, the growth-inhibitor theory might better have weathered this criticism. But two of his most talented assistants left him in the latter twenties, and their subsequent work on cytoplasmic inheritance failed to sustain his theory. As Lehmann's assistant from 1922 to 1929, for example, Joachim Schwemmle had initially defended *Kernmonopol* in his work on *Epilobium*. Toward the end of this period, however, he became interested in von Wettstein's work.[3] On acquiring the chair of botany at Erlangen in 1930, he gradually shifted his work to cytoplasmic inheritance in *Oenothera,* moving steadily closer to von Wettstein's position (see below). Schwemmle's contemporary at Tübingen, Friedrich Oehlkers, was also Lehmann's assistant, from 1922 to 1925 (Marquardt 1974). As professor of botany in Freiburg a decade later, he began to work on cytoplasmic inheritance, adopting a theoretical position even closer to von Wettstein's, as we shall see. By the late thirties Lehmann was in serious trouble. For one thing, several geneticists— including one from his own institute—announced that they were unable to replicate his experiments (Michaelis 1938; 1939a; Brücher 1938). For another, in 1938 Lehmann was found guilty of various forms of misconduct (including misuse of funds) while serving as head of the Botanical Institute and former dean of the science faculty from 1933 to 1937. Suspended from his academic post, he was allowed to resume teaching in 1943, but protests from the rector and deans prevented him from resuming the directorship of his former institute.[4]

By the 1930s, therefore, the evidence for *Kernmonopol* was rather weak. Why then did Kernmonopolists insist upon attacking the evidence for cytoplasmic inheritance? Some observers have suggested that this was, to a large extent, a dispute between botanists and zoologists. Certainly Oehlkers regarded zoology as a bastion of skepticism (Oehlkers 1949, 110–11), and not without reason: The critics Goldschmidt, Stern, Nachtsheim, Hertwig, and Jollos were zoologists.[5] That zoologists were inclined to be skeptical might have been related to the fact that they were less familiar with plants (in which cytoplasmic inheritance was most readily demonstrated), and there were few known cases of the phenomenon in animals. Sapp has argued that critics rejected the plasmon theory because it meant that no phenotype's inheritance could be understood solely in terms of chromosomal genes (Sapp 1987). And that, in turn, would mean that the Mendelian's or cytologist's skills, while necessary, would no longer be sufficient for understanding in-

3. Schwemmle 1927–28; on Schwemmle's career see Arnold 1981.
4. Deichmann 1991, 335–53.
5. As a protozoologist who took his doctorate with Max Hartmann, Hämmerling is not so obviously identifiable as a zoologist. In "Mein Wissenschaftlicher Lebenslauf," however, he states that zoology played a larger role in his education than did botany.

heritance. Thus critics sought to contain this threat to their expertise by insisting that cytoplasmic inheritance was restricted to a few phenotypes in plants, perhaps merely those affected by chloroplasts.

These kinds of explanations in terms of "professional interests" may be part of the reason why cytoplasmic inheritance was rejected by *Drosophila* cytogeneticists like Stern and Nachtsheim who would have found it relatively difficult to take cytoplasmic inheritance into account. Perhaps this is why Stern seems to have been rather cavalier about the evidence for cytoplasmic inheritance.[6] But professional interests cannot account for most of the critics' skepticism. Lehmann, for example, was a botanist, as were several American critics: E. M. East, George Beadle (before turning to *Drosophila* in the early 1930s), and Edmund Sinnott. And although most of the German critics were zoologists, it is difficult to see how their research skills could have been marginalized by cytoplasmic inheritance. Ironically, Goldschmidt himself had been one of the first to demonstrate cytoplasmic inheritance in animals, showing that the pigmentation of the caterpillar stage, rate of development, and extent of sex differentiation in the gypsy moth were due to stable cytoplasmic factors (Goldschmidt 1933c, 612–14). Other critics had expert knowledge of species hybridization, the method whereby cytoplasmic inheritance was generally demonstrated; Hämmerling was constructing species hybrids in *Acetabularia*, and Paula Hertwig published an extensive monograph on species hybrids in animals (Hertwig 1936). And as we saw in chapter 3, Jollos's dauermodifications bore a very close resemblance to cytoplasmic inheritance. Thus the disciplinary affiliation and professional interests of these critics are of limited use in understanding why, in view of the accumulating evidence for cytoplasmic inheritance during the 1930s and the virtual absence of empirical counterarguments, the critics nonetheless rejected the phenomenon so vigorously.

6. In "Der Kern als Vererbungsträger" (1930b), Stern concentrated on summarizing the evidence for the chromosome theory, turning only in the last three paragraphs to the evidence for cytoplasmic inheritance. His remarkably inadequate treatment of the plasmon theory suggests that he was either unaware of the relevant work by Correns and von Wettstein—hardly likely in view of the fact that he worked in the same institute with them for several years—or that he was rather more interested in the validity of the chromosomal theory than in its biological significance. Even much later, when the evidence for cytoplasmic inheritance had grown substantially, Ernst Caspari recalled that in their discussions of his review paper on cytoplasmic inheritance (Caspari 1948; they were then colleagues at the University of Rochester), Stern repeatedly attempted to account for the data in terms of alternative chromosomal explanations (Caspari interview).

9.2 Revisionist Conceptions of the Plasmon

The "orthodox" form of the plasmon theory, as we saw in chapter 2, was originally outlined by Correns, developed by von Wettstein, and subsequently endorsed by Oehlkers, among others. Renner, Michaelis, and Schwemmle also made extremely important contributions to the arguments for cytoplasmic inheritance, but, as we shall see below, each of the latter three parted company with von Wettstein on one or more of the plasmon's features. These three individuals, therefore, adopted a variety of what I will call "revisionist" positions in the debate over cytoplasmic inheritance, occupying a sizable middle ground between von Wettstein et al. and the Kernmonopolists. Let us consider the central issues in the debate on which these revisionists deviated from orthodoxy:

(i) *What was the plasmon's function?* Oehlkers followed Correns in regarding the plasmon as a "trait-constituting structure."[7] Michaelis, too, assigned the plasmon a regulative role in gene activity, although at first he had distanced himself from Correns's conception (in which genes exerted merely a quantitative effect upon reactions qualitatively rooted in the plasmon). Throughout the 1930s, for example, Michaelis argued that at least some of the plasmon's effects were merely to inhibit or stimulate gene action (1929, 311; 1931, 103; 1932, 96; 1933b, 395; 1937, 287; 1939b, 565). From about 1940, however, Michaelis began to endorse Correns's position: The *kind* of phenotype, he argued, was determined by the plasmon, while the genes merely affected its *degree* (1940a, 1940b). A gene's phenotypic outcome was assigned by the plasmon rather than intrinsic to the gene: A gene's usual biochemical consequences could be redirected by the insertion of a new set of plasmon-controlled reaction systems in the cytoplasm (Michaelis and von Dellingshausen 1942, 375, 424–25). While Schwemmle, too, gave the plasmon a regulatory function, he had much less to say about the mechanics of this process. While citing both Correns's monograph and von Wettstein's review paper (both 1937), he avoided commenting upon their models of plasmon function, and in a 175-page discussion of the relations between nucleus, cytoplasm, and chloroplasts, he devoted a grand total of three sentences to the plasmon's function. His remarks about the chloroplasts' role in trait formation were similarly

---

7. For example, Oehlkers took over Correns's term *Entfaltungsmechanismus* (Oehlkers 1941, 177; 1949, 98).

scanty and undeveloped.[8] Renner was even further from orthodoxy; to my knowledge, he never discussed the plasmon's function. Instead he seems to have been interested in cytoplasmic inheritance simply as a form of extrachromosomal inheritance rather than as a regulatory system central to development and/or evolution.

(ii) *What was the scope of the plasmon's effects?* Based on evidence from a small number of traits in three or four organisms, von Wettstein had postulated plasmon counterparts for *all* genes in *all* organisms. This was a very ambitious claim, since he recognized that most instances of maternally inherited patterns of variegation were ascribable to chloroplasts, and that the latter were also known to affect a wide range of other traits. To justify this claim, therefore, he had to rule out the possibility that those traits which he assigned to the plasmon might actually be due to chloroplasts. But neither of the reasons he gave could be conclusive. On the one hand, he felt that the evidence for a plasmon in animals made the chloroplast theory unlikely, but there were far more reported cases of cytoplasmic inheritance in plants than in animals, and there was no need to assume the same genetic basis in both plants and animals. Worse, the mosses with which von Wettstein worked did not permit him to distinguish the effects of plasmon from chloroplasts in the way that Renner and Schwemmle could with *Oenothera*. Thus von Wettstein's conclusion was necessarily cautious: "Some cases of maternally inherited reciprocal differences in morphological traits might also be conditioned by the chloroplasts. In most cases the differences are probably due to the plasmon" (von Wettstein 1937, 358; cf. 350). Michaelis (and apparently Oehlkers) endorsed the principle that the genetic makeup of all cells consisted of genes, plasmon, and (in plants) chloroplasts, as three independent entities normally in harmony with each other (e.g., Michaelis 1933b, 395, 407–8; Oehlkers 1941, 180). Although in principle Renner agreed (1934, 264–65), in practice he often sought to prove that phenomena attributed by others to a plasmon were more simply explained in terms of Baur's theory of chloroplast distribution. In the mid-thirties, for example, Konrad L. Noack and other botanists were advocating a theory of variegation, similar to that in Correns's classic paper of 1909, which explained patterns of color variation in terms of the distribution through the leaf of a "labile" cytoplasm (in all but name, a plasmon) which affected maternal chlo-

---

8. See Schwemmle 1938, 658 (on the plasmon's function) and 649–50 (on the chloroplasts').

roplasts' function. Where the cytoplasm was "healthy," the chloroplasts would be green, and where it was "diseased," they would become yellow or white.[9] But in every case, Renner objected, the data could be more simply explained by the transfer of paternal chloroplasts through the pollen tube, the differential distribution of maternal and paternal chloroplasts during subsequent cell divisions, and the different responses of these two chloroplast types to their genetic and physiological environments (Renner 1924b; 1934; 1936; 1937).

Both theories had their difficulties. Like Correns's before it, Noack's interpretation had to assume that two kinds of cytoplasm would not mix uniformly in the hybrid cell, an assumption which Correns had admitted was unlikely (Correns 1922). Renner's interpretation, on the other hand, required the transmission of paternal chloroplasts via the pollen tube as well as the existence of "mixed" cells containing both green and white chloroplasts, but neither of these could be demonstrated in many cases. Faced with this apparent stalemate, Renner made his choice:

> How happy we would be if the connection between hereditary determinants and the phenotype were even half as clear as the case in hand! And we are supposed to obscure this clarity by assuming a labile condition of the cytoplasm, simply because we cannot observe the transfer of the chloroplasts via the pollen tube!? That we cannot see this transfer in relation to the facts of variegation means as little as the invisibility of chromosomes in mature gametes in relation to the fact of chromosome number in cell-division. Skepticism is good but in moderation. The development of the chromosome theory shows the wisdom of those who stopped doubting the theory at all cost, once a certain body of evidence had accumulated. It is these believers who have erected the structure of modern genetic theory from experimental evidence, while the doubters have contributed nothing essential.[10]

Although Renner's theory made a testable prediction (that chloroplast transfer through the pollen tube should eventually be observable) which was subsequently confirmed, his choice of theory *at the time* was hardly dictated by the evidence then available.[11] In practice he seems to have

9. For a review of this debate see Oehlkers 1937; 1938a.
10. Renner 1934, 251. When he first advanced this theory in 1924 he also conceded that he could not explain why no zygote cells had yet been found which contained both kinds of plastid (1924b).
11. After coming to the conclusion that the choice between Renner's and Noack's conflicting interpretations was essentially a matter of taste, I discovered that Goldschmidt had earlier drawn the same conclusion, for this debate as well as for the 1909 debate between Baur and Correns (Goldschmidt 1961, 231–32).

accepted the plasmon theory as a fallback position only in those cases where chloroplast explanations seemed implausible. In *Oenothera*, however—the genus he knew best—he felt that all cytoplasmic-inheritance phenomena could be accounted for in terms of chloroplasts (Renner 1929; 1936).

Schwemmle was similarly cautious about invoking the plasmon theory. By the late 1930s he demonstrated that in *Oenothera* some of the maternally inherited traits usually ascribed to a plasmon were in fact due to chloroplasts (Schwemmle 1938, 643ff) and warned of the methodological difficulties in showing conclusively that cytoplasmic-inheritance phenomena were due to structures other than chloroplasts (1941a; 1941b). More generally, he was hesitant to adopt a single conception of the plasmon. When assessing apparently contradictory evidence in *Epilobium*, mosses, and *Oenothera*, he repeatedly emphasized that cytoplasmic inheritance might not have the same properties in all organisms (1927–28, 631; 1935, 185).

(iii) *Was the plasmon's genetic structure stable?* In specially constructed species hybrids in which the chromosomes were derived from the male parent and the cytoplasm derived from the female, any sign that the plasmon's effects had declined over several generations was interpreted (by both sides) as evidence that the plasmon was unstable. Such a decline meant that the disharmonies between nucleus and plasmon from different species would eventually be alleviated by the plasmon adapting, presumably under the influence of nuclear genes.[12] The plasmon's stability was thus the major issue separating proponents from critics of cytoplasmic inheritance, since instability suggested that the effects assigned to the plasmon were merely modifications, ultimately under nuclear control.

Oehlkers seems to have had no doubts about the plasmon's stability (Oehlkers 1941, 180; 1949, 118–19). Schwemmle was close to orthodoxy on this issue, presenting evidence from *Oenothera* that the incompatibility between nucleus and cytoplasm in one case declined over several generations due to adaptation by the *nucleus*, since the plasmon and chloroplasts had remained unaltered (Schwemmle 1938). But once again he was loathe to generalize; one ought not to assume, he wrote,

---

12. In principle the issues of stability and independence are separate; one can imagine self-replicating genetic elements in the cytoplasm, independent of the nucleus, which fail to maintain themselves indefinitely. This was in fact the position eventually adopted by Michaelis (cf. chap. 3). In practice, nevertheless, most participants in the debate treated the two issues as connected.

that the plasmon in *every* organism would prove so stable (1935, 184; 1935–36, 601). Renner vigorously defended the genetic stability of the chloroplasts in the presence of a foreign nucleus, but had little to say about the plasmon's stability (Renner 1922; 1934). Furthest from orthodoxy was Michaelis. Although he routinely emphasized the stability of the plasmon in *Epilobium* (since it had survived unaltered during thirteen generations of backcrossing by 1935), he also acknowledged that the plasmon's effects occasionally declined, interpreting this (initially at least) as the adaptation of the plasmon under foreign nuclear influence (Michaelis 1929, 314; 1931, 103–4; 1932, 98; Michaelis and Wertz 1935). (As we saw in chapter 3, he regarded these rare instances of plasmon instability as theoretically important because they suggested that cytoplasmic inheritance and dauermodifications were related phenomena.) This deviation from orthodoxy provoked a rebuttal from von Wettstein, who warned, "It hardly needs emphasizing how crucial a correct interpretation of this case is" (von Wettstein 1937, 357).

(iv) *Was the plasmon a unitary or particulate structure?* As we saw in chapter 2, Correns's earliest discussions of gene action envisioned some kind of unitary regulatory structure in the cytoplasm, since individual genes appeared to be independent agents, incapable of coordinating their own actions. In subsequent papers on cytoplasmic inheritance, Correns's model was developed more fully. We can, he suggested, envisage the cytoplasm as a machine gun and the genes as cartridges (Correns 1928, 164; 1937, 127–28). The analogy was imperfect, since the gun cannot determine the sequence of cartridges' release; the cartridges' fit with the gun is an all-or-none phenomenon; and cartridges cannot exert a quantitative influence over the operation of the gun. What he sought to emphasize, nevertheless, was that the sequence and timing of genes' action is dependent upon the functioning of another suitably designed structure. His intentions become clear if we compare the machine-gun analogy with the earlier analogy of nuclear heredity as a kaleidoscope. Both analogies portray the genes as relatively small and simple elements which only become effective when ordered by a larger, more complex, stable structure.

Von Wettstein deployed a very similar model in one of his earliest papers on cytoplasmic inheritance (von Wettstein 1928, 48). Rather than portray the cytoplasm-gene relation in terms of "lock and key," with its implication of cytoplasm's passivity and genes' activity, he suggested that we think of an organism's phenotype as like an airplane's flight. The latter requires both external enabling conditions (wind, acceptable weather) and an internal driving force (the engine, fuel), but the most essential factor was probably the design of the aircraft itself. With this analogy von Wettstein put forward the view that genes (i.e., the engine) are the driving force in trait formation (flight), but that the plasmon (aircraft) is both a precondition for gene activity and a structuring influence upon it. In his most extensive discussion of the plasmon's nature, von Wettstein sharply distinguished its structure from that of the genes. In order for development to proceed in a coherent and orderly sequence, the otherwise chaotic and rather violent activity of the genes ("*dieses chaotische Gleichzeitig-Losschlagen*") had to be regulated by a mechanism with the appropriate structure (von Wettstein 1937, 362). Unlike genes or chloroplasts, therefore, the plasmon had a unitary structure; it could not be reduced to a swarm of "plasmagenes," as Winkler had earlier suggested. Unlike the genes with their specific effects upon particular traits, the plasmon had a general and uniform effect upon the entire organism.

Oehlkers declared his support for the plasmon theory "in a von Wettsteinian sense." The plasmon was an independent and stable genetic element which guided trait formation, but unlike genes or chloroplasts, it was inherited as a unitary structure (Oehlkers 1931, 231; 1938a, 304; 1941, 177; 1949, 111). Schwemmle had much less to say about the plasmon's structure, pleading that "we really still know too little about the interaction of nucleus, cytoplasm and chloroplasts" (Schwemmle 1938, 658). Nonetheless he evidently felt obliged in a major work on cytoplasmic inheritance to offer some way of conceptualizing this interaction. In an attempt to explain why two *Oenothera* strains with different cytoplasm (i.e., plasmon) but identical chromosomes and chloroplasts differed in a particular trait, Schwemmle merely noted that "two differently constructed cars, equipped with the same amount of fuel, can traverse different distances" (1938, 658). Though miserly, to say the least, his analogy bears a clear resemblance to the mechanical analogies used earlier by Correns and von Wettstein, in which genes provided the driving force while the plasmon directed their energies. Although reluctant to theorize, therefore, Schwemmle seems to have accepted the orthodox view of plasmon structure.

Renner, as we saw above, devoted more effort to developing chloroplast theories of cytoplasmic inheritance than to plasmon theories. Occasionally he tentatively suggested that the structure termed *plasmon* might be some stable chemical feature of the cytoplasm such as pH or viscosity,[13] but nowhere did he advance a holistic model as had Correns and von Wettstein—quite the contrary! In 1929 Renner stated what he took to be von Wettstein's definition of the plasmon: "The sum of the so far inseparable genetic elements in the cytoplasm" (Renner 1929, 25). Referring back to the paper of von Wettstein's which Renner was citing, however, we find a different emphasis. Throughout this paper von Wettstein refers to the plasmon in the singular as "a genetic element." Although he does not rule out the possibility that the plasmon *may* eventually prove to be an aggregate, he regards it for the present as a single entity: "It may be hoped that closer analysis will succeed in breaking down the unity of the plasmon, just as the existence of individual chromosomal genes can be inferred from segregation patterns. For the time being, however, this kind of analysis of the plasmon is out of the question" (von Wettstein 1927a, 265). A year later von Wettstein inclined more strongly to a unitary conception: "[More-or-less uniform effects of the plasmon on different traits], of course, provide no grounds for inferring that the cytoplasmic heredity has a mosaic structure. Undoubtedly this portion of the hereditary material is always transmitted to the next generation as a unit [*als Ganzes*].... Not composed of individual genes, [the plasmon] uniformly alters the entire organism" (1928a, 198, 206). While von Wettstein actually regarded the plasmon—for lack of evidence to the contrary—as a unitary structure, Renner perceived him as saying that the plasmon was an aggregate of as-yet-inseparable parts. For one the glass was half full; for the other, half empty.

Once again, what is striking about the arguments for either unitary or particulate structures for the plasmon is that they had so little empirical justification. Virtually the only evidence which Correns and von Wettstein could cite in support of their unitary model was the fact that maternally inherited traits did not segregate in subsequent crosses the way Mendelian traits did. Unfortunately, however, this evidence was also consistent with a particulate model: If there were a high degree of redundancy in the system due to large numbers of identical plasmagenes specifying the same trait, segregation of plasmon variants would be exceedingly rare. But as we have just seen, von Wettstein's sympathies were evident in 1928, and by 1937 he no longer took seriously

---

13. Renner 1934, 264. In 1924 he asserted that the plasmon could mutate (independently of genes and plastids), but his attempt to characterize the plasmons of different *Oenothera* strains using physiological or other criteria had so far been unsuccessful (1924b).

the possibility that the plasmon might be particulate. By that time, however, Peter Michaelis had concluded precisely the opposite. In the course of the 1930s his model of the plasmon shifted from a unitary to a particulate model. In the early thirties he had rejected the existence of plasmagenes, contrasting the plasmon's structure with that of individually distinct genes (Michaelis 1931, 104; 1933b, 388). By 1935, however, he was drawing upon Baur's theory of variegation, treating the plasmon as a population of independently replicating genetic elements. This particulate model was developed further in subsequent papers: "At least in *Epilobium*," he wrote, "we have to reckon with the possibility that the plasmon is not a unitary substance but the sum of various genetic units" (Michaelis 1938, 455; cf. 1935a, 497–99; 1942a, 178–81). Despite these conflicting views of the plasmon's structure, neither von Wettstein nor Michaelis criticized—or even acknowledged— each other's interpretation during the late thirties. As Knapp concluded, neither holistic nor particulate models could be ruled out at that time (Knapp 1938b, 424). Confrontation between the two plasmon theorists, I suggest, had to be avoided because neither man could think of an experiment which would resolve the issue.

At least one Kernmonopolist noticed that the orthodox plasmon theory relied to some extent upon *a priori* assumptions: "This rejection [of *Kernmonopol* by plasmon theorists and others] is largely without experimental justification; it rests more upon the need to find a *holistic principle* in the developmental process. It is the disinclination to see organisms as *aggregates* of *hereditary factors* [*Anlagen*] . . . whose complete independence is seen by some as incapable of guaranteeing the organism's unity."[14] With this claim, Paula Hertwig undoubtedly captured a feature of the orthodox plasmon theory. On the other hand, she seems not to have noticed that her fellow Kernmonopolists' rejection of cytoplasmic inheritance was also "largely without experimental justification." And in portraying all plasmon theorists as holists, she overlooked the revisionists: Michaelis's preference for a particulate plasmon was equally *a priori*. Each of the three interpretations of the evidence

---

14. Hertwig 1934, 427, emphasis in original. Two years later her skepticism toward cytoplasmic inheritance was again visible in the asymmetric way in which she discussed the evidence for it in animals. Although she *cites* Kühn's paper on cytoplasmic inheritance, she does not discuss his data or interpretation (namely, that a plasmon exists). And although she presents Goldschmidt's data, showing that nonreciprocal differences in race hybrids of the gypsy moth persisted unaltered for over ten generations, she declines to rule out the possibility that cytoplasm might still be merely the "Reaktionsbasis" for gene-controlled developmental sequences (Hertwig 1936, 115–17).

for cytoplasmic inheritance, as we shall see, was based on a different set of assumptions.

9.3 Models of Cellular Order

In trying to uncover these assumptions, let us focus upon the participants' choice of language. More particularly, let us examine the analogies with which they represented the cell's organization. The central role of analogy in scientific thought is well known (Black 1962; Hesse 1966; Schon 1963). More than simply a didactic tool with which established theories can be made intelligible to learners, analogy serves an essential heuristic purpose in research. Much was gained by thinking of light and sound propagation as wavelike and by thinking of natural selection as a form of plant or animal breeding. Furthermore, Hesse has argued, scientific theories explain by a process of "metaphorical redescription" in which the seemingly unconnected observations in an unfamiliar system become ordered and intelligible through juxtaposition with an older, familiar system of meanings.

But choosing an analogy is problematic. No analogy is *obviously* appropriate for a particular set of puzzling observations. One cannot know in advance which features of the older meaning system will turn out to have corresponding properties in the unfamiliar system. Furthermore, there are always noncorrespondences, even with the best of analogies: Gas molecules may behave in many respects like billiard balls, but no one believes them to be numbered. Finally, it is reasonable to assume that there are normally several potentially useful analogies at the theorist's disposal, each of which displays a partial resemblance to the unfamiliar system. Under these circumstances the theorist has to choose between promising analogies on nonempirical grounds. The criteria used, consciously or otherwise, will necessarily be social in two senses (Barnes 1974). First, it is obvious that a private analogy—perhaps derived from a dream—cannot become incorporated into communicable knowledge unless it embodies experiences shared by other members of the scientific community such that it will be understood by those members. But more significant, even where analogies are collectively understood, they must also embody meanings which are acceptable within the community. Many analogies will meet this criterion by being morally and politically neutral: electricity as a "fluid," gene action as the "translation of information" and so forth. But other analogical candidates—no matter how empirically plausible—may carry objectionable meanings and thus be avoided. Mark Adams has shown, for example, how the term *gene fund* (to describe the genetic composition of a population) was avoided by a sector of the Soviet genetics community in the 1930s because of its capitalist connotations (Adams 1979). And by implying, however unintentionally, action by a higher intelligence, Darwin's use of the term *selection* initially worried some who accepted a naturalistic account of evolution, while appealing to some theists (Young 1985a). These cases demonstrate that it is not easy for the theorist to anticipate how a given analogy will be received by all sectors of the relevant community. In general, however, it is obvious that a scientist will avoid analogies which are known to conflict with the values and experience of his or her sector of the scientific community. It then follows that the analogy actually selected may offer a clue as to what those shared values are.

In trying to make sense of complex organized systems, scientists have often found it useful to draw upon social analogies. In the case of evolutionary theory, for example, it has been argued that Darwin got the idea for the concept of natural selection from reading Malthus's analysis of poverty, and that Darwin's mechanism of speciation was modeled on theories of political economy (Young 1985b; Schweber 1980). Often social analogies have been used in representing relations of hierarchy and control. In eighteenth-century Scottish physiology, for example, the nervous system played a key role: It was seen as a mediator between the organism and its environment, shaping the quality of sensations and integrating body functions. In so doing, it mirrored the civilizing role assumed by lowland gentry and Edinburgh literati in their attempt to assimilate dissident groups (Lawrence 1979). Similarly, historians of embryology have shown how conceptions of the process of fertilization have incorporated assumptions about gender relations (Biology and Gender Study Group 1989). Others have discussed the anthropomorphic intentions behind Spemann's "organizer" concept (Horder and Weindling 1985). Some evidence suggests that various later-nineteenth-century German biologists' theories of the organization of multicelled organisms were patterned on their conceptions of an ideal society.[15] Finally, molecular biologists of the 1960s routinely used a "workshop" model of the cell in which the operon's "blueprints" were transcribed into "messenger" RNA which guided the assembly of peptides on ribosomal "workbenches," the whole process being driven by ATP as the "currency" of the cellular "economy."

In the debate over cytoplasmic inheritance, too, participants often

---

15. Weindling 1981 and 1991.

represented the relationships between nuclear genes and cytoplasm in terms of analogies drawn from the social order. Although both advocates and critics of the phenomenon deployed such analogies, however, they portrayed the cellular order in very different ways. Critics saw the relation between nucleus and cytoplasm as strongly hierarchical, while plasmon theorists emphasized equality and cooperation. The earliest such uses of social analogy known to me emphasized the role of the nucleus. Weismann referred to the "domination" (*Beherrschung*) of the cell by the hereditary substance and compared the nucleus's role in development to a military command structure.[16] Richard Hertwig saw in the cell an absolute monarchy, whose achievements "result from the mass of people, the directives from the monarch."[17] Theodor Boveri deployed a different kind of social analogy, portraying the cytoplasm as the mason and carpenter who act under the architect's instruction (Boveri 1904, 104). At the first meeting of the German Genetics Society in 1921, Hans Nachtsheim used a variant of this analogy: "The cytoplasm is the building material for the chromosomes, and just as the composition of building material is of the greatest importance for the architect, if he wishes to realize his design as he conceived it, so also is the composition of the cytoplasm of the greatest importance for the chromosomes" (Nachtsheim 1922a, 251). Although Nachtsheim did not rule out the possibility of cytoplasmic inheritance in principle and seems to have thought he was underlining the importance of the cytoplasm, his model provoked objections in the subsequent discussion from, among others, Renner, Baur, and Winkler.[18]

By the early 1920s the term *Kernmonopol* (*Monopol* having economic implications) was evidently in common usage, conveying the established view that genes were *the* directive force in trait formation. When Winkler opened the debate over cytoplasmic inheritance in 1923, for

---

16. Weismann 1892, 61. I thank K. A. Parkinson for drawing this reference to my attention.

17. Hertwig, "Die Protozoen und die Zelltheorie," *Archiv für Protistenkunde* 1 (1907): 1–40, cited by Gilbert 1988, 337. For Max Verworn's criticism of the "sole domination" (*Alleinherrschaft*) of the cell by the nucleus, see Boveri 1904, 103 (cf. 113).

18. Nachtsheim 1921. Although Nachtsheim was not a particularly active participant in the debate over cytoplasmic inheritance, it is clear from the marginalia in his reprint collection (Institut für Anthropologie und Humangenetik, Universität Heidelberg) that he made some attempt to keep up with at least von Wettstein's publications on the subject. In 1927 he again briefly endorsed *Kernmonopol* (Nachtsheim 1927, 994). By the late 1940s he seems to have accepted some form of cytoplasmic inheritance, while dismissing the evidence for its structure—whether particulate or unitary—as inconclusive (W. Ludwig to Nachtsheim, 23.1.48, and Nachtsheim to Ludwig, 30.1.48, Nachtsheim papers).

example, he used the term with neither quotation marks nor explanation (Winkler 1924, 238). Although neither the scientific nor the linguistic context demanded it, subsequent critics of cytoplasmic inheritance repeatedly chose hierarchical images of the cell. Goldschmidt stressed that it was the genes which "direct and guarantee" (*leiten und gewährleisten*) the sequence of development, when he could just as well have said that they "determine" (*bestimmen*) or "specify" (*festlegen*) this sequence.[19] Lehmann could have written simply that it is not the cytoplasm alone which "causes" (*verursacht*) maternal inheritance. Instead he wrote that it is not "the cytoplasm's business alone to decide" [*das Plasma allein [hat] nicht ... zu entscheiden*]" (Lehmann 1928, 20). If the plasmon theory were true, Paula Hertwig argued, "then the gene would be dethroned from its ruling position in development and evolution and forced into a secondary role [*von seinem ... beherrschenden Platz entthront*]."[20]

Throughout the 'twenties and 'thirties these hierarchical relations were expressed through the critics' use of the enzyme-substrate analogy to represent the relations between genes and cytoplasm. This analogy commonly portrayed the cytoplasm as the passive object upon which genes acted. In Goldschmidt's opinion, all geneticists agreed "that the protoplasm is the necessary, specific substrate for the orderly sequence of development, but that it is the chromosomal genes which—given a specific cytoplasmic substrate—direct and guarantee the correct sequence."[21]

Michaelis initially used the substrate analogy, but abandoned it during the 1930s as his view of the plasmon approximated more closely to Correns's.[22] Plasmon theorists frequently objected that cytoplasm was

---

19. Goldschmidt 1933a, 1. For another instance of this metaphor see Hertwig 1927, 22.

20. Hertwig 1934, 427. For other uses of "domination" (*Beherrschung*) see Hertwig 1927, 21; Michaelis and von Dellingshausen 1942, 426; Hämmerling 1934b, 262–64.

21. Goldschmidt 1933a, 1. Nearly three decades later his position on cytoplasmic inheritance was little different (Goldschmidt 1961, 209–13). See also Hämmerling 1929, 59, 61, 65.

22. In the early thirties, when Michaelis still saw the plasmon as merely a modifier of gene action, he commonly referred to it as a "substrate" (1931, 103; 1933b, 365; 1935b, 148). But in 1933, when he first began to acknowledge that Correns's conception (that genes quantitatively modify the plasmon's action) could not be ruled out, he began to portray genes and plasmon for the first time as "partners" (1933b, 393, 396). By 1937, voicing the suspicion that the cytoplasm might be more than "just the necessary substrate for nuclear gene action," he again used the partner metaphor (1937, 284 and 287, re-

not "mere" substrate, and proposed a variety of alternative analogies (Oehlkers 1931, 230; Renner and Kupper 1921, 202; von Wettstein 1928b, 48; 1934a, 36). In their earliest statements of the plasmon theory, as we have seen, Correns and von Wettstein opted for mechanical alternatives to the enzyme-substrate analogy: the machine gun and the airplane. Social analogies can also be found in these early papers; genes and cytoplasm were said to be "of equal birth" (*ebenbürtig*) or "equally entitled" (*gleichberechtigt*) (von Wettstein 1928a, 6, 194, 196, 202; see also 1937, 360). But thereafter social analogies largely displaced mechanical ones. Although brief and unelaborated, they always emphasized the egalitarian nature of the gene-cytoplasm relation. Genes and plasmon were described as "partners" or "opponents" (*Gegenspieler*) (von Wettstein 1928a, passim; 1934a, 36) which were "of equal value" (*gleichwertig*) (von Wettstein 1937, 360; Renner 1934, 264) and which in some cases reached a "compromise" in the determination of a trait (*gleich beteiligt zum Kompromisse kommen*) (von Wettstein 1928a, 195). *Kernmonopol* was rejected on the grounds that it placed the genetic "initiative" with the genes (Oehlkers 1949, 99). In one particular strain, von Wettstein remarked, "the gene is indifferent [*gleichgültig*] and the decision [over the characteristics of the phenotype] lies solely with the plasmon [*die Entscheidung allein dem Plasmon zufällt*]."[23] Renner insisted that the chloroplasts were not "subjugated to the commands" of the nucleus (*der Botmässigkeit des Zellkerns ... unterworfen*). If the "harmony" of the cell was to be maintained, genes, chloroplasts, and plasmon "have to take account of each other" (*haben ... aufeinander Rücksicht zu nehmen*) (Renner 1934, 263, 265). Correns went further, declaring that "the chromosomes are in no sense rulers [*Beherrscher*] of the cytoplasm's work; they render it assistance" (*sie sind seine Hilfsmittel*) (Correns 1937, 128, citing V. Grégoire).

The most elaborate social analogy was developed by Michaelis, who preferred it to Correns's machine gun or von Wettstein's airplane:

One could compare the organization of the cell with a large army. The cytoplasm would be represented by the large number of squads, the genome by the

---

spectively), and in the early forties—by which time he had accepted Correns's conception—"partner" had replaced "substrate" altogether (1940a, 188, 219; Michaelis and von Dellingshausen 1942, 419, 426).

23. Von Wettstein 1928a, 194. Once again, he could just as easily have written "the phenotype is determined [*bestimmt*] by the plasmon" or "the effect of the plasmon predominates over that of the gene" (*die Wirkung des Plasmons über die des Gens überwiegt*).

334   Styles of Thought

general-staff. As in the division of the entire army into individual sections, where squads are distributed more or less randomly to various corps or regiments while the generals are assigned according to a well-thought-out plan, so also are cytoplasm and nucleus distributed in cell-division. When the relative importance of the constituent groups is assessed, it is clear that the most important role, in battles as well as trait-formation, is played by the general-staff and genome, respectively.

So far the model looks quite like *Kernmonopol*. But at the time he wrote this paper, Michaelis was beginning to move closer to Correns's view of the plasmon. Accordingly, as this passage continues, we see Michaelis steering a path between von Wettstein's egalitarian model on the one hand and *Kernmonopol*'s strict subjugation of the cytoplasm on the other:

> But neither the squads nor the cytoplasm are simply a machine which mechanically carries out all commands of the superior authority; they are rather an entirely specific and . . . independent part of the whole, and the full use of all capacities is only possible through harmonious cooperation. Just as it is difficult to decide after an important battle whether soldiers or officers are responsible for the success, it may also be difficult to assess the relative importance of cytoplasm and nucleus. . . . Just as a group of officers, placed in an army of different racial composition, may not be able to operate effectively, due to the specific characteristics of the squads, and will be inferior in battle to an organically constituted army, so also the transfer of a genome from one species into the cytoplasm of another will lead to inhibition and disrupted development since cytoplasm and nucleus, like squads and officers, each possess their specific and independent properties [Michaelis 1933b, 407–8].

This military analogy, Michaelis believed, could be usefully extended to a number of other phenomena, including the processes by which genes and plasmon mutate.

The frequency with which social analogies were deployed should not be exaggerated. Apart from von Wettstein and Michaelis, most participants in the debate seem to have used such language in only one or two publications. But on those occasions where recognizable analogies of any kind are used to portray nuclear-cytoplasmic relations, social analogies greatly outnumber mechanical ones and are about as common as the enzyme-substrate analogy. Clearly this pattern of usage was in no sense constrained by the available vocabulary: Language less politically laden than "of equal birth," etc., could just as easily have been used. It would appear, therefore, that social analogies had a particular appeal. That is not to say, of course, that this preference was conscious. Indeed, the fact that social analogies were generally used without quotation marks—and as metaphors rather than as similes—suggests just the opposite. Because all parties to the debate deployed social analogies, the methodological legitimacy of this practice was almost never called into question. In the only challenge known to me, von Wettstein picked up Nachtsheim's analogy of the genes as architect and gave it a subversive twist:

> If the cytoplasm has often been described as building material and the genes as architect, one can even perhaps agree with that portrayal in the sense that the form of Gothic buildings and the shape of Egyptian Syenit-statues owe as much to the stone used as to the architect. How often has the artist had to adapt to his materials! [von Wettstein 1928a, 196].

But what prompted the choice of social analogy? Since German geneticists deployed these analogies during a period of political conflict, some might suppose that this usage was merely an aberration peculiar to German society in crisis. The fact that the Germans' use of social analogies seems to have peaked in 1933 and 1934 looks consistent with this interpretation (e.g., Goldschmidt 1933a; Michaelis 1933b; Renner 1934; Hertwig 1934; von Wettstein 1934a). But in fact the Germans were not the only ones to use social analogies in representing the cell's organization. As early as 1917 L. C. Dunn had rejected cytoplasmic inheritance in favor of "the governance of the chromosomes over development" (Dunn 1917, 299). And in the 1930s the embryologist E. E. Just characterized the cell as a federal polity, consisting of a self-organizing populace (the cytoplasm) and a government (the nucleus) whose task was to remove obstacles to popular action (Gilbert 1988, 332–34). In response to the evidence for cytoplasmic inheritance which was accumulating in the United States during the 1940s, George Beadle wrote that one particular gene's function was to "supervise" the replication of cytoplasmic particles, and that cytoplasm was complementary to the nucleus: "Each has its specific duties in the overall economy of the organism."[24] Other English-speaking geneticists chose explicitly political analogies. Nuclear genes, Cyril Darlington wrote, "predominate in the government of heredity as well as in the government of the cell" (Darlington 1944, 164). Tracy Sonneborn shifted the emphasis: "The gene is not in exclusive control of heredity. It operates not in an ivory tower but in the organic unity of the cell, and the rest of the cell has some voice in the directions that are given" (Sonneborn 1950a, 39). His student, D. L. Nanney, developed the analogy further. *Kernmonopol* placed "master molecules" in the chromosomes:

24. Beadle 1949, 223; cf. 230. For genes as "supervisors" see Beadle 1948, 727.

All other cellular constituents are considered relatively inconsequential except as obedient servants of the masters. . . . [A preferable view of heredity] . . . contains the notion of checks and balances in a system of biochemical reactions. In contrast to the totalitarian government by "master molecules," [this alternative] government is a more democratic organisation. . . . It appears unlikely that the role of genes in development is to be understood so long as the genes are considered as dictatorial elements in the cellular economy [Nanney 1957, 136, 162].

Perceptions of the cell's resemblance to society, therefore, were by no means peculiar to Germany nor to the 1930s.

Since the German analogies discussed above appeared in research papers rather than in textbooks or in lectures for a lay audience, their use cannot be written off as merely a didactic or rhetorical device, bearing little relation to the way in which geneticists actually conceptualized the cellular order. Whether these analogies were chosen for their heuristic potential, however, also seems doubtful. Michaelis clearly felt that the military analogy held such promise, and he took the trouble to elaborate it in some detail. But it cannot be overlooked that all other social analogies remained mere words or phrases; their users seem to have been unconcerned to develop them as research instruments.[25] There may be good reason for this. Perhaps there were so few social analogies in the research literature after 1934 because articulating them at that time would have required explicit discussion of political systems which might have attracted unwelcome attention. Many of the participants to this debate would have been keen to avoid such attention. As victims of the civil service law of April 1933, Stern and Jollos emigrated and were soon free of Nazi persecution, but once in the United States their research interests shifted to other problems. Goldschmidt remained until 1936 but had to keep a low profile, as did Oehlkers because his wife was Jewish. Hertwig's and Hämmerling's political sympathies made them vulnerable after 1933, and Nachtsheim had declined to join any National Socialist organizations. Michaelis was working in an institute full of Nazi-sympathisers, and Renner and von Wettstein had already acquired reputations for fighting political appointments in their respective institutions. Lehmann, on the other hand, was in credit with the Party, but that ran out abruptly in 1938. This left only Schwemmle, but

25. The distinction in German between *Metapher* and *Analogie* is useful here; the former denotes a casual perception of similarity of indeterminate scope, while the latter refers to the systematic attempt to pin down similarity relations. I thank W. F. Kümmel for pointing this out to me.

of all the geneticists interested in cytoplasmic inheritance, Schwemmle was the least inclined to theorize.

Nonetheless, even undeveloped analogies may have been useful as a form of conceptual shorthand, enabling geneticists to organize existing data on nuclear-cytoplasmic relations into a system which could be readily visualized and communicated. Under more favorable political circumstances, such a shorthand might have provided the raw material for more elaborated, heuristically promising models. But once the Nazi threat had subsided, German research on cytoplasmic inheritance declined. Correns and von Wettstein were dead by 1945, and by 1950 Schwemmle, Oehlkers, and Renner had largely abandoned this field, leaving only Michaelis to carry on the tradition. If one wanted to assess social analogies' heuristic potential, therefore, one would have to look at their use in other countries, such as France and the United States, where Ephrussi and Sonneborn made major breakthroughs (Sapp 1987).

9.4 The Cell as Political Microcosm

Whatever the function of these analogies may have been, we are left with explaining why participants chose such *different* social analogies to represent nuclear-cytoplasmic relations. Since advocates of cytoplasmic inheritance stressed "equality" while critics referred to "domination," the obvious explanation is that geneticists perceived the cell as some kind of political battleground and that advocates and critics endorsed different political ideals. But if so, what kinds of political issues were perceived to be at stake in the cell?

Some have suggested that the differing conceptions of nuclear-cytoplasmic relations reflected different views of gender relations, where the nucleus represented the male and the cytoplasm, the female (Biology and Gender Study Group 1989). While this explanation may make sense of debates over cytoplasmic inheritance at other times and places, I find no evidence for it in the German debate. Sapp argues that the political language used to describe nuclear-cytoplasmic relations reflected the users' perceptions of the power relations between embryologists and geneticists, as well as between nuclear and cytoplasmic geneticists, but offers no evidence for this claim (Sapp 1987, xv, 100, 116, 197, 228, 234). Alternatively, references to "democracy" and "totalitarianism" in some American discussions of cytoplasmic inheritance, as well as the egalitarian language frequently used by orthodox plasmon theorists, might seem to suggest that participants perceived this as a debate over the distribution of power between rulers (nucleus) and ruled (cytoplasm). Gilbert presents evidence along these lines, arguing that the black American embryologist E. E. Just's emphasis upon the embryological importance of the cytoplasm mirrored his defense of social and political equality for blacks, while Goldschmidt's advocacy of *Kernmonopol* reflected his monarchism (Gilbert 1988). One difficulty with this thesis is that on occasions Just seems to have been notably more sympathetic to monarchism than to equality (Manning 1983, 191–93), and Goldschmidt's monarchist views, as we have seen, were shared by plasmon theorists such as Renner and von Wettstein. But a more serious difficulty with this theory is that if it were generally correct, one would expect orthodox plasmon theorists to be keener democrats than were the critics. As we saw in chapter 7, however, this was not true. If anything, the reverse was the case: Hämmerling was favorably disposed toward social democracy, and Stern and Hertwig were liberals. No comparable degree of left-liberal engagement is known for any of the plasmon theorists, and von Wettstein and Renner were conservative nationalists. Moreover, on reflection it is clear that the plasmon theorists' concern was not to challenge the concept of a hierarchical relationship between the cell's genetic elements and its *cytoplasm*. That kind of radical challenge was more likely to be mounted by embryologists. Instead, the plasmon theorists sought to widen the definition of what comprised the cell's *genetic elements,* while leaving intact the view that the *non*genetic portion of the cytoplasm played an essentially passive role in development. Translated into the political idiom, this hardly constituted a call to redistribute power between elite and masses, but merely to broaden membership of the elite.

This points toward a third interpretation which better accords with the political outlook of participants in the debate. Although plasmon theorists and critics cannot be distinguished along a conventional left-right political spectrum, what does distinguish the two groups is their willingness to embrace the traditional mandarin role. From tables 7.2 and 7.3 it is evident that all of the orthodox plasmon theorists (and some of the revisionists) are identifiable as comprehensive thinkers who endorsed the conventional mandarin role. The critics of cytoplasmic inheritance, in contrast, are predominantly pragmatic thinkers and "outsiders" to the academic mainstream: Stern, Nachtsheim, and Hertwig. As the son of a surveyor, a member of right-wing radical and anti-Semitic organizations from the 1920s, and a propagandist for the Nazi cause among biologists from 1933—all of which were highly unusual among professors of his generation—the critic Lehmann was also clearly an outsider.[26] According to this interpretation, therefore, the participants' concern with the relationship between the nucleus and alleged genetic elements in the cytoplasm reflected their views on the role of academics in politics. The nucleus was perceived as the cell's government, while cytoplasmic heredity served as the cell's mandarinate. Just as the mandarinate called repeatedly for intellectual synthesis, both to stem the fragmenting effects of specialization inside the university and to provide the ideological glue for a reconstituted *Gemeinschaft* outside it, so also did orthodox plasmon theorists picture the plasmon as a holistic structure which would coordinate the actions of individual genes in the developmental process. The debate over cytoplasmic inheritance, like its political counterpart, focused upon the problem of order. Would the exercise of political power alone be sufficient, or would social harmony also require a new *Weltanschauung*, which mandarins were so keen to provide? Should politicians alone shape the development of German society, or only in collaboration with academics?

However simplistic this "microcosm theory" might seem, it is remarkably consistent with several features of the debate. For example, harmony was a central issue for plasmon theorists, as for mandarins more generally. Within the cell, genes, as autonomous determinants, were regarded as incapable (on their own) of creating and sustaining the orderly development of the organism. From the mandarins' point of view, similarly, different political parties in a coalition government (of which there were many during the Weimar Republic) were incapable of carrying out an integrated package of policies which would be in the best interests of the nation. Consider the role which von Wettstein assigned to the plasmon in evolution: as a mediator which reestablished organismic harmony in the aftermath of disruptive random gene mutations (chapter 3). One is reminded of a standard critique of democratic systems: that elections produce sudden shifts of government

---

26. On Lehmann see *Wer Ist's?* 1935; Adam 1977, 30–31. For an example of his public support for National Socialism after 1933, see Lehmann 1936b. In a (vain) attempt to gain admission to the Party after having been ousted from various National Socialist organizations in 1938, Lehmann claimed to have been "practically the only Tübingen professor who took part in every Nazi gathering" (Lehmann to Gaugeschäftsführer Baumert, n.d., Lehmann file, Berlin Document Center).

As we saw in chapter 8, Hämmerling is a borderline case: difficult to classify as either outsider or mandarin. As critics but clear-cut comprehensive thinkers, however, Goldschmidt and Jollos are anomalous for my theory and will be discussed below.

which undermine the possibility of long-term consistency or efficacy in policy-making. The plasmon and the mandarinate were to foster harmony between different sections of the cellular and political systems. And its stability in the presence of a foreign nucleus meant that the plasmon was an independent genetic force within the cell, reflecting the mandarins' self-understanding as a force "above" party politics.[27]

Although the critics of cytoplasmic inheritance acknowledged that the Mendelian chromosome theory had as yet nothing to say about the processes by which genes specified the temporal and spatial order of development, the maintenance of harmony within the cell was simply not an issue for them. In the political arena, analogously, the never-ending clash of interests in the *Reichstag* was not likely to be a source of alarm for outsider academics (at least for the democrats among them), but merely an inevitable feature of a pluralist social order. Moreover, from their point of view, power sharing between nucleus and plasmon must have looked uncomfortably like the political role claimed by the mandarinate. Since most of the Kernmonopolists were democrats, they may have been understandably wary of giving academics—the great majority of whom were hostile to the republic—a share in the governing process. The critics' unease about the political connotations of the plasmon theory would have been magnified by the fact that geneticists in the United States—in the eyes of most German academics, the epitome (for better or worse) of a modern and democratic country—were overwhelmingly critical of the claims for cytoplasmic inheritance. At first sight, Lehmann, as an enthusiastic National Socialist, appears to be the odd one out. But Nazis, too, insisted that academics should defer totally to the new regime in the process of *Gleichschaltung* (forced alignment). While Nazis and democrats had utterly different conceptions of how political power was legitimately *acquired*, therefore, they agreed nonetheless that there was no role for the German mandarinate in the governing process. (Returning to the level of the cell, all critics of cytoplasmic inheritance accordingly emphasized the instability of cytoplasmic effects in the presence of a foreign nucleus.)

Although this interpretation of the debate accounts reasonably well for the positions taken by orthodox plasmon theorists and critics, the

---

27. The reliance upon species hybrids to demonstrate cytoplasmic inheritance may have encouraged comprehensives, at least, to see the cell as a political microcosm: The coexistence in the hybrid cell of the cytoplasm from one species with what was routinely called a "foreign" nucleus from another may have resonated with most mandarins' alienation from Weimar governments.

revisionists look like a troublesome category. As we saw in section 2, they deviated from the orthodox position in a variety of ways, yet two of them (Renner and Michaelis) were comprehensives. This apparent anomaly dissolves, however, once one recognizes that no social stratum is altogether homogeneous. Although classifiable as comprehensives, Renner and Michaelis were in fact marginal figures within the mandarinate in some respects, and from that point of view their deviation from the orthodox plasmon theory makes sense.

Consider, for example, Otto Renner, who had little to say about the nature of the plasmon except that he assumed it would eventually be revealed as a particulate structure. At first sight his breadth of botanical knowledge, his demonstrable *Bildung,* and his aversion to "politics" clearly identify him as a mandarin. When one looks more closely, however, it is evident that some of his intellectual predilections distanced him from the mainstream of the mandarin tradition. For Renner had no time for holism and was generally suspicious of metaphysics. His biographers stress his critical frame of mind, the emphasis he placed on experiment rather than interpretation, and his brusque rejection of "speculation" or "mystical" points of view.[28] Sensing that the last of these was gaining ground in Germany after 1933, he spoke out bluntly: "There are those in Germany today who would like to block the use of chemical and physical approaches in biology and who prefer a 'holistic perspective' to the analytical one [*das Unterscheiden*] which has made the exact sciences great" (Renner 1935, 179). But whoever expects genetics to make a practical contribution, he continued, will want to make sure that biology remains analytical.

Renner's aversion to holism also comes across in his assessment of Goethe's contributions to botany. While paying tribute to Goethe's skill as a naturalist and his physiological approach to plant development, Renner was skeptical about Goethe's search for an underlying unity in nature. To be sure, these "mystical" and "obscure" (*verklärende*) notions made Goethe a poet of genius, but they detracted from his stature as a scientist (Renner 1949, 107–8, 114–15, 120). How different was Alfred Kühn's evaluation of Goethe (cf. chap. 7)! Renner enjoyed reading Kühn on Goethe, but confessed, "I cannot be as tolerant of romantic *Naturphilosophie* as you are."[29]

28. Oehlkers 1961; Stubbe 1962; von Frisch 1960–61; Bünning 1944; Kühn 1960a; Butterfass 1961; Mägdefrau 1961; see also Renner's *Antrittsvorlesung* at Munich in 1948 (Renner 1961a).

29. Renner to Kühn, 25.10.55, Kühn papers, Heidelberg.

Renner's methodological credo as a scientist is spelled out in the speech he gave upon assuming the chair of botany at Munich in 1948. While accepting the harmony of the organism, he insisted that this was better explained in terms of the mechanistic relations among the parts than by invoking some kind of formative holistic force.[30] He likened the task of the geneticist to that of the chemist who analyzes compounds in terms of constituent elements (Renner 1930, 6–7); his test strains were "useful reagents" (1922, 237). This inclination to mechanistic reductionism nicely corresponds to the way in which he dealt with cytoplasmic inheritance: The emphasis was almost entirely upon chloroplasts as a particulate form of cytoplasmic inheritance rather than upon von Wettstein's holistic plasmon.

That Renner could not accept the holistic assumptions so common among his professorial colleagues may well have been simply an idiosyncratic feature of his personality. Without an exhaustive biography of him, we cannot be sure. Another explanation, however, needs to be considered, if not for Renner as an individual, then at least for that category of academics of similar background. For Renner was a Catholic from a lower-middle-class family. Although educated at a classical grammar school, he would not have been able to afford to attend a university had not his brothers (both lawyers) agreed to finance him (von Frisch 1960–61, 131; Butterfass 1961). Thus Renner grew up in a family which was materially deprived by mandarin standards and in a religious denomination whose members had been singled out for attack by Bismarck. This background, I suggest, may be one of the reasons why Renner's embrace of the mandarin tradition (and thus of the orthodox plasmon theory) was less than total.

Like Renner, Julius Schwemmle seems to have been favorably disposed to the Correns/von Wettstein conception in principle, but had very little to say about its structure or function and declined to comment on the scope of its action. As we have seen, Schwemmle seems to have felt no need to relate his results to the major theoretical issues concerning development or evolution and was extraordinarily reluctant to make generalizations, even when they were confined to the properties of cytoplasmic inheritance. One reason for this rather timorous, unmandarin attitude toward theorizing may lie in the fact that, because he was the son of a primary school teacher, a university education was neither socially nor financially straightforward for him. His social marginality to the educated middle class helps to explain why

---

30. Reprinted as Renner 1961a, 110–11.

Schwemmle—unlike most professors—joined a Party organization. Although "anything but a Nazi," he evidently hoped the SA would ameliorate the university's ivory-tower elitism.[31]

If Schwemmle's and Renner's location within the professoriate was that of the ambivalent newcomer to the mandarin tradition, Michaelis's was that of a born mandarin who moved steadily away from that tradition over the course of his career. The son of a portrait painter and a philosophy professor's daughter whose salon in Munich attracted prominent figures, Michaelis grew up in cultivated surroundings, developing talent as a pianist and artist as well as interests in art and literature.[32] Despite this typically mandarin background, however, Michaelis was sent to a modern secondary school. Later, after an unhappy period as Renner's assistant at Jena, he spent the years 1927–33 at Stuttgart's *Technische Hochschule* before joining Baur at the KWI for Breeding Research. Although this move entailed abandoning his attempt to habilitate in Stuttgart, there is no evidence that Michaelis was especially interested in a conventional university career. At the KWI for Breeding Research he enjoyed rapid promotion, becoming a department head in 1935 (with four assistants under him by 1939–40) and a tenured member of the Kaiser-Wilhelm Society (*Wissenschaftliches Mitglied*) in 1941.[33] This position allowed him essentially to devote forty years to one phenomenon in one class of organisms. Specialization on this scale may well have spoiled his chances of getting a chair or even just the title of professor.[34]

---

31. The quote is from Hämmerling to Stern, 20.5.46 (similarly, Hämmerling to Stern, 3.2.47), Stern papers. Schwemmle explained his decision to join the SA in a letter to Stern, 4.4.47, Stern papers. The Party seems to have appreciated Schwemmle's anti-elitist attitude. When asked about his political suitability for the chair of genetics at the German University in Prague, the functionary responsible declared no objections, observing that "he is an open and honorable man devoid of social snobbery [*der keinerlei Standesdünkel besitzt*]" (Sicherheitsdienst des Reichsführers SS to Gauleitung Franken, 24.1.41, Schwemmle file, Berlin Document Center).

32. W. Stubbe 1987; Rothe interview; Kuckuck interview.

33. Rothe interview; W. Stubbe 1987.

34. In 1954 the director of the (by then renamed) Max-Planck Institute for Breeding Research took up references to establish whether Michaelis should be given the professorial title at the University of Cologne. While the American referees (Stern, Caspari, Sonneborn, M. Demerec, and Franz Schrader) were unequivocally favorable, most of the Germans' recommendations (as well as Boris Ephrussi's) were ambivalent. Apart from several referees' doubts about the correctness of Michaelis' theoretical interpretations (voiced by H. Stubbe, Renner, and Oehlkers), there was concern that his research area was so restricted. Nearly all German referees noted the "remarkable perseverance" with which Michaelis had pursued a single problem over a thirty-year period. Observing that Michaelis enjoyed a reputation among foreign geneticists, Oehlkers remarked that "the

But after the war Michaelis seems to have had little interest in a university appointment. At one point he was on the verge of receiving a call but was unenthusiastic, since he felt it would take him away from his research. During the 1950s he did some teaching in genetics and general botany at the neighboring University of Cologne, but found it time-consuming and unrewarding.[35] Michaelis's career trajectory can thus be represented as a progressive shift away from the mandarin mainstream of the professoriate toward the periphery. The revisionists' deviation as a group from the orthodox view of the plasmon theory, therefore, corresponds to their marginality with respect to the mandarin tradition, thus lending support to the microcosm theory.

One of the problems faced by any sociological treatment of theory choice is that the number of scientists whose choices can be studied is usually rather small. The *audience* for theory choice, however, goes beyond those relatively few individuals who are actively involved in any given debate, thus enabling us to test the microcosm theory further. There are two such wider audiences in the debate over cytoplasmic inheritance whose reception of the plasmon theory would be of interest: (a) those geneticists on the periphery of the debate who were sufficiently attracted to make an isolated contribution but who otherwise devoted little or none of their research to cytoplasmic inheritance, and (b) biologists further afield who had nothing to do with cytoplasmic inheritance but who expressed an interest in the outcome of the debate.

One geneticist in the former category is Alfred Kühn, who publicized the plasmon theory through many editions of his textbooks but made only one empirical contribution to the debate. As we have seen, Kühn's holistic sympathies are visible in his admiration for Goethe, his striving for an integrative perspective in science, and his interest in form, whether in art or in the patterning of a flour moth's wing. To what extent, however, are these sympathies evident in his choice of concept or theory? In most respects Kühn's work on developmental genetics reveals him to be an unexceptional reductionist (chapter 2). Organ formation was complex, but by using mutants in relevant genes, one could break it down into constituent processes (Kühn 1934b). The

---

Americans in particular have no objections to narrowness [*Einseitigkeit*]." Richard Harder seems to have tried to counteract the impression of Michaelis's narrowness by devoting as much attention in his letter of recommendation to Michaelis's five early ecological papers as he did to the one hundred later papers on cytoplasmic inheritance which made Michaelis's reputation. In the event, the professorial title was not conferred. (I thank Prof. Georg Michaelis for allowing me to see this correspondence.)

35. Rothe interview; Helga Völkl to me, 3.5.88.

search for the physicochemical basis of the induction process, similarly, he felt to be entirely reasonable (Kühn 1934a, 40). Nor was he shy about describing development as a "mechanical" process (e.g., Kühn 1934b, 370; 1937a, 452). This "segmented" character of Kühn's thought should not be surprising. No scientist of holistic persuasion is going to deny that *some* problems are more fruitfully approached in reductionist fashion (any more than a reductionist would deny the existence of apparently emergent phenomena whose properties cannot be deduced from current knowledge of their parts). But while all theories are underdetermined to some extent by evidence, ontological preferences are likely to be decisive where—as in the case of cytoplasmic inheritance— there is a relatively large amount of latitude for alternative interpretations. And when it came to the plasmon theory, Kühn was reluctant to define the plasmon simply as an aggregate of particles. In his only empirical paper on cytoplasmic inheritance, he defined it as "the genetic element" in the cytoplasm (Kühn 1927, 418). His only other published reference to the structure of the plasmon was more explicit:

> There is no comparable way to analyze the plasmon [like segregation analysis of Mendelian genes]. The cytoplasm always appears as a uniform... mass. We are aware of no process through which individual elements in the cytoplasm could be separated as genes are in meiosis. If the plasmon is made up of constituents related to the formation of particular traits, they should not be thought of as individual *particles* like genes but as individual *substances* which are distributed throughout the cytoplasm.[36]

This passage suggests that Kühn's conception of the plasmon was far closer to von Wettstein's than to Michaelis's, a conclusion borne out by my discussions of this issue with several of Kühn's students.[37] Therefore, insofar as Kühn was a mandarin who endorsed the orthodox conception of the plasmon, his position accords with the microcosm theory.

Kühn's counterpart among the revisionists is perhaps Edgar Knapp. An assistant with Correns and then von Wettstein at the KWI for Biology from 1932, Knapp moved in 1936 to the KWI for Breeding Research to head its Department of Mutation Research. After publishing in various areas of botany between 1930 and 1933, he began to

---

36. Kühn 1937a, 444; emphasis in original. A very similar passage appears in his 1938a, 108.
37. Viktor Schwartz to me, September 1984; Hans Piepho to me, 2.10.82; Georg Melchers to me, 9.11.82. Although a student and assistant of von Wettstein's in Göttingen and Berlin, Melchers was in close contact with Kühn from 1943, when the heavy bombing of Berlin forced much of the work at the KWI for Biology to be transferred to Hechingen.

concentrate on cytogenetics and mutation genetics at the KWI for Biology. To my knowledge, none of his work from then until war's end addressed either developmental or evolutionary issues.[38] Moreover, since his biographers claim neither scientific breadth nor cultural interests for him, he is best classified as a geneticist of the pragmatic variety. As the son of a primary school teacher and graduate of a modern secondary school, furthermore, he was a typical outsider. His sole contribution to the literature on cytoplasmic inheritance was a review paper written at the KWI for Breeding Research, where he became acquainted with the work of Michaelis (Knapp 1938b). After considering the merits of both particulate and holistic models of the plasmon, Knapp concluded that there was no need to draw a distinction between stable forms of maternal inheritance (e.g., cytoplasmic inheritance) and unstable ones (e.g., predetermination or dauermodifications), since both could be accommodated by either model under appropriate conditions. Except for the fact that he declined to say whether he thought the holistic or particulate model of the plasmon to be more likely, Knapp's position was virtually identical to that which Michaelis had held since 1935. As the microcosm theory predicts, therefore, Knapp's revisionist position in the debate corresponds to his distance from the mandarin tradition.

Erwin Baur's relationship to the debate is rather similar to Knapp's. Apart from an early classic paper on plastids' role in variegation (cf. chap. 2), which provided the model for some of Renner's and Michaelis's subsequent work, Baur rarely discussed the plasmon theory, either in print or in private.[39] An outsider in every respect, he fits well in the revisionist category alongside Knapp and Schwemmle.[40]

The second audience for the debate over cytoplasmic inheritance—passive but interested onlookers—constitutes an enormous and largely unexplored territory. Were one to study this group systematically, one might begin by looking first for those individuals who explicitly de-

38. This claim is based on Knapp 1935a; 1935b; 1936a; 1936b; 1937; 1938a; 1944.
39. According to his student and assistant Hans Stubbe, Baur never expressed a view on the plasmon theories of Correns, von Wettstein, or Michaelis between 1927 and his death in 1933 (Stubbe to me, 20.12.82). The only occasion, to my knowledge, on which Baur publicly expressed a view on the plasmon theory was in the 1930 edition of his genetics textbook. While he used the term *plasmon* and praised von Wettstein's work, it is not at all clear that he was using the term in von Wettstein's sense (1911, [1930] 226, 236, 262–67, 269).
40. Hans Winkler and Richard Harder also had a peripheral involvement in cytoplasmic inheritance (cf. chap. 2), and might thus be useful in testing the microcosm theory.

fended either reductionism or holism in biology, and then checking to see which of them expressed a view on the plasmon theory. So far I have examined only one instance of this kind: the holist Richard Woltereck. Although he believed the fundamental hereditary substance ("matrix"; cf. chap. 3) to be distributed throughout the nucleus *and* cytoplasm, its holistic structure and its function—to impose order on genes' activity—closely resembled von Wettstein's concept of the plasmon.[41] I can find no substantial discussion of cytoplasmic inheritance in his work, but it is worth noting that the only work in this field which he cites is von Wettstein's, and at one point he suggests that the matrix consists of "genome" and "plasmon."[42] Finally, his remark that the matrix is essential in order to "direct" (*dirigieren*, normally used in orchestral or managerial contexts) gene activity accords well with the social analogies used by orthodox plasmon theorists.[43]

9.5 Conclusion

In this chapter I have argued that the conflicting interpretations of cytoplasmic inheritance advanced in this debate by plasmon theorists, revisionists, and critics were empirically underdetermined. Each group's standpoint can only be explained, therefore, in terms of its willingness to make particular background assumptions. The widespread use by participants of social analogies in representing the relations between nucleus and cytoplasm suggests that these background assumptions were ultimately political in derivation. Consistent with this microcosm theory is the fact that the three groups differ in their academic self-understanding in ways which correspond to their conceptions of nuclear-cytoplasmic relations. In the dispute between Kernmonopolists

---

41. Woltereck 1931b, 307; 1934, 190. Indeed, Hertwig pointed out the resemblance between Woltereck's and von Wettstein's interpretations (1934, 427).

42. See the authors cited in Woltereck's 1931b and 1932. That he would not have been happy with Michaelis's particulate view of the plasmon seems clear from Woltereck 1924, 297.

43. Woltereck 1931b, 276. His judgments on the theory of natural selection are also sometimes expressed in a political idiom. At one point he dismisses the theory because its account of the emergence of new traits represents a "petit-bourgeois shift of emphasis from the creative toward the useful" (1931a, 237). Later in the same paper he criticizes the idea of random mutation as a source of variability which is undirected by the environment: "Plants and animals, past and present, did not emerge as rootless cosmopolites [*beziehungslose Kosmopoliten*] but were shaped a thousandfold by all conceivable habitats" (250).

and orthodox plasmon theorists, the theory explains why the two sides interpreted the evidence from nonreciprocal species hybrids differently and chose contrasting social analogies in order to represent nuclear-cytoplasmic relations. In the dispute *among* advocates of cytoplasmic inheritance over the plasmon's properties, the microcosm theory explains why revisionists were more hesitant than orthodox plasmon theorists to make general claims about the plasmon's function, stability, and scope, or to assign a holistic structure to the plasmon.

Obviously the fact that I have grouped these geneticists taking part in the debate in three categories does not mean that all members of a given category held identical views on cytoplasmic inheritance. On the contrary, there is individual variation within each category—especially among revisionists—which the microcosm theory does not claim to explain. As with any sociological theory of knowledge, the microcosm theory attempts only to account for common tendencies in the cognitive stances adopted by a group. In order to do this, it has to focus heavily on a small number of exemplary individuals *without* thereby claiming to explain each *individual*'s stance. In seeking to understand the peculiarities of an individual's thought, therefore, the appropriate level of analysis is biography, not social or institutional history.

Notwithstanding this caveat, the microcosm theory is not without its problems. For one thing, both the revisionists and the critics seem to be socially heterogeneous. As it stands, for example, the revisionist category includes not just figures marginal to the mandarin mainstream like Renner or Michaelis, but also outsiders like Schwemmle (or Knapp and Baur). Why these latter (but not other outsiders, such as Stern, Nachtsheim, or Hertwig) should have adopted a revisionist position is thus unclear. Furthermore, the group of critics includes not only outsiders but also two apparent mandarins (Goldschmidt and Jollos). Whether these anomalous individuals might be reclassifiable if we had more detailed information about their academic role conceptions, or whether they would become less significant with larger sample sizes, is impossible to say at present.

More generally, the one-to-one correspondence between natural and social order which the microcosm theory posits is vaguely worrying. Any student of the sociology of knowledge knows that thought almost never emerges from social life in such an unmediated fashion. How can such a simple theory possibly be true? On the other hand, as we saw in section 3, for the scientist struggling to understand a complex natural system, social structures *are* sometimes complicated enough to provide useful models. And if late-nineteenth-century German anatomists were conceptualizing *multi*cellular organization as a "cell-state," it is hardly surprising that others would soon apply related metaphors to *intracellular* organization. If cognitive structure rarely reflects social structure in simple fashion, perhaps the exceptions occur during historical periods when intellectuals are especially preoccupied with the problem of social order and, above all, their own place in it. In Germany, for example, this might explain not only why theories of the cell-state were common in the late nineteenth century, but also why the vast majority of social analogies in the debate over cytoplasmic inheritance were used between 1928 and 1934. In the American context, similarly, it might explain why, in the midst of the depression, Walter B. Cannon drew upon social analogies in portraying the homeostatic properties of the human body. His solution to the social crisis of the time, conversely, called for a series of regulatory agencies—manned by experts—which would stabilize the economy, just as the autonomic nervous system maintained the "organic economy" (Cross and Albury 1987).

In the final analysis the microcosm theory will stand or fall on its ability to cope with new material. At present the theory applies only to a historically specific social context. While social analogies were also deployed by geneticists elsewhere, it is an open question whether some version of the microcosm theory could account for subsequent disputes over cytoplasmic inheritance in the United States or France. If, as seems likely, the vast majority of American academics at this time would have been better described as outsiders than as mandarins, then it is certainly interesting that cytoplasmic inheritance was almost universally rejected by American geneticists before the 1950s. Similarly, if we assume that German mandarins' traditional claim to power sharing would have lost all credibility during the reestablishment of democracy in western Germany after 1945, we might expect corresponding shifts in conceptions of the plasmon. There is in fact some evidence that von Wettstein's version of the theory had lost support by about 1950, not least from Oehlkers, who by then was conceding that some plasmon elements were unstable and that particulate models were promising.[44] Finally, al-

44. After the war Lehmann claimed that von Wettstein's theory had lost advocates (Lehmann to Tracy Sonneborn, 17.7.49, Sonneborn papers). Though Lehmann was an interested party, Ernst Caspari had the same impression (Caspari to me, 5.7.83). In an interview, Gertrud Linnert, one of Oehlkers's students, told me that Oehlkers abandoned the model of the plasmon as a unitary structure after the war, in response to the successful particulate models advanced by Ephrussi and Sonneborn. The discussion of cytoplasmic

though the microcosm theory was designed to explain a debate over the cell's organization, its explanatory scope ought to be far wider. Since it predicts that holistic concepts held greater appeal for mandarins than for outsiders, the theory should be applicable to academic controversies in many German disciplines at this time.

---

inheritance in Oehlkers 1952 (219–20, 233, 242–43) seems to confirm her recollection. By 1950 Kühn, too, was attributing cytoplasmic inheritance to "self-replicating structures . . . ('plasmagenes')" (Kühn 1934e [1950], 100).

Conclusion

The character of the German genetics community after 1900 reflected in microcosm the problems and processes of German modernization. As the pace of industrialization accelerated during the latter half of the nineteenth century, the sons of the newer commercial and industrial strata began to enter higher education in larger numbers. By background and education less committed to the ideals of classical *Bildung* which dominated the nineteenth-century university, they tended to opt for higher education of a more modern, sometimes explicitly vocational, kind: in colleges of engineering, agriculture, or commerce, but gradually, too, in certain university disciplines. Few of those who chose to make a career in the academic community could subscribe to the traditional self-understanding of the "mandarin" whose dedication to pure scholarship and the nation's well-being qualified him as spiritual adviser to the German state. Having little in common with the educated middle class which dominated both the professoriate and the top civil service, these social outsiders embraced the more limited role of "expert," placing their specialized knowledge—rather than wisdom—at the public's disposal.

Although dominated initially by mandarins, the genetics community began to accommodate substantial numbers of outsiders after the First World War. At first the newcomers worked exclusively in institutions of a more modern cast, such as the Berlin Agricultural College. By the 1920s, however, they were also to be found in more traditionally organized research institutions such as the KWI for Biology, and by 1930 a few had acquired chairs in universities. Regardless of institutional location, however, outsiders were inclined to approach matters scientific, cultural, and political from a markedly different perspective than did the mandarins. The latter defined the boundaries of their discipline very broadly, so as to encompass the central problems of biological theory at the turn of the century. In keeping with the *Bildungsideal,* they were not only wary of the trend toward specialization in the sciences, but also aspired—sometimes with remarkable success—to a high degree of cultivation in the arts and humanities. Because of the all-embracing character of the mandarins' interests, combined with their traditional self-understanding as apolitical culture-bearers, I have called their style of thought "comprehensive." The outsiders, in contrast, were altogether indifferent to this ideal, as to the metaphysics upon which it was based. Their narrow conception of the domain of genetics as a discipline, their unself-conscious embrace of specialization, their lack of interest in philosophy or high culture, and their willingness to align themselves with the sectional interests of particular political organizations—all bespeak what I have called the "pragmatic" style of thought.

Since mandarins and outsiders rarely addressed the same genetic problems or competed for the same scarce resources, their differences of perspective seldom came into conflict. The debate over cytoplasmic inheritance, however, was of interest to both factions, especially because the central issue—the nature of the relations between chromosomal genes and the cytoplasm—bore an uncanny resemblance to one of the professoriate's major political preoccupations: their own relationship to elected governments during the Weimar Republic. One particular theory of cytoplasmic inheritance (the plasmon theory) portrayed the relations between nucleus and cytoplasm in a manner similar to the form of political order endorsed by mandarins but shunned by outsiders. Since the available evidence bearing on the theory was singularly inconclusive, mandarins and outsiders tended to adopt opposing positions on the plasmon theory in accord with their political predilections.

That, in summary, is the way in which the concept of "styles of thought" can illuminate the history of genetics in Germany. A few historians have complained that my distinction between comprehensive and pragmatic styles of thought is too general, having nothing specifically to do with genetics. It is precisely this generality, however, which makes the comprehensive and pragmatic styles historiographically interesting. For the concept of style can integrate historical analysis in two respects. On the one hand, it alerts the historian to the possibility that features of scientific thought in a given context may be consonant with fundamental assumptions in other cultural sectors. On the other hand, the search for what Mannheim called a style's "basic intention" prompts the historian to look for connections between intellectual history and social history. In this final chapter I want to demonstrate how the approach to stylistic analysis which I have taken is applicable to a wide range of historical phenomena.

If the styles of thought identifiable within the German genetics community represent the cognitive consequences of modernization, similar styles should be evident in other areas of German scholarship during this period. One area which looks promising in this regard embraces those controversies where the opposing camps were committed, broadly speaking, to instrumentalist versus realist philosophies. In his realist crusade against Machian instrumentalism, Max Planck sought to show that the increasingly unified picture of nature which physicists presented was no mere construct of convenience, but a reflection of the unity of the real world. "The first and foremost characteristic of all scientific research [is] the development of a constant world-picture" (Heilbron 1986, 55; cf. 60–61, 65–66, 113, 157–59). Judging from his "unpolitical" outlook, his fears for national unity, his resistance to the introduction of a section for the technical sciences in the Prussian Academy, and his hope that the Academy and similar organizations would preserve the unity of knowledge against the centrifugal effects of specialty formation, Planck would seem to have been the archetypal mandarin. Mach, in contrast, was the outsider; suspicious of metaphysics, he rejected religion and nationalism as reactionary forces and played an active role in the movement for the reform of secondary schools and the extension of popular adult education.[1]

The controversy in Germany over the foundations of mathematics may have had similar political underpinnings. As Herbert Mehrtens has shown, twentieth-century advocates of "modern" mathematics, such as David Hilbert and Felix Hausdorff, resisted attempts to root the validity of mathematical statements in either a real or a metaphysical world (Mehrtens 1990). For Hilbert, "success" (in the form of logical consistency) was the sole judge of truth. Opponents of modernism, such as Henri Poincaré, Felix Klein, and L. E. J. Brouwer, argued that a mathematical truth was one which resonated with the structure of the human mind. Only the mathematician's intuition, therefore, could reveal "the inner harmony of the world." The modernists' reliance upon logical consistency was thus arbitrary, "artificial," and "mechanical." Although few of the protagonists' views on specialization, the aims of scholarship, or cultural and political matters are known (to me), it is worth noting that many adherents of modern mathematics were Jewish and that others were sympathetic to the left.[2]

---

1. Various features of Mach's background, too, mark him as an outsider. More interested as a boy in science and mathematics than in history and the humanities, he left school and was apprenticed to a cabinetmaker for two years, hoping thus to prepare himself to emigrate to the United States in order to escape the clerical-reactionary regime in Austria-Hungary after 1848. From the 1880s and 1890s he opposed the growth of anti-Semitism in the universities, not least because almost all of his closest friends were Jewish (Blackmore 1972). Apart from the Vienna Circle (discussed below), another outsider who endorsed Mach's positivism, at least before 1914, was Einstein: liberal, pacifist, and Jewish.

2. For example, Hausdorff was Jewish, unenthusiastic about German participation in the First World War, and a member of the German Democratic Party from 1919 (Mehr-

That controversies within physics and mathematics in Germany resembled the differences between comprehensive and pragmatic styles of thought in genetics is consistent with the modernization thesis. But the cognitive "fault line" within German disciplines, induced by the upheavals of modernization, did rather more than merely separate instrumentalists from realists. The latter was not a major issue between the two factions within the genetics community, nor does it distinguish what appear to be comparable factions among German biochemists of the period. As Jeffrey Johnson has recently observed, Emil Fischer was an important figure in the modernization of early-twentieth-century German scientific institutions, not least because of his key role in the founding of the Kaiser-Wilhelm Society (Johnson 1990). Fischer regarded this and other forms of collaboration between academe and industry as essential if both German science and German industry were to remain competitive on the international scene. The son of a wealthy manufacturer, Fischer stood "far closer to industry than most German professors" (Johnson 1990, 27). He supported the *Technische Hochschulen* and modern secondary schools in their quest for higher status, and his ability to recognize and exploit commercially valuable academic research made him an attractive consultant to the chemical industry. At the University of Berlin Fischer organized his enormous institute in "industrial" fashion. Doctoral candidates were assigned narrowly defined dissertation topics, each filling one small slot in Fischer's overall research program, and the most important work done by his research assistants was often published under Fischer's name alone. A high proportion of his students and assistants subsequently entered industrial research and administration. In the political sphere Fischer's stance through most of the First World War was "moderate" rather than "annexationist," and after the war he joined the German Democratic Party. In private life he took no interest in literature, music, or art and was contemptuous of scientists who made what he regarded as pretentious excursions into philosophy (Feldman 1973; Fruton 1985). Thus Fischer perfectly exemplifies what I have called the pragmatic style of thought.

As Joseph Fruton has shown, Fischer's contemporary and fellow protein chemist Franz Hofmeister was a different sort of man (Fruton

---

tens 1980). Athough a gentile, Hilbert was close to the German Democratic Party, and he supported the appointments of women and socialists at the University of Göttingen (Dahms 1987a; Rowe 1986; Norbert Schappacher, "Mathematiker und Politik an der Universität Göttingen, 1918–1934," paper presented at the International Congress for History of Science, Munich, 1989).

1985). The son of a respected doctor, Hofmeister ran his laboratory at Strassburg along more egalitarian lines. He suggested problems to his junior colleagues rather than assigning them, and although he helped to prepare their work for publication, he rarely put his own name on their papers. Furthermore, in contrast with the Fischer school's concerted attack upon the problem of protein structure, Hofmeister's group published not only on protein structure but also on a remarkable variety of other biochemical topics. As one of his students (L. J. Henderson) observed, Hofmeister was "not merely a chemist"; he appreciated the complexities of biological problems and thus the difficulties in using physicochemical methods to solve them (cited in Fruton 1985, 325–26). A much higher proportion of Hofmeister's students and assistants made careers in academic life or medicine than did Fischer's, and far fewer entered industry, perhaps because Hofmeister appears to have had no contacts with industry. Of Hofmeister's life outside science, Fruton observes only that he enjoyed painting and listening to music. Hofmeister, it would appear, stood in relation to Fischer as Alfred Kühn did to Erwin Baur.

Since styles of thought are the product of historically specific social conditions, there are no grounds for assuming that the comprehensive or pragmatic styles will be found in all times and places. On the other hand, to the extent that other countries were undergoing similar structural changes from the late nineteenth century, the modernization theory—no doubt with modifications—may be worth exploring in those national contexts as well. There are indications, for example, that Britain might be promising in this respect. In his study of the biometrician-Mendelian controversy among British geneticists, Donald MacKenzie advanced something like a modernization theory (MacKenzie 1981). For Karl Pearson, Darwinism and biometry were central to a vision of gradual social progress championed by a new professional middle class. In contrast, William Bateson's Mendelian and antiselectionist preferences reflected an antiutilitarian conservatism closer to the establishment. Although the Pearson-Bateson divide does not map neatly onto the split within the German genetics community, Pearson's positivist philosophy of science would have found favor with several of the pragmatic thinkers, and his vision of a eugenic and technocratic social order are reminiscent of Baur. In addition, Bateson's views were remarkably close to those of the comprehensive style.[3] For Bateson the key prob-

3. Both Coleman (1970) and MacKenzie (1981) characterize Bateson's style of thought as "conservative," following Mannheim's usage in his 1953.

lems of genetics lay in development and evolution, where he deployed various holistic concepts. He was critical of narrow specialization in biology, and apart from his aforementioned interests in art, literature, and music, he staunchly defended the value of classical education. Furthermore, antiutilitarian sentiments are evident in his views on the proper conduct of science as well as in his hostility to industrialization, and he is said to have become disillusioned with party politics.[4] More generally, although the conception of the scholar's role at Oxford and Cambridge during the late nineteenth century was quite different from that of the German professor, these two academic communities certainly shared an aversion to industry and to narrowly specialized scholarship. No such hostility was evident at the English "civic universities," located in northern industrial cities, whose relationship to industry was similar to that of the *Technische Hochschulen* and whose academics' class backgrounds were most probably closer to those of the German outsiders.[5] Recent studies of those geneticist/plant breeders who were working at provincial agricultural institutions in Britain, for example, reveal a pattern of role identity, problem choice, and class background resembling those of the outsiders.[6]

The same tension between center and periphery may have been present within the American genetics community. In chapter 5 I suggested that although there were signs that a pragmatic style of thought was more common among American geneticists, there were also a few Americans who closely resembled the German comprehensives. A search for pragmatic and comprehensive styles might thus prove equally fruitful within the American community. The obvious first step would be a comparison of those geneticists working in the elite private universities with those working in the agricultural faculties of state universities. A casual prosopographical survey of these two groups suggests that those working in agricultural institutions were more likely to come from farm backgrounds, more likely to work in transmission genetics,

4. On Bateson's approach to genetics and his sympathies for holism, see Coleman 1970. For his criticisms of specialization, see Bateson 1922 and Harvey 1985. On his belief in the value of a classical education, see B. Bateson 1928, 47–49, 444–48; MacKenzie 1981. On his anti-utilitarianism, see Olby 1989b, 499; MacKenzie 1981. On his disillusionment with party politics, see MacKenzie 1981, 148.

5. On late Victorian Oxford and Cambridge, see Rothblatt 1968 (e.g., 86–93, 250); Engel 1983 (e.g., 233); Geison 1978. On science at Oxbridge versus the civic universities, see Sanderson 1972.

6. See Palladino's 1990 and his "Between Craft and Science: Plant-Breeding, Mendelian Theory, and the Universities in Britain, 1900–1920," forthcoming in *Technology and Culture*.

and less likely to have "high cultural" interests than were those working in private universities.[7] Reading the obituaries of men in this group (e.g., R. A. Emerson, E. M. East, Marcus Rhoades, Karl Sax, George Shull, or J. T. Patterson), I am struck by their similarities with Erwin Baur.[8] Conversely, several of those Americans who worked on developmental and/or evolutionary genetics and displayed a "comprehensive" interest in high culture were employed for most of their careers at private universities.[9]

Nonetheless, as it stands, this hypothesis is probably too crude. For one thing, the best-known transmission geneticists also worked at private universities or research institutes (e.g., Columbia, California Institute of Technology, and the Carnegie Institution's Department of Genetics at Cold Spring Harbor) while evolutionary genetics was pursued at some state universities (e.g., at California/Berkeley by Babcock and Clausen and at Texas by Patterson and Stone).[10] For another, the groups at several institutions were hardly uniform. Morgan may have been a Baur look-alike, but Columbia also made room for the likes of Dunn, Dobzhansky, and Francis J. Ryan.[11] Cold Spring Harbor accommodated not only geneticists with broad interests, such as Oscar Riddle and A. F. Blakeslee, but also the transmission geneticist M. Demerec.[12] At the moment, the data necessary for systematically testing this hy-

7. My sources were the obituaries published primarily in *Genetics* and *Biographical Memoirs of the National Academy of Sciences (U.S.)* between 1940 and 1985, some of which are cited in chap. 4, n. 1.

8. Rudorf 1955a (on Shull), Swanson 1976 (on Sax), Peterson and Peterson 1973 and Dempsey 1973 (on Rhoades), Rhoades 1949 and Beadle 1950b (on Emerson), Painter 1965 (on Patterson), and Jones 1945 (on East). Bateson judged the "corn men" as harshly as he did Baur: "They are pathetic in their simplicity, knowing nothing whatever outside genetics. They are some 150 machines for grinding out genetics" (G8g-1, 1.1.22, Bateson papers, cited in Harvey 1985). One survey suggests that contemporary agricultural scientists within American universities are segmented in terms of social background, education, institutional location, and problem choice (Busch and Lacy 1983, esp. chap. 3).

9. Sewall Wright worked at the University of Chicago, L. C. Dunn and Th. Dobzhansky at Columbia, H. S. Jennings and Raymond Pearl at Johns Hopkins, and Edmund Sinnott at Barnard College and Yale (Whaley 1983).

10. On Babcock see Stebbins 1958; on Clausen see Jenkins 1967; on Stone see Crow 1980.

11. In "Reminiscences of Morgan" (typescript, dated 16.8.67, Sturtevant papers), Sturtevant refers to Morgan's intense hostility to anything which smacked of "metaphysics." For Bateson's dismissal of Morgan as "uncultured" and "narrow-minded," see Allen 1978, 195, 275; Cock 1983, 54–55. On Ryan see Ravin 1976.

12. On Blakeslee see Sinnott 1959; on Riddle see Corner 1974; on Demerec see the biographical essays in *Advances in Genetics* 16 (1971).

pothesis are simply not available. With few exceptions, little is known about two crucial sections of the American genetics community: those working in developmental and evolutionary genetics and those employed in the agricultural faculties of state universities.[13]

Apart from its utility in charting the cognitive consequences of modernization, in Germany as elsewhere, stylistic analysis may illuminate the more general process by which scientists select research problems. To take the most extreme case, faced with a choice between important problems and those which look soluble, why do scientists choose differently? To what extent are they willing to wrestle with complexity at the cost of quick publication? How far are they prepared to go in recasting the original problem in order to render it more soluble? As we saw in chapter 2, Alfred Kühn was far more reluctant than George Beadle to abandon the problem of differentiation in order to study the mechanism of gene action. Students of development have often parted company on just this issue, as one developmental geneticist has recently observed:

> That genes act in ontogenesis as part of a complex system had been recognised by nearly all writers after Weismann's time. However, many of the early (and later) authors fail to acknowledge expressly the complexity which such systems might imply, or the dynamics of interplay between their components; and those who paid tribute to the incredible degree of complexity involved in biological pattern formation took recourse to vitalism. . . . Consequently, most research was aimed at analysing individual components of the developing system to the greatest possible depth—the molecular structure of the gene or the biochemistry of the primary inducer, for example—but not at analysing the principles by which the system as a whole would perform its marvellous feats [Sander 1985, 385].

How might such differences of research strategy be explained? In chapter 7 I suggested that the reason why many geneticists of comprehensive persuasion were interested in the fine arts is that they were fascinated with the holistic properties of form. Trying to convey the essence of development to a general academic audience, Hans Spemann chose a metaphor which emphasized the connection with art: "In animal development nature proceeds just like an artist who creates a draw-

---

13. On agricultural geneticists see Kimmelman (1987)—who discusses developments at Cornell, Wisconsin, and California/Berkeley up to 1915—and Fitzgerald's (1990) account of work at the University of Illinois before 1940. The only American developmental or evolutionary geneticists who have so far attracted historians' attention are Sewall Wright (Provine 1986), Th. Dobzhansky (e.g., Provine 1981 and Beatty 1987), and Tracy Sonneborn (Sapp 1987).

ing or sculpture—indeed, like any organizer who disposes of a given material, be it living or nonliving."[14] Among his listeners was Johannes Holtfreter, who nearly abandoned biology in the 1920s for the life of a painter. For Holtfreter, "this speech was a kind of messianic revelation which touched me deeply. How pleased I was to hear that the artist and the dumb little embryo have much in common, particularly the urge to create!"[15] Holtfreter may not have been the only member of Spemann's institute at Freiburg during the 1920s to relish this parallel; many of the other students and assistants there were also interested in the arts.[16] Indeed, serious engagement with the arts was quite common among interwar students of development, both in Germany and elsewhere. Conrad Waddington wrote a book on modern art; Edmund Sinnott was a painter and sculptor of ability; Theodor Boveri was a gifted artist and musician; and William Bateson was a connoisseur and collector of fine art.[17] One wonders whether these figures' aesthetic concerns are shared by those molecular biologists who have taken up the problem of development more recently. Or are the latter better seen as pragmatic thinkers who, like Beadle, have preferred to redefine the problem in order to make it more manageable?

Finally, what are the prospects for a stylistic analysis of national differences in the sciences? Although my aim in part 1 of this book was to characterize the "German approach" to genetics, that project was abandoned in part 2 in favor of a more limited task: the stylistic analysis of intra-German differences. While I have not analyzed the American genetics community in any detail, the evidence suggests that, however distinctive when viewed from afar, "the" style of American genetics was no more homogeneous than its German counterpart. Though evidently dominated by quasi-"outsiders," the American genetics community seems to have included a number of mandarinlike members. "National styles" in science, therefore, are probably best conceptualized rather like "national character" in psychology or sociology; while there is no denying that certain behavioral habits are more common in one country

14. Spemann, rector's address at Freiburg (1923), cited by Holtfreter, "A New Look at Spemann's 'Organizer,'" undated typescript, ca. 1985.
15. Holtfreter, "A New Look at Spemann's 'Organizer.'"
16. Fritz Baltzer, Viktor Hamburger, Hilde Mangold, and Fritz Süffert were similarly inclined (Holtfreter interview; Holtfreter 1968, xiv; Hamburger 1984; Kühn 1946).
17. Robertson 1977; Whaley 1983; Baltzer 1967; Holtfreter to K. Ishihara, 22.7.81 (an autobiographical sketch kindly shown to me by Professor Holtfreter); B. Bateson 1928.

than another, the difference is largely a statistical one.[18] Nevertheless, I would not want to discourage those interested in pursuing national styles, and I hope that comparative analysis of "local" styles within one country—such as that developed in part 2—can provide analytical tools which are useful for understanding national differences. Looking beyond the history of genetics, for example, there are indications that the categories "mandarin" and "outsider" may illuminate some cases where physical theories have enjoyed a different reception in Germany and the United States. Consider the reception of quantum mechanics in the United States from the late 1920s. Unlike their European counterparts, American theoretical physicists were unperturbed by the philosophical problems which the Copenhagen interpretation seemed to pose: the abstract and unvisualizable nature of microphysical reality or the inseparability of subject and object.[19] The Americans' relaxed attitude stemmed from their broadly empiricist conception of scientific method (of which Bridgeman's operationalism was one version), in which the role of theory was to describe rather than to explain (Schweber 1986, Cartwright 1987). From this point of view, mathematical theories were perfectly acceptable, however unvisualizable the relationships which they portrayed. This standpoint bears a marked resemblance to that nononsense attitude to metaphysics which characterized the pragmatic minority within the Germany genetics community and which (to judge from the evidence presented in chapter 5) was probably quite common among American geneticists as well.[20]

Turning to the "realist" side of the debate, John Heilbron has offered an explanation for the attempts by Niels Bohr and his followers to turn the Copenhagen interpretation into a general philosophy of life (Heilbron 1985). Heilbron's answer is both psychological and asymmetric: Something peculiar to these physicists' mental makeup made them "susceptible" to Bohr's philosophy. Against the backdrop of the traditional *Bildungsideal*, however, the behavior of Bohr et al. hardly looks peculiar; it might instead be seen as one of countless attempts by mandarins to derive spiritual guidance from scholarly synthesis. And if the approach which I have taken in this book is correct, we ought to be just as concerned to explain why American physicists were less susceptible

18. Gerald Geison treats national styles in much the same way in *Michael Foster* (1978, 16–17).
19. Americans "ignored complementarity and came to grips with uncertainty without wincing" (Heilbron 1985, 205). On Heisenberg et al.'s efforts to defend the new quantum mechanics against the charge of unvisualizability, see Forman 1984.
20. On T. H. Morgan as an empiricist, see Roll-Hansen 1987b.

to the Bohr virus and why the Vienna Circle was positively allergic to it. The Vienna Circle's left-liberal political sympathies and its suspicion of metaphysical philosophy support the alternative explanation: that the Vienna Circle's criticism of Bohr represented the hostility of academic outsiders toward a form of mandarin ideology.[21]

I hope that historians working on other disciplines and contexts will find stylistic analysis of use in explaining cognitive change. But many will be equally concerned with the epistemological implications of this kind of analysis. More particularly, what can one say about the relative success of comprehensive and pragmatic styles of thought as research strategies? If one accepts Nobel Prizes as a measure of success, the answer in genetics seems clear: The prizes awarded to Morgan (1933), Muller (1946), and Beadle (1958)—but to none of their German counterparts—suggest that the pragmatic strategy was more productive. While several historians of genetics evidently agree,[22] I prefer to reserve judgment on this matter. There are several ways of defining success. Although there is no doubt that the pragmatic strategy produced more results more quickly and has opened up whole new areas of research, it has done so by steering clear of important and difficult problems, as the comparison of Kühn and Beadle demonstrates. Moreover, the comprehensive perspective also opened up a new and important research area—cytoplasmic inheritance—some twenty to thirty years before it was generally accepted in the United States. The dangers awaiting those historians keen to evaluate the science of the past are well illustrated by the fate of Richard Goldschmidt's book *The Material Basis of Evolution*. On publication in 1940, it was probably regarded by most geneticists as the work of an eccentric, albeit a clever one. Dobzhansky, however, regarded it as essential reading for evolutionists, and by the early 1980s some advocates of "punctuated equilibrium" had concluded that the

---

21. Hahn, Neurath, and Carnap 1973; on the Vienna Circle's criticisms of Bohr, see Heilbron 1985, 217–19.

22. For Garland Allen's assessment, see the concluding section of chapter 2. Marsha Richmond states that, compared with the work of the Morgan school, "no major breakthroughs destined to revamp our understanding of the nature of heredity came from Germany" during the period 1900 to 1933 (Richmond 1986, 6). The only exceptions which she acknowledges are the works of Wilhelm Weinberg and Theodor Boveri, neither of which were included in my analysis. Discussing the Americans' decision early this century to separate the study of heredity and development, Richard Burian has written that "on average, among the best scientists, those who found a useful way of restricting their attention to narrower problems fared best" ("Disciplinary Specialization and the American 'Dis-solution' of the Impasse between Development and Heredity in the First Half of the Twentieth Century," paper given at Gif-sur-Yvette, 14 May 1987, p. 8).

book was important enough to justify reprinting (Goldschmidt 1940 [1982]).

More generally, it would be interesting to see stylistic analysis applied to the architects of the evolutionary synthesis.[23] From the available secondary literature on Dobzhansky, Wright, Huxley, and Rensch, it is my impression that they are better described as comprehensive than as pragmatic thinkers. In addition to the wide range of biological areas in which Julian Huxley was knowledgeable, for example, he published works of politics, philosophy, and poetry, believing that the most valuable intellectual work was synthetic. Dobzhansky, similarly, was a naturalist/geneticist who published on humanism and philosophy; was interested in art, music, history, and literature; and preferred grand projects and big generalizations to narrow specialization in science. As we have seen, the breadth of Sewall Wright's contributions to genetics is one reason why he was admired by leading geneticists in Germany, but his lifelong interest in philosophy—more particularly, philosophy of an antireductionist kind—may have been another (Provine 1986; Wright 1964). Finally, like many of his interwar contemporaries, Bernhard Rensch was interested in expressionist painting, philosophy, and the history of Near and Far Eastern cultures, seeking to develop "as all-embracing a world-picture as possible" (Rensch 1979, 53).

It is by no means clear, therefore, that the pragmatic style of thought is more productive *tout court;* different kinds of problems may require different mental sets to solve them. By and large, it seems to me that evaluative judgments about success are better left to scientists themselves than to historians. Whatever judgment the biological community may eventually render on this issue, there is at least one methodological prescription which does, I think, emerge from the comparative historical analysis in this book. It is a straightforwardly pluralist one. If each style of thought offers certain advantages, illuminating some natural phenomena while obscuring others, the best policy is to make room for both comprehensive and pragmatic thinkers in every discipline.

Lastly, for those who share my conviction that good historical work ought to possess significance which is not "merely" historical, I want to consider briefly how the foregoing analysis impinges upon an issue of general concern within the academic community. The literature on the sociology of intellectuals has routinely distinguished between two kinds of intellectuals whose self-understanding corresponds roughly to that

---

23. I have explored this issue further in "Metaphysical foundations of the evolutionary synthesis," forthcoming.

which I have called the mandarin and the outsider. Stuart Hughes, for example, refers to "intellectuals" versus "mental technicians" (1962); Florian Znaniecki to the "sage" versus the "technologist" (1968, chap. 2); and Lewis Coser to the "intellectual" versus the "technician-expert" (1965, 185–86). Edward Shils implicitly makes a similar distinction when he suggests that the bureaucratization of intellectual roles has brought about the "deeper illiteracy of the professional classes" (1972, 78). Whatever their terminology, all of these authors were worried by signs of the growing domestication of intellectuals. The evident price for a comfortable life in academe or comparable institutions was the replacement of wisdom by analytical skill; intellectuals were seen to be losing their appetite for social criticism.[24] All of these authors' choice of terminology, however, suggests that, consciously or not, they see this split within the intelligentsia in *disciplinary* terms, tending to place scholars from the arts and humanities in the former category and scientists and technologists in the latter. In his analysis of the process of domestication, especially in social science scholarship, J. P. Nettl escapes this temptation to some extent. Recognizing that "technicians" are to be found in all disciplines, he recasts the traditional dichotomy in a useful way, distinguishing between "intellectuals" and "academics." Nevertheless, even Nettl seems to assume that scientists belong exclusively in the latter category (e.g., Nettl 1970, 67, 69, 126–27).

It should by now be clear that this view is ahistorical. From the late nineteenth century, many German academics—both scientists and others—were addressing precisely the issues which Nettl ascribes to "intellectuals": a concern with ends rather than means and a readiness to reflect upon and reorder existing cultural traditions, rather than simply adding to them. As important as Nettl's distinction is for the post-1945 era, therefore, it needs to be tempered by the realization that scientists' self-understanding, in Germany at least, has shifted markedly over the last century and was probably very different throughout this period from that in the United States. That is not to say, however, that the mandarin mentality has by now disappeared altogether from the German academic community. This remains an empirical question, and an intriguing one at that. The heated debate in Germany during the late 1980s over the meaning of the Third Reich and, more generally, over the relevance of the history of that period for the creation of a "German national identity" suggests that the mandarin outlook is alive and well,

---

24. It is probably significant that all of the works in question were written before 1968.

at least among some German historians.[25] In the English-speaking world, too, there is little doubt that the split within the intelligentsia, discussed by Nettl and others, has been visible within the universities, especially since the late 1960s. For those who wish to understand why these distinctive role conceptions persist (or decline, as the case may be), the approach developed in this book offers one starting point.

25. On the persistence and transformation of the mandarin role among West German academics in the arts and humanities since 1945, see Brunkhorst 1987.

Archives Consulted

Abbreviations

MPGA: Archiv zur Geschichte der Max-Planck Gesellschaft, Berlin
APS: American Philosophical Society Library, Philadelphia
SBPK: Staatsbibliothek Preussischer Kulturbesitz, Berlin
CIT: Millikan Library, California Institute of Technology

American Society of Naturalists papers (APS)
William Bateson papers (John Innes Institute, Norwich)
Erwin Baur letters (Sammlung Darmstädter, SBPK)
Erwin Baur letters (von Luschan papers, SBPK)
Erwin Baur papers (MPGA)
George Beadle papers (CIT)
Berlin Document Center
A. F. Blakeslee papers (APS)
Ernst Caspari papers (APS)
Ralph Cleland papers (Lilly Library, Indiana University)
Carl Correns letters (Carl Steinbrinck papers, SBPK)
Carl Correns papers (MPGA)
C. B. Davenport papers (APS)
Max Delbrück papers (CIT)
M. Demerec papers (APS)
Th. Dobzhansky papers (APS)
L. C. Dunn papers (APS)
Ernst Ehlers papers (Niedersächsische Staats- und Universitätsbibliothek, Göttingen)
Emil Fischer papers (Bancroft Library, University of California, Berkeley)
Richard Goldschmidt letters (Julian Huxley papers, Rice University)
Richard Goldschmidt papers (Bancroft Library, University of California, Berkeley)
Göttingen University Archive
Hans Grüneberg papers (Contemporary Medical Archive Centre, Wellcome Institute for History of Medicine, London)
Ross G. Harrison papers (Yale University)
International Education Board papers (Rockefeller Archive Center)
H. S. Jennings papers (APS)
E. E. Just papers (Howard University)
Kaiser-Wilhelm Gesellschaft papers, Abt I: Generalverwaltung (Rep IA), KWI für Biologie (Rep 8), KWI für Hirnforschung (Rep 21), KWI für Züchtungsforschung (Rep 51), all in MPGA.

Archives Consulted

Hans Kappert papers (SBPK)
Alfred Kühn papers (MPGA)
Alfred Kühn letters (Institut für Geschichte der Medizin, Universität Heidelberg)
Alfred Kühn letters (Archive of the Naples Zoological Station)
Alfred Kühn letters (Otto Koehler papers, Universität Freiburg)
F. R. Lillie papers (Marine Biological Laboratory, Woods Hole, Massachusetts)
Ernst Mayr papers (SBPK)
T. H. Morgan papers (CIT)
H. J. Muller papers (Lilly Library, Indiana University)
Hans Nachtsheim papers (MPGA)
Notgemeinschaft der Deutschen Wissenschaft papers (Bundesarchiv, Koblenz)
Klaus Pätau papers (in family possession)
Preussische Akademie der Wissenschaften papers (former Akademie der Wissenschaften der DDR, Berlin)
Otto Renner letters (Bayerische Staatsbibliothek, Munich)
Rockefeller Foundation papers (Rockefeller Archive Center)
Elisabeth Schiemann papers (MPGA)
Reinhold von Sengbusch papers (MPGA)
George Shull papers (APS)
Society for the Protection of Science and Learning papers (Bodleian Library, Oxford University)
Tracy M. Sonneborn papers (Lilly Library, Indiana University)
Hans Spemann papers (Senckenbergische Bibliothek, Frankfurt)
Emmy Stein papers (MPGA)
Curt Stern papers (APS)
Curt Stern letters (Ernst Mayr papers, APS)
A. H. Sturtevant papers (CIT)
Fritz von Wettstein letters (Correns papers, MPGA)
Richard Woltereck letters (Hermann Hesse papers, Schiller Nationalmuseum/ Deutsches Literaturarchiv, Marbach am Neckar)
Zentrales Staatsarchiv (ZStA), Merseburg: Preussisches Kultusministerium (Rep 76), Preussisches Ministerium für Landwirtschaft . . . (Rep 87B), Nachlass Schmidt-Ott (Rep 92).

Interviews

Ingrun Anton-Lamprecht, 26 April 1983
Charlotte Auerbach, 21 November 1986
G. H. Beale, 24 November 1986
Ernst Caspari, 23 September 1981
Albrecht Egelhaaf, 3 May 1983
Wolf-Dietrich Eichler, 10 June 1983
Raphael Falk, June 1989
Bentley Glass, 24 July 1984
Cornelia Harte, 1 May 1983
Hermann Hartwig, 1 May 1983
Johannes Holtfreter, 23 August 1981; 10 and 11 August 1985
Hans Kalmus, 26 April 1984
Hermann Kuckuck, 24 March 1983
Gertrud Linnert, 28 August 1987
Hans Marquardt, 25 September 1981
Ernst Mayr, 30 May 1984
Georg Melchers, 28 September 1981
Ursula Philip, 15 December 1986
Hans Querner, 4 June 1984
Hans Ross, 2 May 1983
Lotte Rothe, 2 May 1983
Hans-Georg Schweiger, 25 April 1983
Berthold Schwemmle, 14 May 1988
Joseph Straub, 3 May 1983
W. Stubbe, 4 May 1983
Friedrich Vogel, 19 February 1988

Bibliography

Note: *ZIAV* = *Zeitschrift für induktive Abstammungs- und Vererbungslehre*
Abel, Wolfgang. 1980. Edgar Knapp, 1906–1978. *Berichte und Mitteilungen der Max-Planck-Gesellschaft* 3:20–23.
Adam, Uwe D. 1977. *Hochschule und Nationalsozialismus: Die Universitat Tübingen im Dritten Reich.* Tübingen: J. C. B. Mohr.
Adams, Mark. 1968. The founding of population genetics: Contributions of the Chetverikov school, 1924–1934. *Journal of the History of Biology* 1:23–39.
———. 1979. From "gene fund" to "gene pool": On the evolution of evolutionary language. *Studies in the History of Biology* 3:241–85.
———. 1980a. Sergei Chetverikov, the Kol'tsov Institute, and the evolutionary synthesis. In Mayr and Provine (eds.), *The Evolutionary Synthesis*, 242–78.
———. 1980b. Severtsov and Schmalhausen: Russian morphology and the evolutionary synthesis. In Mayr and Provine (eds.), *The Evolutionary Synthesis,* 193–225.
Adorno, Theodor. 1969. Scientific experiences of a European scholar in America. In Donald Fleming and Bernard Bailyn (eds.), *The Intellectual Migration: Europe and America, 1930–1960* (Cambridge, Mass: Harvard University Press), 338–70.
Allen, Garland. 1969. Hugo de Vries and the reception of the "mutation theory." *Journal of the History of Biology* 2:55–87.
———. 1974. Opposition to the Mendelian-chromosome theory: The physiological and developmental genetics of Richard Goldschmidt. *Journal of the History of Biology* 7:49–92.
———. 1975. *Life Science in the Twentieth Century.* New York: John Wiley.
———. 1978. *Thomas Hunt Morgan: The Man and His Science.* Princeton: Princeton University Press.
———. 1979a. The transformation of a science: T. H. Morgan and the emergence of a new American biology. In A. Oleson and J. Voss (eds.), *The Organization of Knowledge in Modern America, 1860–1920* (Baltimore: Johns Hopkins University Press), 173–211.
———. 1979b. Naturalists and experimentalists: The genotype and the phenotype. *Studies in History of Biology* 3:179–209.
———. 1979c. The rise and spread of the classical school of heredity, 1910–1930: Development and influence of the Mendelian chromosome theory. In N. Reingold (ed.), *The Sciences in the American Context: New Perspectives* (Washington, D.C.: Smithsonian Institution Press), 209–28.
———. 1983. T. H. Morgan and the influence of mechanistic materialism on the development of the gene concept, 1910–1940. *American Zoologist* 23:829–43.

———. 1985. T. H. Morgan and the split between embryology and genetics, 1910–1935. In Horder, Witkowski, and Wyllie (eds.), *A History of Embryology*, 113–46.

———. 1986. The Eugenics Record Office at Cold Spring Harbor, 1910–1940. *Osiris* 2:225–64.

Anderson, T. F. 1975. J. Schultz. *Biographical Memoirs of the National Academy of Sciences (U.S.)* 47:393–424.

Appel, Toby. 1988. Organizing biology: The American Society of Naturalists and its "affiliated societies," 1883–1923. In Rainger, Benson, and Maienschein (eds.), *The American Development of Biology*, 87–120.

Arnheim, R. 1972. Max Wertheimer and Gestalt psychology. In Robert Boyers (ed.), *The Legacy of German Refugee Intellectuals* (New York: Schocken), 97–102.

Arnold, C.-G. 1981. Julius Schwemmle, 1894–1979. *Berichte der Deutschen Botanischen Gesellschaft* 94:749–56.

Asen, Johannes. 1955. *Gesamtverzeichnis des Lehrkörpers der Universität Berlin, 1810–1945*, vol. 1 Leipzig: O. Harrassowitz.

Ash, Mitchell. 1980. Academic politics in the history of science: Experimental psychology in Germany, 1879–1941. *Central European History* 13:255–86.

———. 1984. Disziplinentwicklung und Wissenschaftstransfer: Deutschsprachige Psychologen in der Emigration. *Berichte zur Wissenschaftsgeschichte* 7:207–26.

Autrum, H. 1965. A. Kühn zum 80. Geburtstag. *Die Naturwissenschaften* 52:173.

———. 1969. A. Kühn. *Jahrbuch der Bayerischen Akademie der Wissenschaften*, 263–66.

Ayala, Francisco J. 1985. Theodosius Dobzhansky. *Biographical Memoirs of the National Academy of Sciences (U.S.)* 55:163–214.

Baker, W. K., T. G. Gregg, J. V. Neel, and H. D. Stalker. 1975. Warren Spencer (1898–1969). *Genetics* 79:1–6.

Ball, Margaret M. 1937. *Post-war German-Austrian Relations: The Anschluss Movement, 1918–1936*. Stanford: Stanford University Press.

Baltzer, Fritz. 1918. Theodor Boveris Lehrtätigkeit. In W. C. Röntgen (ed.), *Erinnerungen an Theodor Boveri* (Tübingen: Mohr), 38–66.

———. 1967. *Theodor Boveri: Life and Work of a Great Biologist, 1862–1915*. Berkeley: University of California Press.

Barfurth, Dietrich. 1910. Wilhelm Roux zum 60. Geburtstag. *Archiv für Entwicklungsmechanik* 30:vii–xxxvii.

———. 1924. Wilhelm Roux: ein Nachruf. *Archiv für mikroskopische Anatomie und Entwicklungsmechanik* 104:i–xxii.

Barkin, Kenneth D. 1971. Fritz Ringer's *The Decline of the German Mandarins*. *Journal of Modern History* 43:276–86.

Barnes, Barry. 1974. *Scientific Knowledge and Sociological Theory*. London: Routledge and Kegan Paul.

Barthelmess, Alfred. 1952. *Vererbungswissenschaft*. Freiburg/Munich: Karl Alber.

Bateson, Beatrice. 1928. *William Bateson, FRS, Naturalist.* Cambridge: Cambridge University Press.
Bateson, William. 1922. Evolutionary faith and modern doubts. *Science* 55:55–61.
Bauer, Hans. 1966. Hans Bauer. In W. Böhm and G. Paehlke (eds.), *Forscher und Gelehrte* (Stuttgart: Battenberg), 187–88.
———. 1963. Karl Belar, 1895–1931. In Hugo Freund and A. Berg (eds.), *Geschichte der Mikroskopie (Band I—Biologie)* (Frankfurt a.M.: Umschau), 111–19.
Baumgardt, D. 1965. Looking back on a German university career. *Yearbook of the Leo Baeck Institute* 10:239–65.
Baur, Erwin. 1908–09. Das Wesen und die Erblichkeitsverhältnisse der "Varietates albomarginatae hort." von Pelargonium zonale. *ZIAV* 1:330–51.
———. 1911. *Einführung in die experimentelle Vererbungslehre* Berlin: Borntrager; 3.–4. Auflage 1919; 5.–6. Auflage 1922; 7.–11. Auflage 1930.
———. 1912a. [Review of R. Semon, "Der Stand der Frage nach der Vererbung erworbener Eigenschaften," 1910]. *ZIAV* 6:244–47.
———. 1912b. [Review of R. Semon. *Das Problem der Vererbung erworbener Eigenschaften,* 1912]. *ZIAV,* 8:337–39.
———. 1913. Die Frage nach der Vererbung erworbener Eigenschaften im Lichte der neuen experimentellen Forschung mit Pflanzen. *Archiv für Soziale Hygiene* 8:117–30.
———. 1921. *Die wissenschaftlichen Grundlagen der Pflanzenzüchtung: Ein Lehrbuch für Landwirte, Gärtner, und Forstleute.* Berlin: Borntrager; 5th ed. 1924.
———. 1924. Untersuchungen über das Wesen, die Entstehung, und Vererbung der Rassenunterschiede bei A. majus. *Bibliotheca Genetica* 4:1–170.
———. 1925. Die Bedeutung der Mutation für das Evolutionsproblem. *ZIAV* 37:107–15.
———. 1928. Die Möglichkeit eines gesetzlichen Schutzes von Neuzüchtungen. *ZIAV* supp. 1:399–401.
———. 1931. Evolution. *Journal of the Royal Horticultural Society* (London) 56:176–82.
———. 1932. Artumgrenzung und Artbildung in der Gattung Antirrhinum, Sektion Antirrhinastrum. *ZIAV* 63:256–302.
———. 1932b. Müssen wir wirklich Forschungsanstalten und Hochschulen schliessen? *Deutsche Allgemeine Zeitung,* 10 Feb. 1932.
———. 1932c. Der Untergang der Kulturvölker im Lichte der Biologie. *Volk und Rasse* 7:3–19.
———. 1932d. Wie lässt sich die fortschreitende Degeneration der Kulturvölker aufhalten? *Fortschritte der Gesundheitsfürsorge* 6:1–3.
———. 1933. Carl Correns. *Forschungen und Fortschritte* 9:120.
Baur, Erwin, Eugen Fischer, and Fritz Lenz. 1921. *Grundlagen der menschlichen Erblichkeitslehre und Rassenhygiene.* Munich: Lehmann; 4th ed. 1932.
Beach, Mark. 1968. Professional versus professorial control of higher education. *Educational Record* 49:263–73.

Bibliography

Beadle, George. 1937. The development of eye-colors in Drosophila as studied by transplantation. *American Naturalist* 71:120–26.

———. 1939. Physiological aspects of genetics. *Annual Reviews of Physiology* 1:41–62.

———. 1941. Genetic control of production and utilisation of hormones. In *Proceedings of the Seventh International Genetical Congress*, 58–62.

———. 1945. Biochemical genetics. *Chemical Reviews* 37:15–96.

———. 1946a. Genes and the chemistry of the organism. *American Scientist* 34:31–53.

———. 1946b. The gene. *Proceedings of the American Philosophical Society* 90:422–31.

———. 1948. Physiological aspects of genetics. *Annual Reviews of Biochemistry* 17:727–52.

———. 1949. Genes and biological enigmas. In G. A. Baitsell (ed.), *Science in Progress*, 6th series (New Haven: Yale University Press), 184–248.

———. 1950a. Chemical genetics. In L. C. Dunn (ed.), *Genetics in the Twentieth Century* (New York: Macmillan), 221–25.

———. 1950b. R. A. Emerson, 1873–1947. *Genetics* 35:1–3.

———. 1963. *Genetics and Modern Biology*. Philadelphia: American Philosophical Society.

———. 1966. Biochemical genetics: Some recollections. In J. Cairns, G. Stent, and J. B. Watson (eds.), *Phage and the Origins of Molecular Biology* (Cold Spring Harbor, NY: Cold Spring Harbor Laboratory of Quantitative Biology), 23–32.

———. 1974. Recollections. *Annual Reviews of Biochemistry* 43:1–13.

———. 1977. Genes and chemical reactions in Neurospora (1958). In *Nobel Lectures in Molecular Biology, 1933–1975* (New York: Elsevier), 51–63.

Beadle, George, R. L. Anderson, and Jane Maxwell. 1938. A comparison of the diffusible substances concerned with eye color development in Drosophila, Ephestia, and Habrobracon. *Proceedings of the National Academy of Sciences* 24:80–85.

Beadle, George, and Boris Ephrussi. 1935. Transplantation in Drosophila. *Proceedings of the National Academy of Sciences* 21:642–46.

Beadle, George, and E. L. Tatum. 1941. Genetic control of biochemical reactions in Neurospora. *Proceedings of the National Academy of Sciences* 27:499–506.

Beale, G. H. 1982. Tracy Morton Sonneborn, 1905–1981. *Biographical Memoirs of Fellows of the Royal Society* 28:537–74.

Beatty, John. 1987. Dobzhansky and drift: Facts, values, and chance in evolutionary biology. In Lorenz Krüger, Gerd Gigerenzer, and Mary S. Morgan (eds.), *The Probabilistic Revolution*, vol. 2: *Ideas in the Sciences* (Cambridge, Mass: MIT Press), 271–311.

Becker, C. H. 1919. *Gedanken zur Hochschulreform*. Leipzig: Quelle und Meyer.

———. 1930. *Das Problem der Bildung in der Kulturkrise der Gegenwart*. Leipzig: Quelle und Meyer.

Becker, Heinrich. 1987. Von der Nahrungssicherung zu Kolonialträumen: Die landwirtschaftlichen Institute im Dritten Reich. In H. Becker, H.-J. Dahms, and Cornelia Wegeler (eds.), *Die Universität Göttingen unter dem Nationalsozialismus* (Munich: Saur), 410–36.

Beer, J. J. 1958. Coal tar dye manufacture and the origins of the modern industrial research laboratory. *Isis* 49:123–31.

Belar, Karl. 1924. [Review of T. J. Stomps, *Erblichkeit und Chromosomen*]. *ZIAV* 33:373–75.

———. 1925. Chromosomen und Vererbung: Eine Antwort an Herrn Fick. *Die Naturwissenschaften* 13:717–23.

Ben-David, Joseph. 1968. *Fundamental Research and the Universities*. Paris: OECD.

———. 1968–69. The universities and the growth of science in Germany and the United States. *Minerva* 7:1–35.

———. 1971. *The Scientist's Role in Society*. Englewood Cliffs, N.J.: Prentice-Hall.

Ben-David, Joseph, and Avraham Zloczower, 1962. Universities and academic systems in modern societies. *European Journal of Sociology* 3:45–84.

Bendix, Reinhard. 1967. Tradition and modernity reconsidered. *Comparative Studies in Society and History* 9:292–346.

Bennett, Dorothea. 1977. L. C. Dunn and his contribution to T-locus genetics. *Annual Reviews of Genetics* 11:1–12.

Benson, Keith. 1979. *W. K. Brooks (1848–1908): A Case Study in Morphology and the Development of American Biology*. Ph.D. diss., Oregon State University.

———. 1985. American morphology in the late nineteenth century: The Biology department at Johns Hopkins University. *Journal of the History of Biology* 18:163–205.

———. 1988. From museum research to laboratory research: The transformation of natural history into academic biology. In Rainger, Benson, and Maienschein (eds.), *The American Development of Biology*, 49–83.

Benz, Wolfgang. 1983. Emil J. Gumbel: Die Karriere eines deutschen Pazifisten. In Ulrich Walberer (ed.), *10. Mai 1933: Bücherverbrennung in Deutschland und die Folgen* (Frankfurt: Fischer Taschenbuch), 160–98.

Bertalanffy, Ludwig. 1927. Studien über theoretische Biologie. *Biologisches Zentralblatt* 47:210–42.

———. 1928. Kritische Theorie der Formbildung. *Abhandlungen zur theoretischen Biologie* no. 27.

Beyerchen, Alan. 1988. On the stimulation of excellence in Wilhelmian science. In Jack Dukes and J. Remak (eds.), *Another Germany: A Reconsideration of the Imperial Era* (Boulder, Colorado: Westview), 139–68.

Biology and Gender Study Group. 1989. The importance of feminist critique for contemporary cell biology. In Nancy Tuana (ed.), *Feminism and Science* (Bloomington: Indiana University Press), 172–87.

Black, Max. 1962. *Models and Metaphors*. Ithaca, N.Y.: Cornell University Press.

Blackmore, John. 1972. *Ernst Mach: His Work, Life, and Influence*. Berkeley: University of California Press.

Bledstein, Burton. 1976. *The Culture of Professionalism*. New York: Norton.
Bloor, David. 1978. Polyhedra and the abominations of Leviticus. *British Journal for the History of Science* 11:245–72.
Bluhm, Agnes. 1933. Nachruf auf C. Correns. *Eugenik* 3:49–52.
Bock, K.-D. 1972. *Strukturgeschichte der Assistentur: Personalgefüge, Wert- und Zielvorstellungen in der deutschen Universität des 19. und 20. Jahrhunderts*. Düsseldorf: Bertelsmann.
Böhm, Wolfgang. 1988. Geschichte des Landwirtschaftsstudiums in Deutschland. *Berichte über Landwirtschaft* 66:1–36.
Bonner, Thomas N. 1963. *American Doctors and German Universities*. Lincoln: University of Nebraska Press.
Borscheid, Peter. 1976. *Naturwissenschaft, Staat, und Industrie in Baden (1848–1914)*. Stuttgart: Ernst Klett.
Boveri, Theodor. 1904. *Ergebnisse über die Konstitution der chromatischen Substanz des Zellkerns*. Jena: G. Fischer.
———. 1918. Zwei Fehlerquellen bei Merogonieversuchen und die Entwicklungsfähigkeit merogonischer und partiell-merogonischer Seeigelbastarde. *Archiv für Entwicklungs-Mechanik* 44:417–71.
Bowler, Peter. 1983. *The Eclipse of Darwinism: Anti-Darwinian Evolution Theories in the Decades around 1900*. Baltimore: Johns Hopkins University Press.
Brink, R. A. 1941. Victor Jollos, 1887–1941. *Science* 94:270–72.
vom Brocke, Bernhard. 1980. Hochschul- und Wissenschaftspolitik in Preussen und im deutschen Kaiserreich, 1882–1907: Das "System Althoff." In Peter Baumgart (ed.), *Bildungspolitik in Preussen zur Zeit des Kaiserreichs* (Stuttgart: Klett-Cotta), 9–118.
vom Bruch, Rüdiger. 1980. *Wissenschaft, Politik, und öffentliche Meinung: Gelehrtenpolitik im Wilhelminischen Deutschland, 1890–1914*. Husum: Matthiesen.
———. 1988. Historiker und Nationalökonomen im Wilhelminischen Deutschland. In K. Schwabe (ed), *Deutsche Hochschullehrer als Elite, 1815–1945* (Boppard: Boldt), 105–50.
Brücher, H. 1938. Die reziprok-verschiedene Art- und Rassenbastarde von Epilobium und ihre Ursachen, I: Die Nichtbeteiligung von Hemmungsgenen. *ZIAV* 75:298–340.
Bruford, W. H. 1975. *The German Tradition of Self-Cultivation*. Cambridge: Cambridge University Press.
Brunkhorst, Hauke. 1987. *Der Intellektuelle im Land der Mandarine*. Frankfurt: Suhrkamp.
Brush, Stephen G. 1967. Thermodynamics and history. *Graduate Journal* 7:477–565.
Bünning, Erwin. 1944. Otto Renner, 60 Jahre. *Flora* N.F. 37:vii–x.
———. 1977. Fifty years of research in the wake of Wilhelm Pfeffer. *Annual Reviews of Plant Physiology* 28:1–22.
Burchardt, Lothar. 1975. *Wissenschaftspolitik im Wilhelminischen Deutschland: Vorgeschichte, Gründung, und Aufbau der Kaiser-Wilhelm Gesellschaft*. Göttingen: Vandenhoeck und Ruprecht.

———. 1980. Professionalisierung oder Berufskonstruktion? *Geschichte und Gesellschaft* 6:326–48.
———. 1988. Naturwissenschaftliche Universitätslehrer im Kaiserreich. In K. Schwabe (ed.), *Deutsche Hochschullehrer als Elite, 1815–1945* (Boppard: Boldt), 151–214.
Burian, R., J. Gayon, and D. Zallen. 1988. The singular fate of genetics in the history of French biology, 1900–1940. *Journal of the History of Biology* 21:357–402.
Burkhardt, Richard. 1981. On the emergence of ethology as a scientific discipline. *Conspectus of History* 1:62–81.
Burnham, J. C. 1973. Herbert Spencer Jennings. *Dictionary of Scientific Biography* (New York: Scribners) 7:98–100.
Busch, Alexander. 1959. *Geschichte des Privatdozentens*. Stuttgart: Enke.
Busch, Lawrence, and William Lacy. 1983. *Science, Agriculture, and the Politics of Research*. Boulder, Colorado: Westview.
Butenandt, Adolf. 1981. *Das Werk eines Lebens (Band II: Wissenschaftspolitische Aufsätze, Ansprachen, und Reden)*. Göttingen: Vandenhoeck und Ruprecht.
Butenandt, A., W. Weidel, and E. Becker. 1940. Kynurenin als Augenpigmentbildung auslösendes Agens bei Insekten. *Die Naturwissenschaften* 28: 363–64.
Butterfass, Theodor, 1961. Otto Renner, 1883–1960. *Mitteilungen des Vereins für Naturwissenschaft und Mathematik in Ulm* 26:IX–XII.
———. 1980. Edgar Knapp: 1906 bis 1978. *Berichte der Deutschen Botanischen Gesellschaft* 93:505–515.
Cahan, David. 1982. Werner Siemens and the origin of the Physikalisch-Technisches Reichsanstalt, 1872–1887. *Historical Studies in the Physical Sciences* 12:253–83.
Caneva, Kenneth. 1978. From galvanism to electrodynamics: The transformation of German physics and its social context. *Historical Studies in the Physical Sciences* 9:63–159.
———. 1981. What should we do with the monster? Electromagnetism and the psychosociology of knowledge. *Sociology of the Sciences* 5:101–31.
Carlson, E. A. 1966. *The Gene: A Critical History* Philadelphia: W. B. Saunders.
———. 1981. *Genes, Radiation, and Society: The Life and Work of H. J. Muller*. Ithaca, N.Y.: Cornell University Press.
———. 1983. [Review of Richard Goldschmidt, *The Material Basis of Evolution*, and L. K. Piternick (ed.), *Richard Goldschmidt: Controversial Geneticist and Creative Biologist*]. *Isis* 74:294–96.
Carson, Hampton L. 1980. A provocative view of the evolutionary process. In Leonie Piternick (ed.), *Richard Goldschmidt: Controversial Geneticist and Creative Biologist* (Basel: Birkhäuser), 24–26.
Cartwright, Nancy. 1987. Philosophical problems of quantum theory: The response of American physicists. In L. Krüger, G. Gigerenzer, and Mary S. Morgan (eds.), *The Probabilistic Revolution*, vol. 2: *Ideas in the Sciences* (Cambridge: MIT Press), 417–35.

Caspari, Ernst. 1933. Über die Wirkung eines pleiotropen Gens bei der Mehlmotte Ephestia kühniella Z. *Archiv für Entwicklungsmechanik der Organismen* 130:353–81.

———. 1948. Cytoplasmic inheritance. *Advances in Genetics* 2:1–66.

———. 1957. K. Henke, developmental biologist. *Science* 125:1076.

———. 1960. Richard Goldschmidt. *Genetics* 45:1–5.

———. 1964. H. B. Goodrich (1887–1963). *Genetics* 50:12–13.

———. 1980. An evaluation of Goldschmidt's work after twenty years. In Leonie Piternick (ed.), *Richard Goldschmidt: Controversial Geneticist and Creative Biologist* (Basel: Birkhäuser), 19–23.

Castle, W. E. 1950. The beginnings of Mendelism in America. In L. C. Dunn (ed.), *Genetics in the Twentieth Century* (New York: Macmillan), 59–76.

Caullery, Maurice. 1922. *Universities and Scientific Life in the United States*, trans. J. H. Woods and E. Russell. Cambridge: Harvard University Press.

Chargaff, Erwin. 1978. *Heraclitean Fire: Sketches of a Life before Nature*. New York: Rockefeller University Press.

Churchill, F. B. 1973. Chabry, Roux, and the experimental method in nineteenth century embryology. In R. N. Giere and R. S. Westfall (eds.), *Foundations of Scientific Method: The Nineteenth Century* (Bloomington: Indiana University Press), 161–205.

———. 1980. The modern evolutionary synthesis and the biogenetic law. In Mayr and Provine (eds.), *The Evolutionary Synthesis*, 112–22.

Cleland, Ralph. 1966. Otto Renner, 1883–1960. *Genetics* 53:1–6.

Coben, Stanley. 1979. American foundations as patrons of science: The commitment to individual research. In N. Reingold (ed.), *The Sciences in the American Context: New Perspectives* (Washington, D.C.: Smithsonian Institution Press), 229–47.

Cock, A. G. 1983. William Bateson's rejection and eventual acceptance of the chromosome theory. *Annals of Science* 40:19–59.

Cole, L. J. 1940. Introduction. *American Naturalist* 74:193–97.

Coleman, William. 1970. Bateson and chromosomes: Conservative thought in science. *Centaurus* 15:228–314.

———. 1971. *Biology in the Nineteenth Century*. New York: Wiley.

Collins, Harry. 1985. *Changing Order: Replication and Induction in Scientific Practice*. London/Beverly Hills: Sage.

Conze, Werner, and Jürgen Kocka. 1985. Einleitung. In Conze and Kocka (eds.), *Bildungsbürgertum im 19. Jahrhundert, Teil I: Bildungssystem und Professionalisierung in internationalen Vergleichen* (Stuttgart: Klett-Cotta), 9–28.

Corner, G. W. 1974. Oscar Riddle, 1877–1968. *Biographical Memoirs of the National Academy of Sciences* 45:427–65.

Correns, Carl. 1901. Die Ergebnisse der neuesten Bastardforschungen für die Vererbunsglehre. *Berichte der Deutschen Botanischen Gesellschaft* 19:71–94.

———. 1904. Experimentelle Untersuchungen über die Entstehung der Arten auf botanischem Gebiet. *Archiv für Rassen- und Gesellschaftsbiologie* 1:27–52.

———. 1908–09. Vererbungsversuche mit blass(gelb)-grünen und buntblättrigen Sippen bei Mirabilis, Urtica, und Lunaria. *ZIAV* 1:291–329.
———. 1909. Zur Kenntnis der Rolle von Kern und Plasma bei der Vererbung. *ZIAV* 2:331–40.
———. 1915. Antrittsrede. *Sitzungsberichte der Preussischen Akademie der Wissenschaften*, 499–501.
———. 1921. Die ersten 20 Jahre Mendelscher Vererbungslehre. In Carl Neuberg (ed.), *Festschrift der Kaiser-Wilhelm Gesellschaft zu ihrem 10. Jahres Jubiläum* (Berlin: Springer), 42–49.
———. 1922. Vererbungsversuche mit buntblättrigen Sippen. VI: Einige neue Fälle von Albomaculatio. VII: Über die peraurea-Sippe des Urtica urens. *Sitzungsberichte der Preussischen Akademie der Wissenschaften, phys-math Klasse*, 460–86.
———. 1928. Über nichtmendelnde Vererbung. *ZIAV* Supp. Band 1:131–68.
———. 1937. (Ed. F. von Wettstein.) Nicht mendelnde Vererbung. *Handbuch der Vererbungswissenschaft* IIH:1–132.
Coser, Lewis. 1965. *Men of Ideas: A Sociologist's View*. New York: Free Press.
Cravens, Hamilton. 1977. The role of universities in the rise of experimental biology. *Science Teacher* 44:33–37.
———. 1978. *The Triumph of Evolution*. Philadelphia: University of Pennsylvania Press.
Crosland, Maurice. 1977. History of science in a national context. *British Journal for the History of Science* 10:95–113.
Cross, Stephen, and W. R. Albury. 1987. Walter B. Cannon, L. J. Henderson, and the organic analogy. *Osiris* 3:165–92.
Crow, James F. 1980. Wilson Stone (1907–1968). *Biographical Memoirs of the National Academy of Sciences (U.S.)* 52:451–70.
Dahms, H.-J. 1987a. Einleitung. In H. Becker, H.-J. Dahms, and C. Wegeler (eds.), *Die Universität Göttingen unter dem Nationalsozialismus* (Munich: Saur), 15–60.
———. 1987b. Aufstieg und Ende der Lebensphilosophie: Das Philosophische Seminar der Universität Göttingen zwischen 1917 und 1950. In H. Becker, H.-J. Dahms, and C. Wegeler (eds.), *Die Universität Göttingen unter dem Nationalsozialismus* (Munich: Saur), 169–99.
Dahrendorf, Ralf. 1967. *Democracy and Society in Germany*. Garden City, N.Y.: Doubleday.
Danbom, D. B. 1986. The agricultural experiment station and professionalization: Scientists' goals for agriculture. *Agricultural History* 60:246–55.
Danziger, Kurt. 1979. The social origins of modern psychology. In A. Buss (ed.), *Psychology in Social Context* (New York: Irvington), 27–45.
Darlington, C. D. 1944. Heredity, development, and infection. *Nature* 154:164–69.
Deak, Istvan. 1968. *Weimar Germany's Left-Wing Intellectuals*. Berkeley: University of California Press.

Deichmann, Ute, 1991. *Biologen unter Hitler: die Vertreibung der jüdischen Biologen und die biologische Forschung in Deutschland, 1933–1945* Ph.D. diss., Universität zu Köln.

Dembowski, Jan. 1925. Zur Kritik der Faktoren- und Chromosomenlehre. *ZIAV* 41:216–47.

Dempsey, Ellen. 1973. Random observations on a distinguished professor. *Theoretical and Applied Genetics* 43:97–100.

Dingler, Hugo. 1930. Das Privatdozententum. In Michael Doeberl and Otto Scheel et al. (eds.), *Das Akademische Deutschland,* vol. 3 (Berlin: Weller), 205–18.

Dobzhansky, Th. 1937. *Genetics and the Origin of Species.* New York: Columbia University Press. Published in translation as *Die genetischen Grundlagen der Artenbildung.* Jena: G. Fischer, 1939.

———. 1940a. Speciation as a stage in evolutionary divergence. *American Naturalist* 74:312–21.

———. 1940b. Catastrophism versus evolutionism [review of Goldschmidt, *The Material Basis of Evolution*]. *Science* 92:356–58.

———. 1978. L. C. Dunn. *Biographical Memoirs of the National Academy of Sciences* 49:79–106.

———. 1980a. Morgan and his school in the 1930s. In Mayr and Provine (eds.), *The Evolutionary Synthesis,* 445–52.

———. 1980b. The birth of the genetic theory of evolution in the Soviet Union in the 1920s. In Mayr and Provine (eds.), *The Evolutionary Synthesis,* 229–42.

———. n.d. The reminiscences of Th. Dobzhansky. Columbia University Oral History Project, typescript.

Doerry, Martin. 1986. *Übergangsmenschen: Die Mentalität der Wilhelminer und die Krise des Kaiserreichs,* 2 vols. Weinheim/Munich: Juventa.

Dolby, Alex. 1977. The transmission of two new scientific disciplines from Europe to North America in the late nineteenth century. *Annals of Science* 34:287–310.

Döring, Herbert. 1975. *Der Weimarer Kreis: Studien zum politischen Bewusstsein verfassungstreuer Hochschullehrer in der Weimarer Republik.* Meisenheim: Anton Hain.

Douglas, Mary (ed.). 1982. *Essays in the Sociology of Perception.* London: Routledge and Kegan Paul.

Duhem, Pierre. 1954. *The Aim and Structure of Physical Theory.* Princeton: Princeton University Press.

Dunn, L. C. 1917. Nucleus and cytoplasm as vehicles of heredity. *American Naturalist* 51:286–300.

———. 1965a. W. E. Castle, 1867–1962. *Biographical Memoirs of the National Academy of Sciences (U.S.)* 38:33–80.

———. 1965b. *A Short History of Genetics.* New York: McGraw-Hill.

———. n.d. The reminiscences of L. C. Dunn. Columbia University Oral History Project, typescript.

Dupree, A. Hunter. 1957. *Science in the Federal Government: A History of Policies and Activities to 1940*. Cambridge: Harvard University Press.
Dürken, Bernhard. 1928. *Lehrbuch der Experimentalzoologie*, 2d ed. Berlin: Bornträger.
Dürken, Bernhard, and Hans Salfeld. 1921. *Die Phylogenese*. Berlin: Bornträger.
East, E. M. 1934. The nucleus-plasma problem. *American Naturalist* 68:289–303, 402–39.
Eicher, Eva. 1987. Ernst W. Caspari: Geneticist, teacher, and mentor. *Advances in Genetics* 24:xv–xxix.
Eichler, Wolf-Dietrich. 1982. Zum Gedenken an N. W. Timofeeff-Ressovsky (1900–1981). *Deutsche Entomologische Zeitschrift* N.F. 29:287–91.
Eisler, Colin. 1969. *Kunstgeschichte* American style. In D. Fleming and B. Bailyn (eds.), *The Intellectual Migration: Europe and America, 1930–1960* (Cambridge: Harvard University Press), 544–629.
Elias, Norbert. 1978. *The Civilizing Process: The Development of Manners*. New York: Urizen.
Emerson, Stirling. 1971. A. H. Sturtevant. *Annual Reviews of Genetics* 5:1–4.
———. 1973. E. G. Anderson. *Genetics* 74:19–20.
Engel, A. J. 1983. *From Clergyman to Don: The Rise of the Academic Profession in Nineteenth Century Oxford*. Oxford: Clarendon.
Epstein, Fritz T. 1960. Wartime activities of the SS-Ahnenerbe. In Max Beloff (ed.), *On the Track of Tyranny* (London: Vallentine, Mitchell), 77–95.
Escherich, K. 1944. *Leben und Forschen: Kampf um eine Wissenschaft*. Berlin: Wissenschaftliche Verlagsgesellschaft.
Eulenburg, Franz. 1908. *Der "Akademische Nachwuchs": Eine Untersuchung über die Lage und die Aufgaben der Extraordinarien und Privatdozenten*. Leipzig/Berlin: Teubner.
Fankhauser, G. 1972. Memories of great embryologists: Reminiscences of Fritz Baltzer, Hans Spemann, F. R. Lillie, R. G. Harrison, and E. G. Conklin. *American Scientist* 60:46–55.
Federley, Harry. 1930. Weshalb lehnt die Genetik die Annahme einer Vererbung erworbener Eigenschaften ab? *ZIAV* 54:20–50.
Feinberg, B., and R. Kasrils (eds.). 1973. *Bertrand Russell's America: His Transatlantic Travels and Writings*, vol. 1 (1896–1945). London: Allen and Unwin.
Feldman, Gerald. 1973. A German scientist between illusion and reality: Emil Fischer, 1909–1919. In I. Geiss and B.-J. Wendt (eds.), *Deutschland in der Weltpolitik des 19. und 20. Jahrhunderts* (Düsseldorf: Bertelsmann), 341–62.
———. 1987. The politics of *Wissenschaftspolitik* in Weimar Germany: A prelude to the dilemmas of twentieth-century science policy. In Charles S. Maier (ed.), *Changing Boundaries of the Political* (Cambridge: Cambridge University Press), 255–85.
———. 1990. The private support of science in Germany, 1900–1933. In R. vom Bruch and R. A. Müller (eds.), *Formen ausserstaatlicher Wissenschaftsförderung im 19. und 20. Jahrhundert* (Stuttgart: Franz Steiner), 87–111.

von Ferber, Christian. 1956. *Die Entwicklung des Lehrkörpers der deutschen Universitäten und Hochschulen, 1864–1954*. Göttingen: Vandenhoeck und Ruprecht.

Fick, Rudolf. 1925. Bemerkungen über einige Vererbungslehren. *Die Naturwissenschaften* 13:524–29.

Fischbeck, Gerhard. 1979. Ehrenpromotion . . . Prof. Dr. Hermann Kuckuck. *Jahrbuch 1979 der Technischen Universität München*.

Fischer, Eugen. 1933. Erwin Baur. *Zeitschrift für Ethnologie* 65:390–93.

Fischer, Peter. 1985. *Licht und Leben: Ein Bericht über Max Delbrück*. Konstanz: Universitätsverlag.

Fitzgerald, Deborah. 1990. *The Business of Breeding: Hybrid Corn in Illinois, 1890–1940*. Ithaca, N.Y.: Cornell University Press.

Flexner, Abraham. 1930. *Universities: American, English, German*. London: Oxford University Press.

Fliess, Gerhard. 1959. *Die politische Entwicklung der Jenaer Studentenschaft 1918–1930* Ph.D. diss.: Universität Jena.

Forman, Paul. 1971. Weimar culture, causality, and quantum theory, 1918–1927: Adaptation by German physicists and mathematicians to a hostile intellectual environment. *Historical Studies in the Physical Sciences* 3:1–115.

———. 1984. *Kausalität, Anschaulichkeit,* and *Individualität,* or how cultural values prescribed the character and lessons ascribed to quantum mechanics. In Nico Stehr and Volker Meja (eds.), *Society and Knowledge* (New Brunswick, N.J.: Transaction Books), 333–47.

Forman, Paul, John Heilbron, and Spencer Weart. 1975. Physics circa 1900: Personnel, funding, and productivity of the academic establishments. *Historical Studies in the Physical Sciences* 5:1–185.

Foster, Michael. 1899. Integration in science. *Naturalist* (Hull) (July), 209–23.

Freye, H.-A. 1965. Valentin Haecker (1864–1927) und die Phänogenetik. *Zoologischer Anzeiger* 174:401–10.

Fricke, Dieter. 1964. *Julius Schaxel, 1887–1943: Leben und Kampf eines marxistischen deutschen Naturwissenschaftlers und Hochschullehrers*. Leipzig: Urania.

Frisby, David. 1985. *Fragments of Modernity: Theories of Modernity in the Work of Simmel, Kracauer, and Benjamin*. Cambridge: Polity Press.

von Frisch, Karl. 1960–61. Otto Renner. *Orden Pour le Merite für Wissenschaften und Künste: Reden und Gedenkworte* 4:129–37.

———. 1967. *A Biologist Remembers*, trans. L. Gombrich. London: Pergamon.

Fruton, Joseph. 1985. Contrasts in scientific style: Emil Fischer and Franz Hofmeister; their research groups and their theory of protein structure. *Proceedings of the American Philosophical Society* 129:313–53.

———. 1990. *Contrasts in Scientific Style: Research Groups in the Chemical and Biochemical Sciences*. Philadelphia: American Philosophical Society.

Gaffron, Hans. 1969. Resistance to knowledge. *Annual Reviews of Plant Physiology* 20:1–40.

Galtung, Johan. 1981. Structure, culture, and intellectual style: An essay com-

paring saxonic, teutonic, gallic, and nipponic approaches. *Social Science Information* 20:817–56.
Gay, Peter. 1974. *Weimar Culture: The Outsider as Insider.* Harmondsworth: Penguin.
Geiger, Roger. 1986. *To Advance Knowledge: The Growth of American Research Universities, 1900–1940.* New York: Oxford University Press.
Geiger, Theodor. 1932. *Die soziale Schichtung des deutschen Volkes.* Stuttgart: F. Enke.
Geison, Gerald. 1978. *Michael Foster and the Cambridge School of Physiology.* Cambridge: Cambridge University Press.
Geith, K. 1924. Experimentell-systematische Untersuchungen in der Gattung Epilobium. *Botanisches Archiv* 6:123–39.
Gerth, H. H., and C. Wright Mills (eds.). 1970. *From Max Weber: Essays in Sociology.* London: Routledge and Kegan Paul.
Gilbert, Scott. 1988. Cellular politics: Ernest Everett Just, Richard B. Goldschmidt, and the attempt to reconcile embryology and genetics. In R. Rainger, K. Benson, and J. Maienschein (eds.), *The American Development of Biology,* 311–46.
von Gizycki, R., and F. Pfetsch. 1975. Die Gesellschaft deutscher Naturforscher und Ärzte: Bildung von Sektionen und Abspaltung von Gesellschaften. In F. Pfetsch (ed.), *Innovationsforschung als multidisziplinäre Aufgabe* (Göttingen: Vandenhoeck und Ruprecht), 101–53.
Gluecksohn-Waelsch, S. 1974. L. C. Dunn, 1893–1974. *Genetics* 77:99–100.
———. 1983. Fifty years of developmental genetics. *Transactions of the New York Academy of Sciences* 41:243–51.
Glum, Friedrich. 1928. Die Kaiser-Wilhelm Gesellschaft zur Förderung der Wissenschaften. In Adolf von Harnack (ed.), *Handbuch der Kaiser-Wilhelm Gesellschaft zur Förderung der Wissenschaften* (Berlin: Reimar Hobbing), 11–37.
Goldschmidt, Richard. 1911. *Einführung in die Vererbungswissenschaft.* Leipzig/Berlin: W. Engelmann; subsequent eds. in 1913, 1919, 1923, and 1928.
———. 1916. Theodor Boveri. *Science* 43:263–70.
———. 1924a. Untersuchungen zur Genetik der geographischen Variation. *Archiv für mikroskopische Anatomie und Entwicklungsmechanik* 101:92–337.
———. 1924b. Einige Probleme der heutigen Vererbungswissenschaft. *Die Naturwissenschaften* 12:769–71.
———. 1927. *Physiologische Theorie der Vererbung.* Berlin: Springer.
———. 1929. Experimentelle Mutation und das Problem der sogenannten Parallelinduktion: Versuche an Drosophila. *Biologisches Zentralblatt* 49:437–48.
———. 1933a. Protoplasmatische Vererbung. *Scientia* (Feb. 1933), 1–6.
———. 1933b. Some aspects of evolution. *Science* 78:539–47.
———. 1933c. Untersuchungen zur Genetik der geographischen Variation VII. *Archiv für Entwicklungsmechanik* 130:562–615.

———. 1934. The influence of the cytoplasm upon gene-controlled heredity. *American Naturalist* 68:5–23.

———. 1938. *Physiological Genetics*. New York: McGraw-Hill.

——— 1940. *The Material Basis of Evolution*. New Haven, Conn.: Yale University Press; Yale reprint, ed. Stephen J. Gould, 1982.

———. 1950. Fifty years of genetics. *American Naturalist* 84:313–39.

———. 1954. Different philosophies of genetics. *Science* 119:703–10.

———. 1956. *Portraits from Memory: Recollections of a Zoologist*. Seattle: University of Washington Press.

———. 1960. *In and Out of the Ivory Tower*. Seattle: University of Washington Press.

———. 1961. *Theoretische Genetik*. Berlin: Akademie-Verlag.

Golinski, Jan. 1990. The theory of practice and the practice of theory: Sociological approaches in the history of science. *Isis* 81:492–505.

Gombrich, Ernst. 1968. Style. In D. L. Sills (ed.), *International Encyclopedia of the Social Sciences* (New York: Macmillan), 353–61.

Gould, Stephen Jay. 1977. *Ontogeny and Phylogeny*. Cambridge: Harvard University Press.

———. 1983. The hardening of the modern synthesis. In M. Grene (ed.), *Dimensions of Darwinism*, 71–93.

Gramsci, Antonio. 1971. The intellectuals. In *Selections from the Prison Notebooks of Antonio Gramsci*, ed. Quintin Hoare and G. N. Smith (London: Lawrence and Wishart), 5–23.

Grasse, G. (ed.). 1972. *Alfred Kühn zum Gedachtnis* (= 5. *Biologisches Jahresheft 1972*). Iserlohn: Verband Deutscher Biologen.

Grau, C., W. Schlicker, and L. Zeil (eds.). 1979. *Die Berliner Akademie der Wissenschaften in dem Zeitalter des Imperialismus, Bd. 3 (1933–1945)*. Berlin: Akademie-Verlag.

Graven, Hubert. 1930. Gliederung der heutigen Studentenschaft nach statistischen Ergebnissen. In M. Doeberl and Otto Scheel et al. (eds.), *Das Akademische Deutschland*, vol. 3 (Berlin: Weller), 317–49.

Green, Martin. 1974. *The von Richthofen Sisters: The Triumphant and the Tragic Modes of Love*. London: Weidenfeld and Nicolson.

Gregor-Dellin, Martin. 1988. Unscharfe Welten: Einige Abweichungen in den Lebensläufen Heisenbergs. *Frankfurter Allgemeine Zeitung*, 16 Jan. 1988.

Grene, Marjorie (ed.). 1983. *Dimensions of Darwinism*. Cambridge: Cambridge University Press.

Grossbach, Ulrich. 1988. Der Weg zur allgemeinen Biologie: Entwicklungsbiologie und Genetik in Göttingen. *Göttinger Universitätsschriften* series A, vol. 13:85–97.

Haberlandt, G. 1933. Gedächtnisrede auf C. Correns. *Sitzungsberichte der preussischen Akademie der Wissenschaften, Phys-Math Kl.*, Sonderausgabe, 3–10.

Habermas, Jürgen. 1971. The intellectual and social background of the German university crisis. *Minerva* 9:422–28.

Haecker, Rudolf. 1965. Das Leben von Valentin Haecker. *Zoologischer Anzeiger* 174:1–22.
Haecker, Valentin. 1911. *Allgemeine Vererbungslehre*. Braunschweig: Vieweg; subsequent eds. in 1912 and 1921.
———. 1918. *Entwicklungsgeschichtliche Eigenschaftsanalyse (Phänogenetik)*. Jena: G. Fischer.
———. 1922. Einfach-mendelnde Merkmale. *Genetica* 4:195–234.
———. 1923. Einige Aufgaben der Phänogenetik. *Studia Mendeliana* (Brünn), 78–91.
———. 1925a. Aufgaben und Ergebnisse der Phänogenetik. *Bibliographia Genetica* 1:93–314.
———. 1925b. *Pluripotenzerscheinungen: Synthetische Beiträge zur Vererbungs- und Abstammungslehre*. Jena: G. Fischer.
———. 1926. Phänogenetisch gerichtete Bestrebungen in Amerika. *ZIAV* 40:232–38.
———. 1927. *Goethes Morphologische Arbeiten und die neuere Forschung*. Jena: G. Fischer.
Hagemann, R. 1984. Professor H. Stubbe zum 80. Geburtstag... *Wissenschaftliche Zeitung der Universität Halle*, Math-Naturw. Reihe, vol. 33:95–99.
Hagen, Joel. 1984. Experimentalists and naturalists in twentieth century botany: Experimental taxonomy, 1920–1950. *Journal of the History of Biology* 17:249–70.
Hahn, H., O. Neurath, and R. Carnap. 1973. Wissenschaftliche Weltauffassung: Der Wiener Kreis (1929). English reprint. Dordrecht: Reidel.
Hamburger, Viktor. 1980a. Embryology and the modern synthesis in evolutionary theory. In Mayr and Provine (eds.), *The Evolutionary Synthesis*, 97–112.
———. 1980b. Evolutionary theory in Germany: A comment. In Mayr and Provine (eds.), *The Evolutionary Synthesis*, 303–8.
———. 1984. Hilde Mangold, co-discoverer of the organizer. *Journal of the History of Biology* 17:1–11.
———. 1988. *The Heritage of Experimental Embryology*. Oxford: Oxford University Press.
Hämmerling, Joachim. 1929. Dauermodifikationen. *Handbuch der Vererbungswissenschaft* IE:1–65.
———. 1934a. Entwicklungs-physiologische und genetische Grundlagen der Formbildung bei der Schirmalge *Acetabularia*. *Die Naturwissenschaften* 22:829–36.
———. 1934b. Die Bedeutung des Zellkerns für die Lebensvorgänge. *Forschungen und Fortschritte* 10:262–64.
———. 1935. Über Genomwirkungen und Formbildungsfähigkeit bei *Acetabularia*. *Archiv für Entwicklungsmechanik* 132:424–62.
Harder, Richard. 1927. Zur Frage nach der Rolle von Kern und Protoplasma im Zellgeschehen und bei der Übertragung von Eigenschaften. *Zeitschrift für Botanik* 19:337–407.

———. 1928. Die Rolle des Zellplasmas bei der Übertragung von Eigenschaften. *Medizinische Welt* 40:1485–87 and 44:1629–30.

———. 1929. Über den Anteil des Kerns und des Plasmas an die Vererbung. *Unterrichtsblätter für Mathematik und Naturwissenschaft* 35:11–18.

von Harnack, Adolf (ed.). 1928. *Handbuch der Kaiser-Wilhelm Gesellschaft zur Förderung der Wissenschaften*. Berlin: Reimar Hobbing.

Harris, Henry. 1982. Joachim Hämmerling, 1901–1980. *Biographical Memoirs of Fellows of the Royal Society* 28:111–24.

Harrison, R. G. 1937. Embryology and its relations. *Science* 85:369–74.

Hart, James M. 1878. *German Universities: A Narrative of Personal Experience* ... New York: Putnams.

Hartmann, Max. 1934a. Erwin Baur. *Die Naturwissenschaften* 22:258–60.

———. 1934b. Der Naturforscher Correns: Ein Porträt. *Geistige Arbeit: Zeitung aus der wissenschaftlichen Welt* (20 Feb.), 12.

———. 1939. Eröffnungsansprache ... bei der 13. Jahresversammlung der Deutschen Gesellschaft für Vererbungswissenschaft. *ZIAV* 76:11–13.

Harvey, R. D. 1985. The William Bateson letters at the John Innes Institute. *Mendel Newsletter* no. 25:1–11.

Harwood, Jonathan. 1984. The reception of Morgan's chromosome theory in Germany: Inter-war debate over cytoplasmic inheritance. *Medizinhistorisches Journal* 19:3–32.

———. 1985. Geneticists and the evolutionary synthesis in inter-war Germany. *Annals of Science* 42:279–301.

———. 1986. Ludwik Fleck and the sociology of knowledge. *Social Studies of Science* 16:173–87.

———. 1987. National styles in science: Genetics in Germany and the United States between the world wars. *Isis* 78:390–414.

———. 1989. Genetics, eugenics, and evolution. *British Journal for the History of Science* 22:257–65.

Hawkins, Hugh. 1979. University identity: The teaching and research functions. In Alexandra Oleson and John Voss (eds.), *The Organization of Knowledge in Modern America, 1860–1920* (Baltimore: Johns Hopkins University Press), 285–312.

Heberer, Gerhard (ed.). 1943. *Die Evolution der Organismen: Ergebnisse und Probleme der Abstammungsgeschichte*. Jena: G. Fischer.

———. 1964. Valentin Haecker—Klassiker der Genetik. *Frankfurter Allgemeine Zeitung*, 15.9.64.

Heilbron, John. 1985. The earliest missionaries of the Copenhagen spirit. *Revue d'histoire des sciences* 38:194–230.

———. 1986. *The Dilemmas of an Upright Man: Max Planck as Spokesman for German Science*. Berkeley: University of California Press.

Heimans, J. 1978. Hugo deVries and the gene theory. In Eric Forbes (ed.), *The Human Implications of Scientific Advance* (Edinburgh: Edinburgh University Press), 469–80.

Hendry, John. 1980. Weimar culture and quantum causality. *History of Science* 18:155–80.
Henke, Karl. 1928. Moderne Malerei. In Dr. Erasmus (ed.), *Geist der Gegenwart: Formen, Kräfte, und Werte einer neuen deutschen Kultur* (Stuttgart: Stuttgarter Verlagsinstitut), 213–78.
———. 1930. Wissenschaftliche Erziehung in den Naturwissenschaften. *Handbuch der Pädagogik* 3:384–407.
———. 1955. Forschung und Lehre heute; zum 70. Geburtstag von Alfred Kühn. *Die Naturwissenschaften* 42:193–99.
Henning, Eckart, and Marion Kazemi. 1988. *Chronik der Kaiser-Wilhelm-Gesellschaft zur Förderung der Wissenschaften*. Berlin: Archiv zur Geschichte der Max-Planck-Gesellschaft.
Herf, Jeffrey. 1984. *Reactionary Modernism: Technology, Culture, and Politics in Weimar and the Third Reich*. Cambridge: Cambridge University Press.
Herter, Konrad. 1979. *Begegnungen mit Menschen und Tieren: Erinnerungen eines Zoologen, 1891–1978*. Berlin: Duncker und Humblot.
Hertwig, Paula. 1927. Partielle Keimesschädigungen durch Radium und Röntgenstrahlen. *Handbuch der Vererbungswissenschaft*, vol. 3.
———. 1934. Probleme der heutigen Vererbungslehre. *Die Naturwissenschaften* 22:425–30.
———. 1936. Artbastarde bei Tieren. *Handbuch der Vererbungswissenschaft* IIB:1–140.
———. 1950. Hans Kappert zum 60. Geburtstag. *Zeitschrift für Pflanzenzüchtung* 29:133–34.
———. 1956. Elisabeth Schiemann zum 75. Geburtstag. *Zeitschrift für Pflanzenzüchtung* 36:129–32.
———. 1962. Hans Stubbe zum 60. Geburtstag. *Biologisches Zentralblatt* 81:1–4.
Herz, J. 1923. Die experimentelle Vererbungslehre. *Die Naturwissenschaften* 11:833–42.
Hesse, Mary. 1966. *Models and Analogies in Science*. South Bend, Ind.: Notre Dame University Press.
Hirsch, Walter. 1975. The autonomy of science in totalitarian societies: The case of Nazi Germany. In K. Knorr, H. Strasser, and H. G. Zilian (eds.), *Determinants and Controls of Scientific Development* (Dordrecht: Reidel), 343–66.
Hoch, Paul. 1983. The reception of central European refugee physicists of the 1930s: USSR, UK, USA. *Annals of Science* 40:217–46.
Hoffmann, W. 1971. Nachruf auf Gustav Becker. *Zeitschrift für Pflanzenzüchtung* 65:89–94.
Höfler, K. 1949. Hans Winkler. *Almanach der Österreichischen Akademie der Wissenschaften*, 247–50.
Hofstadter, Richard. 1963. *Anti-Intellectualism in American Life*. New York: Random House.

Hofstadter, Richard, and Walter Metzger. 1955. *The Development of Academic Freedom in the United States.* New York: Columbia University Press.
Holtfreter, Johannes. 1968. Address in honor of Viktor Hamburger. *Developmental Biology* supp. 2:ix–xx.
Holton, Gerald. 1973. *Thematic Origins of Scientific Thought.* Cambridge: Harvard University Press.
———. 1978. *The Scientific Imagination.* Cambridge: Cambridge University Press.
———. 1986. *The Advancement of Science and Its Burdens.* Cambridge: Cambridge University Press.
Hondelmann, W. 1984. The lupin—ancient and modern crop plant. *Theoretical and Applied Genetics* 68:1–9.
Horder, T. J., and P. Weindling. 1985. Hans Spemann and the organiser. In Horder, Witkowski, and Wylie (eds.), *A History of Embryology,* 183–242.
Horder, T. J., J. A. Witkowski, and C. C. Wylie (eds.). 1985. *A History of Embryology.* Cambridge: Cambridge University Press.
Horowitz, Horman. 1974(?). Neurospora and the beginnings of molecular genetics. *Neurospora Newsletter* no. 20:4–6 (copy in George Beadle papers).
———. 1979. Genetics and the synthesis of proteins. *Annals of the New York Academy of Sciences* 325:253–66.
Horton, Robin. 1971. African traditional thought and western science. In Michael F. D. Young (ed.), *Knowledge and Control* (London: Collier-Macmillan), 208–66.
Hughes, H. Stuart. 1952. *Oswald Spengler: A Critical Estimate.* New York: Scribner's.
———. 1959. *Consciousness and Society: The Reorientation of European Social Thought, 1890–1930.* London: MacGibbon and Kee.
———. 1962. Is the intellectual obsolete? In *An Approach to Peace, and Other Essays.* New York: Atheneum.
Hull, David L. 1973. *Darwin and His Critics: The Reception of Darwin's Theory of Evolution by the Scientific Community.* Chicago: University of Chicago Press.
Hunecke, Volker. 1979. Der "Kampf ums Dasein" und die Reform der technischen Erziehung im Denken Alois Riedlers. In Reinhard Rürup (ed.), *Wissenschaft und Gesellschaft: Beiträge zur Geschichte der T.H./T.U. Berlin, 1879–1979,* vol. 1. Berlin/Heidelberg: Springer.
Hünemörder, C., 1987. Zur Geschichte der Botanik in Hamburg bis 1945. *Berichte der Deutschen Botanischen Gesellschaft* 100:215–32.
Irwin, M. R., and R. W. Cumley. 1940. Speciation from the point of view of genetics. *American Naturalist* 74:222–31.
Jablonski, W. 1928. [Review of R. Hertwig, *Abstammungslehre und neuere Biologie*]. *Archiv für Rassen- und Gesellschaftsbiologie* 20:176–78.
Jacobs, Natasha. 1989. From unity to unity: Protozoology, cell theory, and the new concept of life. *Journal of the History of Biology* 22:215–42.

Jahn, Ilse, Rolf Löther, and Konrad Senglaub. 1982. *Geschichte der Biologie.* Jena: G. Fischer.
Jarausch, Konrad. 1982. *Students, Society, and Politics in Imperial Germany.* Princeton: Princeton University Press.
——— (ed.). 1983. *The Transformation of Higher Learning, 1860–1930.* Stuttgart: Klett-Cotta.
———. 1984. *Deutsche Studenten, 1800–1970.* Frankfurt: Suhrkamp.
———. 1988. The universities: An American view. In J. R. Dukes and J. Remak (eds.), *Another Germany: A Reconsideration of the Imperial Era* (Boulder, Colorado: Westview), 181–206.
Jastrow, I. 1930. Kollegiengelder und Gebühren. In Michael Doeberl and Otto Scheel et al. (eds.), *Das Akademische Deutschland,* vol. 3 (Berlin: A. Weller), 277–84.
Jastrow, Joseph. 1906. The academic career as affected by administration. *Science* 23:561–74.
Jenkins, J. A., 1967. Roy E. Clausen (1891–1956). *Biographical Memoirs of the National Academy of Sciences (U.S.)* 39:37–54.
Jennings, H. S. 1941. Raymond Pearl. *Genetics* 26.
———. 1943. Raymond Pearl (1879–1940). *Biographical Memoirs of the National Academy of Sciences (U.S.)* 22:295–347.
Jensen, Paul. 1919. Physiologische Bemerkungen zur Vererbungs- und Entwicklungslehre. *Die Naturwissenschaften* 9:519–24.
Johannsen, Wilhelm. 1909. *Elemente der exakten Erblichkeitslehre.* Jena: G. Fischer; subsequent eds. 1913 and 1926.
———. 1922. Hundert Jahre Vererbungsforschung. *Verhandlungen der Gesellschaft deutscher Naturforscher und Ärzte* 87:70–104.
———. 1923. Some remarks on units in heredity. *Hereditas* 4:133–41.
Johannson, Ivar. 1961. L. J. Cole. *Genetics* 46:1–4.
Johnson, Jeffrey. 1985. Academic chemistry in imperial Germany. *Isis* 76:500–524.
———. 1990. *The Kaiser's Chemists: Science and Modernization in Imperial Germany.* Chapel Hill: University of North Carolina Press.
Jollos, Victor. 1922. *Selektionslehre und Artbildung.* Jena: G. Fischer.
———. 1924. Untersuchungen über Variabilität und Vererbung bei Arcellen. *Archiv für Protistenkunde* 49:307–74.
———. 1930. Studien zum Evolutionsproblem I: Über die experimentelle Hervorrufung und Steigerung von Mutanten bei D. melanogaster. *Biologisches Zentralblatt* 50:541–54.
———. 1931a. Genetik und Evolutionsproblem. *Zoologischer Anzeiger* supp. 5:252–95.
———. 1931b. Die experimentelle Auslösung von Mutanten und ihre Bedeutung für das Evolutionsproblem. *Die Naturwissenschaften* 19:171–77.
———. 1933. Vererbung. *Neue Rundschau* (Dec.), 796–819.
———. 1934. Inherited changes produced by heat-treatment in D. melanogaster. *Genetica* 16:476–94.

———. 1935a. Studien zum Evolutionsproblem II: Dauermodifikationen und "plasmatische Vererbung" und ihre Bedeutung für die Entstehung der Arten. *Biologisches Zentralblatt* 55:390–436.

———. 1935b. Sind Dauermodifikationen—"Schwachmutationen" und der "Parallelismus von Modifikationen und Mutanten" eine Stütze für das lamarckistische Prinzip? *ZIAV* 69:418–25.

———. 1939. Grundbegriffe der Vererbungslehre: Insbesondere Mutation, Dauermodifikation, Modifikation. *Handbuch der Vererbungswissenschaft* ID: 1–106.

Jones, D. F. 1921. Meeting of the Geneticists Interested in Agriculture. *Science* 53:429–31.

———. 1939. E. M. East. *Genetics* 24.

———. 1945. E. M. East. *Biographical Memoirs of the National Academy of Sciences (U.S.)* 23:217–42.

Joravsky, David. 1983. The Stalinist mentality and the higher learning. *Slavic Review* 42:575–600.

Jost, L. 1933–34. [Review of G. Haberlandt, *Erinnerungen, Bekenntnisse, und Betrachtungen*]. *Zeitschrift für Botanik* 26:171–73.

Just, G. 1921. Die Gründungsversammlung der Deutschen Gesellschaft für Vererbungswissenschaft. *Aus der Natur* 18:211–16, 255–64.

Kaelble, Hartmut, 1973. Sozialer Aufstieg in Deutschland, 1850–1914. *Vierteljahrsschrift für Sozial- und Wirtschaftsgeschichte* 60:41–71.

Kappert, Hans. 1978. *Vier Jahrzehnte miterlebte Genetik*. Berlin/Hamburg: Parey.

Kargon, Robert H. (ed.). 1974. *The Maturing of American Science*. Washington, D.C.: American Association for the Advancement of Science.

———. 1982. *The Rise of Robert Millikan*. Ithaca, N.Y.: Cornell University Press.

Kay, Lily. 1986. *Cooperative Individualism and the Growth of Molecular Biology at the California Institute of Technology, 1928–1953*. Ph.D. diss., Johns Hopkins University.

———. 1989a. Selling pure science in wartime: The biochemical genetics of G. W. Beadle. *Journal of the History of Biology* 22:73–101.

———. 1989b. Beyond the organism: G. W. Beadle's approach to the gene problem. Paper given on 24.6.89 at conference of the Society for the History, Philosophy, and Social Studies of Biology, London, Ontario.

Keller, Evelyn Fox. 1983. *A Feeling for the Organism: The Life and Work of Barbara McClintock*. San Francisco: W. H. Freeman.

Kelly, Reece Conn. 1973. *National Socialism and German University Teachers*. Ph.D. diss., University of Washington.

Kenney, Martin. 1986. *Biotechnology: The University-Industry Complex*. New Haven: Yale University Press.

Kershaw, Ian. 1985. *The Nazi Dictatorship*. London: Edward Arnold.

Kevles, Daniel J. 1979. *The Physicists: The History of a Scientific Community in Modern America*. New York: Vintage.

———. 1980. Genetics in the United States and Great Britain: A review with speculations. *Isis* 71:441–55.

———. 1986. *In the Name of Eugenics: Genetics and the Uses of Human Heredity.* Harmondsworth: Penguin.

Kimmelman, Barbara. 1981. An effort in reductionist sociobiology: The Rockefeller Foundation and physiological genetics, 1930–1942. Unpublished typescript.

———. 1983. The American Breeders' Association: Genetics and eugenics in an agricultural context, 1903–1913. *Social Studies of Science* 13:163–204.

———. 1987. *A Progressive Era Discipline: Genetics at American Agricultural Colleges and Experiment Stations, 1900–1920.* Ph.D. diss., University of Pennsylvania.

———. 1992. Organisms and interests in scientific research: R. A. Emerson's plea for the unique contributions of agricultural genetics. In Adele Clarke and Joan Fujimura (eds.), *The Right Tools for the Job in 20th Century Life Sciences* (Princeton, NJ: Princeton University Press).

Kingsland, Sharon E., 1991. The battling botanist: Daniel Trembly MacDougal, mutation theory, and the rise of experimental evolutionary biology in America, 1900–1912. *Isis* 82:479–509.

Kleinhempel, D. 1985. Rudolf Schick: Begründer der Kartoffelforschung in Gross Lüsewitz. In Akademie der Landwirtschaftswissenschaften der DDR (ed.), *Rudolf Schick—80 Jahre* (Gross Lüsewitz: Institut für Kartoffelforschung), 34–47.

Knapp, Edgar. 1935a. Untersuchungen über die Wirkung von Röntgenstrahlen an den Lebermoos Sphaerocarpus mit Hilfe der Tetraden-Analyse, I. *ZIAV* 70:309–49.

———. 1935b. Zur Frage der genetischen Aktivität des Heterochromatins nach Untersuchungen am X=Chromosom von Sphaerocarpus Donnellii. *Berichte der Deutschen Botanischen Gesellschaft* 53:751–60.

———. 1936a. Zur Genetik von Sphaerocarpus (Tetradenanalytische Untersuchungen). *Berichte der Deutschen Botanischen Gesellschaft* 54:(58)–(69).

———. 1936b. Heteroploidie bei Sphaerocarpus. *Berichte der Deutschen Botanischen Gesellschaft* 54:346–61.

———. 1937. Crossing-over und Chromosomenreduktion. *ZIAV* 73:409–18.

———. 1938a. Über Fragen der Geschlechtsbestimmung nach Untersuchungen an Sphaerocarpus. *Berichte der Deutschen Botanischen Gesellschaft* 56:(36)–(37).

———. 1938b. Über genetisch bedeutsame Zellbestandteile ausserhalb der Chromosomen (eine theoretische Untersuchung). *Biologisches Zentralblatt* 58:411–25.

———. 1943. Eckhard Kuhn. *Berichte der Deutschen Botanischen Gesellschaft* 61:344–49.

———. 1944. Das Problem der Erbsubstanz. *Die Naturwissenschaften* 32:39–47.

Koestler, Arthur. 1975. *The Case of the Midwife Toad*. London: Pan Books.
Kohler, Robert E. 1979. Warren Weaver and the Rockefeller Foundation program in molecular biology. In Nathan Reingold (ed.), *The Sciences in the American Context: New Perspectives* (Washington, D.C.: Smithsonian Institution Press), 249–93.
———. 1982. *From Medical Chemistry to Biochemistry*. Cambridge: Cambridge University Press.
Köhler, Wolfgang. 1953. The scientists from Europe and their new environment. In Franz Neumann et al. (eds.), *The Cultural Emigration: The European Scholar in America* (Philadelphia: University of Pennsylvania Press), 112–37.
Kosswig, Curt. 1959. Paula Hertwig zum 70. Geburtstag. *Biologisches Zentralblatt* 78:671–73.
Krebs, Hans. 1981. *Reminiscences and Reflections*, ed. Anne Martin. Oxford: Oxford University Press.
Kuckuck, Hermann. 1980. Elisabeth Schiemann, 1881–1972. *Berichte der Deutschen Botanischen Gesellschaft* 93:517–37.
———. 1988. *Wandel und Beständigkeit im Leben eines Pflanzenzüchters*. Berlin/Hamburg: Parey.
Kuckuck, Hermann, and Hans Stubbe. 1970. Nachruf auf Rudolf Schick. *Theoretical and Applied Genetics* 40:1–5.
Kühn, Alfred. 1919. *Die Orientierung der Tiere im Raum*. Jena: G. Fischer.
———. 1922. *Grundriss der allgemeinen Zoologie*. Leipzig: Georg Thieme; 3d ed. 1928, 17th ed. 1968.
———. 1926. Ernst Ehlers zum Gedächtnis. *Zeitschrift für wissenschaftliche Zoologie* 128:I–XV.
———. 1927. Die Pigmentierung von *H. juglandis Ashmed*, ihre Prädetermination, und ihre Vererbung durch Gene und Plasmon. *Nachrichten der Gesellschaft der Wissenschaften zu Göttingen, Math-Phys Klasse*, 407–21.
———. 1932. Entwicklungs-physiologische Wirkungen einiger Gene von Ephestia kühniella. *Die Naturwissenschaften* 20:974–79.
———. 1932–33. Goethe und die Naturforschung. *Nachrichten der Gesellschaft der Wissenschaften zu Göttingen*, 47–69.
———. 1934a. Vererbung und Entwicklungs-physiologie. In W. Kolle (ed.), *Wissenschaftliche Woche zu Frankfurt*, vol. 1: "Erbbiologie," 37–48.
———. 1934b. Genetik und Physiologie. *Berichte über die gesamte Physiologie* 81:370–71.
———. 1934c. Genwirkung und Artveränderung. *Der Biologe* 3:217–27.
———. 1934d. Über den biologischen Wert von Mutationsrassen. *Forschungen und Fortschritte* 10:359–60.
———. 1934e. *Grundriss der Vererbungslehre*. Heidelberg: Quelle und Meyer; 2d ed. 1950, 4th ed. 1965.
———. 1935. Karl Heider und ein Entwicklungsabschnitt der Zoologie. *Die Naturwissenschaften* 23:791–96.

———. 1936. Versuch über die Wirkungsweise der Erbanlagen. *Die Naturwissenschaften* 24:1–10.
———. 1937a. Entwicklungs-physiologisch-genetische Ergebnisse an Ephestia kühniella Z. *ZIAV* 73:419–55.
———. 1937b. Kern- und Plasmavererbung. *Züchtungskunde* 12:433–49.
———. 1937c. [Review of O. Schindewolf, *Paläontologie, Entwicklungslehre, und Genetik*]. *Biologisches Zentralblatt* 57:551–53.
———. 1938a. Grenzprobleme zwischen Vererbungsforschung und Chemie. *Berichte der Deutschen Chemischen Gesellschaft* 71:107–14.
———. 1938b. Theodor Boveri. *Genetics* 23:i–vi.
———. 1939a. Über die fruchtbare Zusammenarbeit von Naturwissenschaft und Medizin. *Verhandlungen der Gesellschaft Deutscher Naturforscher und Ärzte*, v–x.
———. 1939b. H. Spemann zum 70. Geburtstag. *Die Naturwissenschaften* 27:425–26.
———. 1941. H. Spemann. *Forschungen und Fortschritte* 17:371.
———. 1942. H. Spemann. *Jahrbuch der Akademie der Wissenschaften in Göttingen*, 99–103.
———. 1946. Fritz Süffert zum Gedächtnis. *Die Naturwissenschaften* 33:161–63.
———. 1947. Fritz von Wettstein zum Gedächtnis. *Jahrbuch der Akademie der Wissenschaften zu Göttingen*, 1–6.
———. 1948a. Zum 70. Geburtstag Richard Goldschmidts am 12. April 1948. *Experientia* 4:239–42.
———. 1948b. Biologie der Romantik. In *Romantik: Ein Zyklus Tübinger Vorlesungen* (Tübingen/Stuttgart: Rainer Wunderlich/Hermann Leins), 215–34.
———. 1950. *Anton Dohrn und die Zoologie seiner Zeit.* Naples: Zoological Station.
———. 1952. Entwicklung und Problematik der Genetik. *Die Naturwissenschaften* 40:65–69.
———. 1955. *Vorlesungen über Entwicklungsphysiologie.* Berlin/Göttingen/Heidelberg: Springer; 2d ed. 1965.
———. 1957a. Karl Henke. *Die Naturwissenschaften* 44:25.
———. 1957b. August Weismann, 1834–1914. In Johannes Vincke (ed.), *Freiburger Professoren des 19. und 20. Jahrhunderts* (Freiburg i.Br.: Eberhard Albert), 191–99.
———. 1959. Alfred Kühn. *Nova Acta Leopoldina* 21:274–80.
———. 1960a. Nachruf auf Otto Renner. *Zeitschrift für Naturforschung* 15B:8.
———. 1960b. [Untitled obituary for Karl Henke]. *Jahrbuch der Akademie der Wissenschaften in Göttingen* (vol. for 1944–1960), 165–67.
Kühn, A., E. Caspari, and E. Plagge. 1935. Über hormonale Genwirkungen bei Ephestia kühniella Z. *Gesellschaft der Wissenschaften zu Göttingen, Nachr. aus der Biologie*, N.F. 2:1–29.

Kühn, Alfred, and K. Henke. [1929.] Genetisch und entwicklungs-physiologische Untersuchungen an der Mehlmotte E. kühnielle Zeller, I–VII. *Abhandlungen der Gesellschaft der Wissenschaften zu Göttingen, Math-Phys Klasse*, N.F. 15, no. 1:1–121.

———. 1932. ibid, VIII–XII, as above, 127–219.

———. 1936. ibid, XIII–XIV, as above, 225–72.

Kuhn, Harald. 1963. *Soziologie der Apotheker*. Stuttgart: G. Fischer.

Kuhn, T. S. 1970. *The Structure of Scientific Revolutions*, 2d ed. Chicago: University of Chicago Press.

Laitko, Hubert, E. Fabian, et al. (eds.). 1987. *Wissenschaft in Berlin: Von den Anfängen bis zum Neubeginn nach 1945*. Berlin.

Lakatos, Imre. 1970. Falsification and the methodology of scientific research programmes. In I. Lakatos and A. Musgrave (eds.), *Criticism and the Growth of Knowledge* (Cambridge: Cambridge University Press), 91–196.

Landauer, Walter. 1924. [Review of G. Just, *Praktische Übungen zur Vererbungslehre*]. *Die Naturwissenschaften* 12:462–63.

Lang, Anton. 1980. Some recollections and reflections. *Annual Reviews of Plant Physiology* 31:1–28.

———. 1987. Elisabeth Schiemann: Life and career of a woman scientist in Berlin. *Englera* 7:17–28.

Laqueur, Walter. 1974. *Weimar: A Cultural History, 1918–1933*. London: Weidenfeld and Nicolson.

Lash, Scott, and S. Whimster (eds.). 1987. *Max Weber: Rationality and Modernity*. London: Allen and Unwin.

Latour, Bruno, and Steve Woolgar. 1979. *Laboratory Life: The Social Construction of Scientific Facts*. London/Beverly Hills: Sage.

Law, John, and Ursula Sharma. 1977. [Extended review of Mary Douglas, *Implicit Meanings*]. *Sociological Review* 25:463–69.

Lawrence, Christopher. 1979. The nervous system and society in the Scottish Enlightenment. In Barry Barnes and Steven Shapin (eds.), *Natural Order: Historical Studies of Scientific Culture* (Beverly Hills and London: Sage), 19–40.

Lebovics, Herman. 1969. *Social Conservatism and the Middle Classes in Germany, 1914–1933*. Princeton, N.J.: Princeton University Press.

Lehmann, Ernst. 1928. Reziprok-verschiedene Bastarde in ihrer Bedeutung für das Kern-Plasma-Problem. *Tübinger Naturwissenschaftliche Abhandlungen* 11:5–39.

———. 1932. Der Anteil von Kern und Plasma an die reziproke Verschiedenheiten von Epilobium Bastarden. *Zeitschrift für Pflanzenzüchtung* A17:157–72.

———. 1936a. Versuch zur Klärung der reziproken Verschiedenheiten von Epilobium-Bastarden I: Der Tatbestand und die Möglichkeit seiner Klärung durch differente Wuchsstoffbildung. *Jahrbuch für wissenschaftliche Botanik* 82:657–68.

---. 1936b. *Wege und Ziele einer deutschen Biologie*. Munich: J. F. Lehmann.
---. 1938. Vererbungslehre, Rassenkunde, und Rassenhygiene. *Der Biologe* 7:306–10.
---. 1938–39. Zur Genetik der Entwicklung in der Gattung Epilobium. *Jahrbuch für wissenschaftliche Botanik* 87:625–41.
Lehmann, Ernst, and R. Beatus. 1931–32. Der Unterricht in der Vererbungswissenschaft an den deutschen Hochschulen. *Der Biologe* 1:89–96.
Lehmann, Ernst, and J. Schwemmle. 1927. Genetische Untersuchungen in der Gattung Epilobium. *Bibliotheca Botanica* 95:1–156.
Lenoir, Timothy. 1988. A magic bullet: Research for profit and the growth of knowledge in Germany around 1900. *Minerva* 26:66–88.
Lepsius, M. R. 1977. Social theories of "modernity" and "modernisation." In R. Koselleck (ed.), *Studien zum Beginn der modernen Welt*. Stuttgart: Klett-Cotta.
Lewis, D., and D. M. Hunt. 1984. Hans Grüneberg, 1907–1982. *Biographical Memoirs of Fellows of the Royal Society* 30:227–47.
Liebersohn, Harry. 1985. The American academic community before the First World War. In W. Conze and J. Kocka (eds.), *Bildungsbürgertum im 19. Jahrhundert: Bildungssystem und Professionalisierung in internationalen Vergleichen*, part 1. Stuttgart: Klett-Cotta.
Linskens, H. 1986. Friedrich Gustav Brieger, 1900–1985. *Berichte der Deutschen Botanischen Gesellschaft* 99:137–43.
Loeb, Jacques. 1916. *The Organism as a Whole*. New York: Putnam.
Ludmerer, Kenneth. 1972. *Genetics and American Society*. Baltimore: Johns Hopkins University Press.
Ludwig, Karl-Heinz. 1979. *Technik und Ingenieure im Dritten Reich*. Königstein/Düsseldorf: Athenäum und Droste.
Ludwig, Wilhelm. 1940. Selektion und Stammesentwicklung. *Die Naturwissenschaften* 28:689–705.
---. 1943. Die Selektionstheorie. In G. Heberer (ed.), *Die Evolution der Organismen*, 479–520.
Lungreen, Peter. 1981. Bildung und Besitz—Einheit oder Inkongruenz in der europäischen Sozialgeschichte? Kritische Auseinandersetzung mit einer These von Fritz Ringer. *Geschichte und Gesellschaft* 7:262–75.
---. 1983. Differentiation in German higher education. In Konrad Jarausch (ed.), *The Transformation of Higher Learning, 1860–1930* (Chicago: University of Chicago Press), 149–79.
---. 1985. Zur Konstituierung des "Bildungsbürgertums": Berufs- und Bildungsauslese der Akademiker in Preussen. In Werner Conze and Jürgen Kocka (eds.), *Bildungsbürgertum im 19. Jahrhundert, Teil I: Bildungssystem und Professionalisierung in internationalen Vergleichen* (Stuttgart: Klett-Cotta), 79–108.
MacKenzie, Donald. 1981. *Statistics in Britain, 1865–1930: The Social Construction of Scientific Knowledge*. Edinburgh: Edinburgh University Press.

MacKenzie, Donald, and B. Barnes. 1975. Biometriker versus Mendelianer; eine Kontroverse und ihre Erklärung. *Kölner Zeitschrift für Soziologie und Sozialpsychologie,* Sonderheft 18 ("Wissenschaftssoziologie"), 165–96.

Macrakis, Kristie. 1986. Wissenschaftsförderung durch die Rockefeller-Stiftung im "Dritten Reich": Die Entscheidung, das KWI für Physik finanziell zu unterstützen, 1934–39. *Geschichte und Gesellschaft* 12:348–79.

Mägdefrau, K. 1961. Otto Renner. *Berichte der Bayerischen Botanischen Gesellschaft* 34:103–13.

Maienschein, Jane. 1981. Shifting assumptions in American biology: Embryology, 1890–1910. *Journal of the History of Biology* 14:89–113.

———. 1983. Experimental biology in transition: Harrison's embryology, 1895–1910. *Studies in the History of Biology* 6:107–27.

———. 1984. What determines sex? A study of converging approaches, 1880–1916. *Isis* 75:457–80.

———. 1985. Preformation or new formation—or neither or both. In Horder, Witkowski, and Wylie (eds.), *A History of Embryology,* 73–108.

———. 1987. Physiology, biology, and the advent of physiological morphology. In Gerald Geison (ed.), *Physiology in the American Context, 1850–1940* (Bethesda: American Physiological Society), 177–93.

———. 1988. Whitman at Chicago: Establishing a Chicago style of biology? In Rainger, Benson, and Maienschein (eds.), *The American Development of Biology,* 151–82.

Mandler, Jean M., and George Mandler. 1969. The diaspora of experimental psychology: The Gestaltists and others. In D. Fleming and B. Bailyn (eds.), *The Intellectual Migration: Europe and America, 1930–1960* (Cambridge: Harvard University Press), 371–419.

Manegold, Karl-Heinz. 1970. *Universität, Technische Hochschule, und Industrie.* Berlin: Duncker und Humblot.

Mangelsdorf, P. C. 1955. G. H. Shull. *Genetics* 40:1–4.

Mannheim, Karl. 1932. [Review of Stuart Rice, ed., *Methods in Social Science*]. *American Journal of Sociology* 38:276–77.

———. 1953. Conservative thought. In *Essays in Sociology and Social Psychology* (London: Routledge and Kegan Paul), 74–164.

———. 1956. The problem of the intelligentsia: An enquiry into its past and present role. In *Essays in the Sociology of Culture* (London: Routledge and Kegan Paul), 91–170.

Manning, Kenneth R. 1983. *Black Apollo of Science: The Life of Ernest Everett Just.* New York: Oxford University Press.

Marquardt, Hans. 1974. Friedrich Oehlkers, 1890–1971. *Berichte der Deutschen Botanischen Gesellschaft* 87:185–92.

Mayr, Ernst. 1940. Speciation phenomena in birds. *American Naturalist* 74:249–78.

———. 1980a. Prologue: Some thoughts on the history of the evolutionary synthesis. In Mayr and Provine (eds.), *The Evolutionary Synthesis,* 1–48.

———. 1980b. Curt Stern. In Mayr and Provine (eds.), *The Evolutionary Synthesis,* 424–29.
———. 1980c. Introduction to section 9. In Mayr and Provine (eds.), *The Evolutionary Synthesis,* 279–84.
———. 1980d. How I became a Darwinian. In Mayr and Provine (eds.), *The Evolutionary Synthesis,* 413–23.
———. 1980e. The role of systematics in the evolutionary synthesis. In Mayr and Provine (eds.), *The Evolutionary Synthesis,* 123–36.
———. 1982. *The Growth of Biological Thought.* Cambridge, Mass.: Belknap Press.
Mayr, Ernst, and W. Provine (eds.). 1980. *The Evolutionary Synthesis: Perspectives on the Unification of Biology.* Cambridge: Harvard University Press.
McClelland, Charles E. 1980. *State, Society, and University in Germany, 1700–1914.* Cambridge: Cambridge University Press.
McMurrich, J. Playfair. 1974. A retrospect (1923). In Robert Kargon (ed.), *The Maturing of American Science* (Washington, D.C.: American Association for the Advancement of Science), 37–48.
Medvedev, Zhores. 1982. N. W. Timofeeff-Ressovsky (1900–1981). *Genetics* 100:1–5.
Mehrtens, Herbert. 1980. *Felix Hausdorff: Ein Mathematiker seiner Zeit.* Bonn: Fachschaftsrat Mathematik und Mathematisches Institut der Universität.
———. 1990. *Moderne, Sprache, Mathematik: Eine Geschichte des Streits um die Grundlagen der Disziplin und des Subjekts formaler Systeme.* Frankfurt: Suhrkamp.
Meijer, Onno. 1985. Hugo deVries no Mendelian? *Annals of Science* 42: 189–232.
Melchers, Georg. 1939. Genetik und Evolution (Bericht eines Botanikers). *ZIAV* 76:229–59.
———. 1953. Fritz von Wettstein (1895–1945). *Mitteilungen aus der Max-Planck-Gesellschaft* no. 6:11–15.
———. 1961. Otto Renner. *Mitteilungen aus der Max-Planck-Gesellschaft (1961),* 38–43.
———. 1972. Hans Stubbe zum 70. Geburtstag. *Theoretical and Applied Genetics* 42:1–2.
———. 1975. Theodor Schmucker, 1894–1970. *Berichte der Deutschen Botanischen Gesellschaft* 88:473–84.
———. 1987a. Ein Botaniker auf dem Wege in die Allgemeine Biologie auch in Zeiten moralischer und materieller Zerstörung. *Berichte der Deutschen Botanischen Gesellschaft* 100:373–95.
———. 1987b. Fritz von Wettstein, 1895–1945. *Berichte der Deutschen Botanischen Gesellschaft* 100:396–405.
Merriam, C. Hart. 1893. Biology in our colleges: A plea for a broader and more liberal biology. *Science* 21:352–55.
Merz, J. T. 1896–1914. *A History of European Thought in the Nineteenth Century,* 4 vols. Edinburgh: Blackwood.

Meyer-Abich, A. 1935. *Krisenepochen und Wendepunkte des biologischen Denkens.* Jena: G. Fischer.

Meyer-Thurow, Georg. 1982. The industrialisation of invention: A case study from the German chemical industry. *Isis* 73:363–81.

Michaelis, Peter. 1929. Über den Einfluss von Kern und Plasma auf die Vererbung. *Biologisches Zentralblatt* 49:302–16.

———. 1931. Die Bedeutung des Plasmas für die Pollenfertilität reziprokverschiedener Epilobium-bastarde. *Berichte der Deutschen Botanischen Gesellschaft* 49:96–104.

———. 1932. Über die Beziehungen zwischen Kern und Plasma bei den reziprok-verschiedenen Epilobium-bastarden. *ZIAV* 62:95–102.

———. 1933a. Entwicklungs-geschichtlich-genetische Untersuchungen an Epilobium I. *ZIAV* 65:1–71.

———. 1933b. Entwicklungs-geschichtlich-genetische Untersuchungen an Epilobium II: Die Bedeutung des Plasmas für die Pollenfertilität des Epilobium luteum-hirsutum Bastards. *ZIAV* 65:353–411.

———. 1935a. Entwicklungs-geschichtliche Untersuchungen an Epilobium III: Zur Frage der Übertragung von Pollenschlauchplasma in die Eizelle und ihre Bedeutung für die Plasmavererbung. *Planta* 23:486–500.

———. 1935b. Entwicklungs-geschichtliche Untersuchungen an Epilobium IV: Der Einfluss des Plasmons auf Verzweigung und Pilzresistenz. *Berichte der Deutschen Botanischen Gesellschaft* 53:143–50.

———. 1937. Untersuchungen zum Problem der Plasmavererbung. *Protoplasma* 27:284–89.

———. 1938. Über die Konstanz des Plasmons. *ZIAV* 74:435–59.

———. 1939a. Keimstimmung und Plasmavererbung. *Jahrbuch für wissenschaftliche Botanik* 88:69–88.

———. 1939b. Über den Einfluss des Plasmons auf die Manifestation der Gene. *ZIAV* 77:548–67.

———. 1940a. Über reziprok-verschiedene Sippenbastarde bei E. hirsutum I: Die reziprok-verschiedene Bastarde der Epilobium hirsutum- Sippe Jena. *ZIAV* 78:187–222.

———. 1940b. Über reziprok-verschiedene Sippenbastarde bei E. hirsutum III: Über die genischen Grundlagen der im Jena-Plasma auftretenden Hemmungsreihe. *ZIAV* 78:295–337.

———. 1941. Die Vererbung. In T. Roemer and W. Rudorf (eds.), *Handbuch der Pflanzenzüchtung* 1:99–149.

———. 1942a. Experimentelle Untersuchungen über die geographische Verbreitung von Plasmon-Unterschieden und der auf diese Unterschiede empfindlichen Gene, sowie deren theoretischen Bedeutung für das Kern-Plasma-Problem. *Biologisches Zentralblatt* 62:170–86.

———. 1942b. Experimentelle Untersuchungen ... VI: In welcher Weise sind an der Manifestation der im Jena-Plasma auftretenden Entwicklungstendenzen die Gene dieser Sippe beteiligt? *ZIAV* 80:454–99.

———. 1949. Prinzipielles und Problematisches zur Plasmavererbung. *Biologisches Zentralblatt* 68:173–95.
———. 1954. Cytoplasmic inheritance in Epilobium and its theoretical significance. *Advances in Genetics* 6:287–401.
———. 1963. Probleme, Methoden, und Ergebnisse der Plasmavererbung. *Die Naturwissenschaften* 50:581–85.
Michaelis, Peter, and M. von Dellingshausen. 1942. Experimentelle Untersuchungen . . . IV: Weitere Untersuchungen über die genischen Grundlagen der extrem stark gestörten Bastarde der E. hirsutum-Sippe Jena. *ZIAV* 80:373–428.
Michaelis, Peter, and E. Wertz. 1935. Entwicklungs-geschichtliche Untersuchungen an Epilobium VI: Vergleichende Untersuchungen über das Plasmon von E. hirsutum, E. luteum, E. montanum, und E. roseum. *ZIAV* 70:138–59.
Miller, Howard S. 1970. *Dollars for Research: Science and Its Patrons in Nineteenth Century America*. Seattle: University of Washington Press.
Mommsen, Wolfgang. 1974. *Max Weber und die deutsche Politik, 1890–1920*, 2d ed. (Tübingen: J. C. B. Mohr).
Monroy, Alberto, and Christiane Groeben. 1985. The "new" embryology at the Zoological Station and at the Marine Biological Laboratory. *Biological Bulletin* 168 (supp.): 35–43.
Morgan, T. H. 1917. The theory of the gene. *American Naturalist* 51:513–44.
———. 1921. *Die stoffliche Grundlage der Vererbung*, trans. H. Nachtsheim. Berlin: Bornträger.
———. 1923a. The modern theory of genetics and the problem of development. *Physiological Reviews* 3:603–27.
———. 1923b. The bearing of Mendelism on the origin of species. *Scientific Monthly* 16:237–47.
———. 1926a. Genetics and the physiology of development. *American Naturalist* 60:489–515.
———. 1926b. William Bateson. *Science* 63:531–35.
———. 1932. The rise of genetics. *Science* 76:261–67, 285–88.
Morgan, T. H., A. H. Sturtevant, H. J. Muller, and C. B. Bridges. 1915. *Mechanism of Mendelian Heredity*. New York: Holt.
Mulkay, Michael. 1972. *The Social Process of Innovation*. London: Macmillan.
Muller, H. J. 1929. The gene as the basis of life. In *Studies in Genetics* (Bloomington: Indiana University Press, 1962), 188–204.
———. 1934. Lenin's doctrines in relation to genetics. Reprinted in Loren Graham, *Science and Philosophy in the Soviet Union* (London: Allen Lane, 1973), 453–69.
———. 1949. E. B. Wilson. *Genetics* 34:1–9.
Nachtsheim, Hans. 1919. Die Analyse der Erbfaktoren bei Drosophila und deren zytologischen Grundlage: Ein Bericht über die bisherigen Ergebnisse der Vererbungsexperimente Morgans und seiner Mitarbeiter. *ZIAV* 20:118–56.

———. 1920. Ergebnisse und Fortschritte der Vererbungswissenschaft im Jahre 1920. *Jahresbericht über die gesamte Physiologie 1920,* 28–51.

———. 1921. Die Gründung der Deutschen Gesellschaft für Vererbungswissenschaft. *Die Naturwissenschaften* 9:844–50, 879–81.

———. 1922a. Kern und Plasma in ihrer Bedeutung für die Vererbung. *ZIAV* 27:249–51.

———. 1922b. Einige Ergebnisse und Fortschritte der Vererbungswissenschaft. *Jahresbericht über die gesamte Physiologie 1922,* 548–58.

———. 1927. Der V. Internationale Kongress für Vererbungswissenschaft. *Die Naturwissenschaften* 15:989–95.

———. 1959. Paula Hertwig, zu ihrem 70. Geburtstag. *Münchener Medizinische Wochenschrift* 101:1798–99.

———. 1966. Hans Nachtsheim. In W. E. Böhm and Gerda Paehlke (eds.), *Forscher und Gelehrte* (Stuttgart: Battenberg,), 190–91.

Nanney, David L. 1957. The role of the cytoplasm in heredity. In W. D. McElroy and H. B. Glass (eds.), *The Chemical Basis of Heredity* (Baltimore: Johns Hopkins University Press), 134–64.

Nauck, E. T. 1937. *Franz Keibel: Zugleich eine Untersuchung über das Problem des wissenschaftlichen Nachwuchs.* Jena: G. Fischer.

———. 1956. *Die Privatdozenten der Universität Freiburg, 1818–1955.* Freiburg i.Br.: Eberhard Albert.

Neel, J. V. 1987. Curt Stern. *Biographical Memoirs of the National Academy of Sciences (U.S.)* 56:443–74.

Nettl, J. P. 1970. Ideas, intellectuals, and structures of dissent. In Philip Rieff (ed.), *On Intellectuals* (Garden City, N.J.: Doubleday), 57–134.

Neumann, Franz. 1953. The social sciences. In Franz Neumann et al. (eds.), *The Cultural Emigration: The European Scholar in America* (Philadelphia: University of Pennsylvania Press), 4–26.

Neumann, Michael. 1987. Über den Versuch ein Fach zu verhindern: Soziologie in Göttingen 1920–1950. In H. Becker, H.-J. Dahms, and Cornelia Wegeler (eds.), *Die Universität Göttingen unter dem Nationalsozialismus* (Munich: Saur), 298–312.

Nicolson, Malcolm. 1989. National styles, divergent classifications: A comparative case study from the history of French and American plant ecology. *Knowledge and Society* 8:139–86.

Noack, K. L. 1925. [Review of Winkler, "Über die Rolle von Kern und Plasma bei der Vererbung," 1924]. *Zeitschrift für Botanik* 17:462–63.

Norton, Bernard. 1973. The biometric defence of Darwinism. *Journal of the History of Biology* 6:283–316.

Nyhart, Lynn. 1986. *Morphology and the German University, 1860–1900.* Ph.D. diss., University of Pennsylvania.

O'Boyle, Lenore. 1968. Klassische Bildung und soziale Struktur in Deutschland zwischen 1800 und 1848. *Historische Zeitschrift* 207:584–608.

Oehlkers, Friedrich. 1927. *Erblichkeitsforschung an Pflanzen: Eine Abriss ihrer Entwicklung in den letzten 15 Jahren.* Dresden/Leipzig: Theodor Steinkopff.

———. 1929–30. [Review of Renner, *Artbastarde bei Pflanzen*, 1929]. *Zeitschrift für Botanik* 22:538–40.

———. 1930. Der Aufbau der Biologie und ihre Stellung im System der Wissenschaften, in *Reichsgründungsfeier der Technischen Hochschule Darmstadt am 18. Januar 1930* (= *Schriften der Hessischen Hochschulen: T.H. Darmstadt* Heft 1), 9–20.

———. 1930–31. [Review of J. Hämmerling, "Dauermodifikationen," 1929]. *Zeitschrift für Botanik* 24:381.

———. 1931. Vererbung. *Fortschritte der Botanik* 1:222–36.

———. 1933–34. [Review of R. Woltereck, *Grundzüge einer allgemeinen Biologie*]. *Zeitschrift für Botanik* 26:27–28.

———. 1937. Vererbung. *Fortschritte der Botanik* 6:288–306.

———. 1938a. Vererbung. *Fortschritte der Botanik* 7:293–312.

———. 1938b. Bastardierungsversuche in der Gattung Streptocarpus Lindley. I: Plasmatische Vererbung und die Geschlechtsbestimmung von Zwitterpflanzen. *Zeitschrift für Botanik* 32:305–93.

———. 1940. Bastardierungsversuche . . . III: Neue Ergebnisse über die Genetik von Wuchsgestalt und Geschlechtsbestimmung. *Berichte der Deutschen Botanischen Gesellschaft* 58:76–91.

———. 1940–41. [Review of Th. Dobzhansky, *Die genetischen Grundlagen der Artbildung*]. *Zeitschrift für Botanik* 36:141–43.

———. 1941. Bastardierungsversuche . . . IV: Weitere Untersuchungen über plasmatische Vererbung und Geschlechtsbestimmung. *Zeitschrift für Botanik* 37:158–82.

———. 1948. Genetik und Zytogenetik der Pflanzen. In A. Kühn and E. Bünning (eds.), *Naturforschung und Medizin in Deutschland, 1939–1946*, Band 53, *Biologie*, Teil II ("FIAT Bericht II") (Wiesbaden: Dieterich'sche Verlagsbuchhandlung), 17–60.

———. 1949. Über Erbträger ausserhalb des Zellkerns. *Berichte der Naturforschenden Gesellschaft zu Freiburg i.Br.*

———. 1952. Neue Überlegungen zum Problem der Ausserkaryotischen Vererbung. *ZIAV* 84:213–50.

———. 1961. Otto Renner. *Berichte der Deutschen Botanischen Gesellschaft* 74:82–94.

———. 1964. Cytoplasmic inheritance in the genus Streptocarpus Lindley. *Advances in Genetics* 12:329–70.

Olby, Robert. 1974. *The Path to the Double Helix*. London: Macmillan.

———. 1985. *Origins of Mendelism*, 2nd ed. Chicago: University of Chicago Press.

———. 1989a. The dimensions of scientific controversy: The biometric-Mendelian debate. *British Journal for the History of Science* 22:299–320.

———. 1989b. Scientists and bureaucrats in the establishment of the John Innes Horticultural Institution under William Bateson. *Annals of Science* 46:497–510.

Oppenheimer, Jane. 1956. [Review of Alfred Kühn, *Vorlesungen über Entwicklungsphysiologie*]. *Quarterly Review of Biology* 31:31–34.
Painter, T. S. 1965. J. T. Patterson (1878–1960). *Biographical Memoirs of the National Academy of Sciences (U.S.)* 38:223–62.
Palladino, Paolo. 1990. The political economy of applied research: Plantbreeding in Great Britain, 1910–1940. *Minerva* 28:446–68.
Panofsky, Erwin. 1953. The history of art. In Franz Neumann et al. (eds.), *The Cultural Emigration: The European Scholar in America* (Philadelphia: University of Pennsylvania Press), 82–111.
Pätau, Klaus. 1939. Die mathematische Analyse der Evolutionsvorgänge. *ZIAV* 76:220–28.
Paul, Diane. 1983. A war on two fronts: J. B. S. Haldane and the response to Lysenkoism in Britain. *Journal of the History of Biology* 16:1–37.
Paul, Diane, and Barbara Kimmelman. 1988. Mendel in America: Theory and practice, 1900–1919. In Rainger, Benson, and Maienschein (eds.), *The American Development of Biology*, 281–310.
Paulsen, Friedrich. 1906. *The German Universities and University Study*. London: Longmans Green.
———. 1921. *Geschichte des gelehrten Unterrichts, Bd. 2 (1740–1892)*, 3rd ed., ed. Rudolf Lehmann. Berlin/Leipzig: deGruyter.
Pauly, Philip. 1984. The appearance of academic biology in late nineteenth century America. *Journal of the History of Biology* 17:369–97.
Penners, A. 1922. Über die Rolle von Kern und Plasma bei der Embryonalentwicklung. *Die Naturwissenschaften* 10:727–33, 761–65.
Peterson, P. A., and S. R. Peterson. 1973. Marcus M. Rhoades. *Theoretical and Applied Genetics* 43:93–96.
Pfetsch, Frank. 1970. Scientific organisation and science policy in Imperial Germany, 1871–1914. *Minerva* 8:557–80.
Pfetsch, Frank, and A. Zloczower (eds.). 1973. *Innovation und Widerstände in der Wissenschaft: Beiträge zur Geschichte der deutschen Medizin*. Düsseldorf: Bertelsmann.
Phelan, Anthony (ed.). 1985. *The Weimar Dilemma: Intellectuals in the Weimar Republic*. Manchester: Manchester University Press.
Pickstone, J. V. 1989. Medicine in industrial Britain: The uses of local studies. *Social History of Medicine* 2:197–203.
———. 1990. A profession of discovery: Physiology in nineteenth-century history. *British Journal for the History of Science* 23:207–16.
Plagge, Ernst. 1936. Bewirkung der Augenausfärbung der rotäugugen Rasse von Ephestia kühniella Z. durch Implantation artfremder Hoden. *Gesellschaft der Wissenschaften zu Göttingen, Math-Phys Klasse, Nachr. aus der Biologie*, N.F. 2:251–56.
Plarre, Werner. 1987. A contribution to the history of the science of heredity in Berlin. *Englera* 7:147–217.
Plate, Ludwig. 1913. *Vererbungslehre*. Leipzig: W. Engelmann. 2d ed., Jena: G. Fischer, 1932.

———. 1926. Lamarckismus und Erbstockhypothese. *ZIAV* 43:88–113.
———. 1932. Genetik und Abstammungslehre. *ZIAV* 62:47–67.
———. 1935. Kurze Selbstbiographie. *Archiv für Rassen- und Gesellschaftsbiologie* 29:84–87.
Plough, H. H. 1954. E. G. Conklin, 1863–1952. *Genetics* 39:1–3.
Plough, H. H., and P. T. Ives. 1934. Heat-induced mutations in Drosophila. *Proceedings of the National Academy of Sciences (U.S.)* 20:268–73.
———. 1935. Induction of mutations by high temperature in Drosophila. *Genetics* 20:42–69.
Pontecorvo, Guido. 1959. *Trends in Genetic Analysis*. New York: Columbia University Press.
Poor, Harold. 1968. *Kurt Tucholsky and the Ordeal of Germany, 1914–1935*. New York: Scribner's.
Proceedings of International Conference on Plant Breeding and Hybridization 1902. *Memoirs of the Horticultural Society of New York* 1 (1904).
Provine, William. 1971. *Origins of Theoretical Population Genetics*. Chicago: University of Chicago Press.
———. 1979. Francis B. Sumner and the evolutionary synthesis. *Studies in History of Biology* 3:211–40.
———. 1980. Introduction to section 1. In Mayr and Provine (eds.), *The Evolutionary Synthesis*, 51–58.
———. 1981. Origins of the "genetics of natural populations" series. In R. C. Lewontin, J. A. Moore, W. B. Provine, and Bruce Wallace (eds.), *Dobzhansky's Genetics of Natural Populations (Numbers 1–43)* (New York: Columbia University Press), 1–76.
———. 1983. The development of Wright's theory of evolution: Systematics, adaptation, and drift. In M. Grene (ed.), *Dimensions of Darwinism*, 43–70.
———. 1986. *Sewall Wright and Evolutionary Biology*. Chicago: University of Chicago Press.
Pyenson, Lewis. 1983. *Neo-Humanism and the Persistence of Pure Mathematics*. Philadelphia: American Philosophical Society.
Querner, Hans. 1975. Beobachtung oder Experiment? Die Methodenfrage in der Biologie um 1900. *Verhandlungen der Deutschen Zoologischen Gesellschaft*, 4–12.
———. 1980a. Die Methodenfrage in der Biologie des 19. Jahrhunderts. *Nachrichten der deutschen Gesellschaft für Geschichte der Medizin, Naturwissenschaft, und Technik* 30:111–26.
———. 1980b. Die Zoologie in der Aera von Eugen Korschelt. *Sudhoffs Archiv* 64:313–29.
Rainger, Ronald. 1988. Vertebrate paleontology as biology: Henry Fairfield Osborn and the American Museum of Natural History. In Rainger, Benson, and Maienschein (eds.), *The American Development of Biology*, 219–56.
Rainger, R., K. Benson, and J. Maienschein (eds.). 1988. *The American Development of Biology*. Philadelphia: University of Pennsylvania Press.

von Rauch, Konrad. 1935. *Cytologisch-embryologische Untersuchungen an Scurrula atropurpurea Dans* . . . Ph.D. diss., Universität Zurich.

———. 1944(?). *Erwin Baur: Gestalt und Wirken eines deutschen Biologen.* Typescript, n.d. (ca. 1944), Erwin Baur papers (MPG-Archiv).

Ratzke, Erwin. 1987. Das Pädagogische Institut der Universität Göttingen: Ein Überblick über seine Entwicklung in den Jahren 1923–1949. In H. Becker, H.-J. Dahms, and Cornelia Wegeler (eds.), *Die Universität Göttingen unter dem Nationalsozialismus* (Munich: Saur), 200–218.

Ravin, A. 1976. F. J. Ryan, 1916–1963. *Genetics* 84:1–25.

Reif, Wolf-Ernst. 1983. Evolutionary theory in German paleontology. In M. Grene (ed.), *Dimensions of Darwinism*, 173–203.

———. 1986. The search for a macroevolutionary theory in German paleontology. *Journal of the History of Biology* 19:79–130.

Reingold, Nathan. 1978. National style in the sciences: The United States case. In Eric Forbes (ed.), *Human Implications of Scientific Advance*, (Edinburgh: Edinburgh University Press), 163–73.

Reinke, J. 1904. Botanik und Zoologie. In W. Lexis (ed.), *Das Unterrichtswesen im Deutschen Reich; Band 1—Die Universitäten* (Berlin: Asher), 280–89.

Remme, Karl. 1926. *Die Hochschulen Deutschlands: Ein Führer durch Geschichte, Landschaft, Studium.* Berlin: Universität Berlin.

Renner, Otto. 1922. Eiplasma und Pollenschlauchplasma als Vererbungsträger bei den Oenotheren. *ZIAV* 27:235–37.

———. 1924a. Vererbung bei Artbastarden. *ZIAV* 33:317–47.

———. 1924b. Die Scheckung der Oenothera-Bastarden. *Biologisches Zentralblatt* 44:309–36.

———. 1929. Artbastarde bei Pflanzen. *Handbuch der Vererbungswissenschaft* IIA:1–161.

———. 1930. Eröffnungsansprache. *Berichte der Deutschen Botanischen Gesellschaft* 48:(2)–(13).

———. 1934. Die pflanzlichen Plastiden als selbständige Elemente der genetischen Konstitution. *Berichte der Sächsischen Akademie der Wissenschaften zu Leipzig, Math-Phys Klasse* 86:241–66.

———. 1935. Hugo deVries. *Der Erbarzt* 2:177–79.

———. 1936. Zur Kenntnis der nicht-mendelnden Buntheit der Laubblätter. *Flora* 130:218–90.

———. 1937. Zur Kenntnis der Plastiden- und Plasma-Vererbung. *Cytologia* Fujii jubilee vol. 2:644–55.

———. 1946. Friedrich Wettstein, Ritter von Westersheim. *Die Naturwissenschaften* 33:97–100.

———. 1944–48. Fritz Wettstein von Westersheim. *Jahrbuch der Bayerischen Akademie der Wissenschaften*, 261–65.

———. 1949. Goethes Verhältnis zur Pflanzenwelt, von Jena gesehen. In *Dem Tüchtigen ist diese Welt nicht stumm: Beiträge zum Goethe-Bild* (Jena: Wilhelm Gronau), 100–120.

———. 1950. Die Situation der Biologie nach 50 Jahren Mendelforschung. *Berichte der Deutschen Botanischen Gesellschaft* 63:(4)–(10).
———. 1959. Botanik. Teil V, in *Geist und Gestalt (Naturwissenschaftliche Biographie: Beiträge zur Geschichte der Bayerischen Akademie der Wissenschaften)*, vol. 2. Munich: C. H. Beck.
———. 1961a. Münchener Antrittsvorlesung 1948. *Berichte der Bayerischen Botanischen Gesellschaft* 34:105–12.
———. 1961b. William Bateson and Carl Correns. *Sitzungsberichte der Heidelberger Akademie der Wissenschaften, Math-Naturwiss. Klasse,* 6. Abhandlung, 159–81.
Renner, Otto, and W. Kupper. 1921. Artkreuzungen in der Gattung Epilobium. *Berichte der Deutschen Botanischen Gesellschaft* 39:201–6.
Rensch, Bernhard. 1979. *Lebensweg eines Biologen in einem turbulenten Jahrhundert.* Stuttgart: G. Fischer.
———. 1980. Historical development of the present synthetic neo-Darwinism in Germany. In Mayr and Provine (eds.), *The Evolutionary Synthesis,* 284–303.
———. 1983. The abandonment of Lamarckian explanations: The case of climatic parallelism of animal characteristics. In M. Grene (ed.), *Dimensions of Darwinism,* 31–42.
Rhoades, M. M. 1949. R. A. Emerson (1873–1947). *Biographical Memoirs of the National Academy of Sciences (U.S.)* 25:313–23.
———. 1956. L. J. Stadler. *Genetics* 41:1–3.
———. 1984. The early years of maize genetics. *Annual Reviews of Genetics* 18:1–29.
Richmond, Marsha. 1986. *Richard Goldschmidt and Sex-determination: The Growth of German Genetics, 1900–1935.* Ph.D. diss., Indiana University.
Riese, Reinhard. 1977. *Die Hochschule auf dem Wege zum wissenschaftlichen Grossbetrieb: Die Universität Heidelberg und das badische Hochschulwesen, 1860–1914.* Stuttgart: Klett.
Ringer, Fritz K. 1969. *Decline of the German Mandarins: The German Academic Community, 1890–1933.* Cambridge: Harvard University Press.
———. 1979. *Education and Society in Modern Europe.* Bloomington: Indiana University Press.
Robertson, A. 1977. Conrad Hal Waddington. *Biographical Memoirs of Fellows of the Royal Society* 23:575–622.
Roegele, O. 1966. Student im Dritten Reich. In *Die Deutsche Universität im Dritten Reich (eine Vortragsreihe der Universität München)* (Munich: Piper), 137–74.
Roll-Hansen, Nils. 1978a. The genotype theory of Wilhelm Johannsen and its relation to plant breeding and evolution. *Centaurus* 22:201–35.
———. 1978b. Drosophila genetics: A reductionist research program. *Journal of the History of Biology* 11:159–210.
———. 1989. Geneticists and the eugenics movement in Scandinavia. *British Journal for the History of Science* 22:335–46.

Romeis, B. 1953. Hermann Stieve. *Anatomischer Anzeiger* 99:401–40.
Rosenberg, Charles. 1976. *No Other Gods: On Science and American Social Thought*. Baltimore: Johns Hopkins University Press.
———. 1980. [Review of B. Barnes and S. Shapin (eds.), *Natural Order*]. *Isis* 71:291–95.
———. 1983. Science in American society: A generation of historical debate. *Isis* 74:356–67.
Rossiter, Margaret. 1975. *The Emergence of Agricultural Science: Justus Liebig and the Americans, 1840–1880*. New Haven: Yale University Press.
———. 1979. The organization of the agricultural sciences. In A. Oleson and J. Voss (eds.), *The Organization of Knowledge in Modern America, 1860–1920* (Baltimore: Johns Hopkins University Press), 211–48.
———. 1982. *Women Scientists in America: Struggles and Strategies to 1940*. Baltimore: Johns Hopkins University Press.
Rothblatt, Sheldon. 1968. *Revolution of the Dons: Cambridge and Society in Victorian England*. Cambridge: Cambridge University Press.
Rowe, David. 1986. Jewish mathematicians at Göttingen in the era of Felix Klein. *Isis* 77:422–49.
Rübensam, Erich. 1982. Laudatio zum 80. Geburtstag Hans Stubbes. In *Hans Stubbe—80 Jahre* (Akademie der Landwirtschaftswissenschaften der DDR), 11–16.
Rudorf, Wilhelm. 1955a. G. H. Shull, 1874–1954. *Berichte der Deutschen Botanischen Gesellschaft* 68a:237–44.
———. 1955b. Hans Kappert zum 65. Geburtstag. *Der Züchter* 25:193–94.
Rudwick, Martin. 1982. Cognitive styles in geology. In Mary Douglas (ed.), *Essays in the Sociology of Perception*, 219–41.
Rüschemeyer, Dietrich. 1976. Partial modernization. In J. J. Loubser et al. (eds.), *Exploration in General Theory in Social Science* (New York: Free Press), 756–72.
———. 1981. Die Nichtrezeption von Karl Mannheims Wissenssoziologie in der amerikanischen Soziologie. In M. R. Lepsius (ed.), *Soziologie in Deutschland und Österreich, 1918–1945* (Opladen: Westdeutscher Verlag), 414–20.
Saha, M. S. 1984. *Carl Correns and an alternative approach to genetics: The study of heredity in Germany between 1880 and 1930*. Ph.D. diss., Michigan State University.
Sander, Klaus. 1985. The role of genes in ontogenesis—evolving concepts from 1883 to 1983 as perceived by an insect embryologist. In Horder, Witkowski, and Wylie (eds.), *A History of Embryology*, 363–95.
Sanderson, Michael. 1972. *The Universities and British Industry, 1850–1970*. London: Routledge and Kegan Paul.
Sapp, Jan. 1987. *Beyond the Gene: Cytoplasmic Inheritance and the Struggle for Authority in Genetics*. New York: Oxford University Press.
Schappacher, Norbert. 1987. Das Mathematische Institut der Universität Göttingen 1929–1950. In H. Becker, H. J. Dahms, and Cornelia Wegeler (eds.),

Die Uniersität Göttingen unter dem Nationalsozialismus (Munich: Saur), 345–73.

Schaxel, Julius. 1916. *Über den Mechanismus der Vererbung.* Jena: G. Fischer.

———. 1919a. *Grundzüge der Theorienbildung in der Biologie.* Jena: G. Fischer: 2d ed. 1922.

———. 1919b. Über die Darstellung allgemeiner Biologie. *Abhandlungen zur theoretischen Biologie* 1.

Scheler, Max. 1980. *Problems of a Sociology of Knowledge,* trans. M. S. Frings, ed. K. W. Stikkers. London: Routledge and Kegan Paul.

Schiemann, Elisabeth. 1934. Erwin Baur. *Berichte der Deutschen Botanischen Gesellschaft* 52:51–114.

———. 1949. Tine Tammes zum Gedächtnis. *Der Züchter* 19:181–84.

———. 1955. Emmy Stein, 1879–1954. *Der Züchter* 25:65–67.

———. 1960. Erinnerungen an meine Berliner Universitätsjahre. In H. Leussink, E. Neumann, and G. Kotowski (eds.), *Studium Berolinense: Aufsätze . . . zur Geschichte der Friedrich-Wilhelms-Universität zu Berlin* (Berlin: deGruyter), 845–57.

———. 1963. Hermann Kuckuck zum 60. Geburtstag. *Zeitschrift für Pflanzenzüchtung* 50:1–8.

Schipperges, H. 1976. *Weltbild und Wissenschaft: Eröffnungsreden zu den Naturforscherversammlungen, 1822–1972.* Hildesheim: H. A. Gerstenberg.

Schleip, W. 1927. Entwicklungsmechanik und Vererbung bei Tieren. *Handbuch der Vererbungswissenschaft* IIIA.

Schlicker, Wolfgang, et al. 1975. *Die Berliner Akademie der Wissenschaften in der Zeit des Imperialismus,* part 2 (1917–1933). Berlin: Akademie-Verlag.

Schlösser, L. A. 1933. Erwin Baur zum Gedächtnis. *Rasse, Volk, und Staat (Rassenhygienisches Beiblatt zum "Völkischen Beobachter"),* December 1933.

———. 1935. Beitrag zu einer physiologischen Theorie der plasmatischen Vererbung. *ZIAV* 69:159–92.

Schmidt, Siegfried, et al. (eds.). 1983. *Alma mater Jenensis: Geschichte der Universität Jena.* Weimar: Boehlaus.

Scholder, Klaus (ed.). 1982. *Die Mittwochs-Gesellschaft: Protokolle aus dem geistigen Deutschland, 1932–1944.* Berlin: Severin und Siedler.

Schon, Donald. 1963. *The Displacement of Concepts.* London: Tavistock.

Schönemann, Friedrich. 1930. Das Verhältnis der deutschen Universitäten zu den Hochschulsystemen des Auslandes: Amerika. In M. Doeberl and Otto Scheel et al. (eds.), *Das Akademische Deutschland,* vol. 3 (Berlin: Weller), 53–59.

Schorske, Carl. 1981. *Fin de Siècle Vienna: Politics and Culture.* Cambridge: Cambridge University Press.

Schroeder-Gudehus, Brigitte. 1972. The argument for the self-government and public support of science in Weimar Germany. *Minerva* 10:537–70.

Schwabe, Klaus. 1969. *Wissenschaft und Kriegsmoral: Die deutschen Hochschullehrer und die politischen Grundfragen des Ersten Weltkrieges.* Göttingen: Musterschmidt.

——— (ed.). 1988. *Deutsche Hochschullehrer als Elite, 1815–1945*. Boppard: Boldt.
Schwanitz, Franz. 1943. Genetik und Evolutionsforschung bei Pflanzen. In G. Heberer (ed.), *Die Evolution der Organismen*, 430–78.
Schweber, S. S. 1980. Darwin and the political economists: Divergence of character. *Journal of the History of Biology* 13:195–289.
———. 1986. The empiricist temper regnant: Theoretical physics in the United States 1920–1950. *Historical Studies in the Physical and Biological Sciences* 17:55–98.
Schweiger, H.-G. 1981. Joachim Hämmerling. *Max-Planck-Gesellschaft: Berichte und Mitteilungen*, Heft 3:24–27.
Schwemmle, Julius. 1927–28. [Review of von Wettstein, "Über plasmatische Vererbung...," 1927]. *Zeitschrift für Botanik* 20:628–31.
———. 1927. Erklärung der reziproken Verschiedenheit der Bastarde zwischen E. parviflorum und E. roseum auf Grund hybridologischen Untersuchungen ihrer Sterilitätserscheinungen. *Bibliotheca Botanica* 95:17–50.
———. 1934. Carl Correns zum Gedächtnis. *Medizinische Klinik* 14.9.34, 1251.
———. 1935. Die Rolle des Plasmas für die Vererbung. *Erbarzt* 2:179–85.
———. 1935–36. [Review of Michaelis and von Dellinghausen]. *Zeitschrift für Botanik* 29:600–601.
Schwemmle, Julius, et al., 1938. Genetische und Zytologische Untersuchungen an Eu-Oenotheren, I–VI. *ZIAV* 75:358–800.
———. 1941a. Plastidenmutationen bei Eu-Oenotheren. *ZIAV* 79:171–87.
———. 1941b. Weitere Untersuchungen an Eu-Oenotheren über die genetische Bedeutung des Plasmas und der Plastiden. *ZIAV* 79:321–35.
———. 1944. Plastiden und Genmanifestation. *Flora* N.F. 37:61–72.
Schwemmle, Julius, and M. Zintl. 1939. Genetische und Zytologische Untersuchungen an Eu-Oenotheren: Die Analyse der Oen. argentinea. *ZIAV* 76:353–410.
Seeberg, Reinhold. 1930. Hochschule und Weltanschauung. In Michael Doeberl and Otto Scheel et al. (eds.), *Das Akademische Deutschland*, vol. 3 (Berlin: Weller), 163–78.
Seier, Hellmut. 1988. Die Hochschullehrerschaft im Dritten Reich. In Klaus Schwabe (ed.), *Deutsche Hochschullehrer als Elite*, 247–95.
Seidel, F. 1936. [Review of R. Woltereck, *Grundzüge einer allgemeinen Biologie*]. *Die Naturwissenschaften* 24:378–79.
Shapin, Steven. 1982. History of science and its sociological reconstructions. *History of Science* 20:157–211.
Shapiro, Meyer. 1953. Style. In A. L. Kroeber (ed.), *Anthropology Today* (Chicago: University of Chicago Press), 287–312.
Shils, Edward. 1972. The traditions of intellectual life: Their conditions of existence and growth in contemporary societies. In *The Intellectuals and the Powers, and Other Essays* (Chicago: University of Chicago Press), 71–94.
———. 1978. The order of learning in the United States from 1865 to 1920: The ascendancy of the universities. *Minerva* 16:159–95.

Shull, George. 1923. The species concept from the point of view of a geneticist. *American Journal of Botany* 10:221–28.
Sierp, Hans. 1924. Die nicht-vererbungswissenschaftliche Arbeiten von Correns. *Die Naturwissenschaften* 12:772–80.
Sinnott, Edmund W. 1959. A. F. Blakeslee (1874–1954). *Biographical Memoirs of the National Academy of Sciences (U.S.)* 33:1–38.
Sinnott, Edmund W., and L. C. Dunn. 1939. *Principles of Genetics*, 3d ed. New York: McGraw-Hill.
Smith, David, and Malcolm Nicolson. 1989. The "Glasgow School" of Paton, Findlay, and Cathcart: Conservative thought in chemical physiology, nutrition, and public health. *Social Studies of Science* 19:195–238.
Snell, G. D. 1975. C. C. Little. *Biographical Memoirs of the National Academy of Sciences (U.S.)* 46:241–64.
Snyder, L. H. 1946. C. B. Davenport. *Genetics* 31.
Sonneborn, T. M. 1948. H. S. Jennings. *Genetics* 33:1–4.
———. 1950a. Partner of the genes. *Scientific American*, Nov. 1950, 30–39.
———. 1950b. The role of the genes in cytoplasmic inheritance. In L. C. Dunn (ed.), *Genetics in the Twentieth Century* (New York: Macmillan), 291–314.
Spatz, Hugo. 1962. Max-Planck Institut für Hirnforschung, Frankfurt am Main. *Jahrbuch der Max-Planck Gesellschaft, 1961*, part 2, 405–21.
Spencer, Warren. 1940. Levels of divergence in Drosophila speciation. *American Naturalist* 74:299–311.
Spranger, Eduard. 1930. Das Wesen der deutschen Universität. In Michael Doeberl and Otto Scheel et al. (eds.), *Das Akademische Deutschland*, vol. 3 (Berlin: Weller), 1–38.
Srb, Adrian. 1974(?). Beadle and Neurospora: Some recollections. *Neurospora Newsletter* no. 20:8–9 (copy in George Beadle papers).
Stebbins, G. Ledyard. 1958. E. B. Babcock, 1877–1954. *Biographical Memoirs of the National Academy of Sciences (U.S.)* 32:50–66.
———. 1980. Botany and the synthetic theory of evolution. In Mayr and Provine (eds.), *The Evolutionary Synthesis*, 139–52.
Stein, Emmy. 1950. Dem Gedächtnis von C. E. Correns: Nach einem halben Jahrhundert der Vererbungswissenschaft. *Die Naturwissenschaften* 37:457–63.
———. 1952. E. Baur. *Mitteilungen der Max-Planck-Gesellschaft, 1952*, 7–11.
Steiner, E. 1982. R. E. Cleland. *Biographical Memoirs of the National Academy of Sciences (U.S.)* 53:121–40.
Stern, Curt. 1928. Fortschritte der Chromosomentheorie der Vererbung. *Ergebnisse der Biologie* 4:206–359.
———. 1929. Erzeugung von Mutationen durch Röntgenstrahlen. *Natur und Museum* 59:577–83.
———. 1930a. Entgegnung auf den Bemerkungen von Franz Weidenreich... *Natur und Museum* 60:133–34.
———. 1930b. Der Kern als Vererbungsträger. *Die Naturwissenschaften* 18:1117–25.

———. 1931. Karl Belar zum Gedächtnis. *Die Naturwissenschaften* 19: 921–23.

———. 1936. Genetics and ontogeny. *American Naturalist* 70:29–35.

———. 1954. George W. Beadle. *Science* 119:229–30.

———. 1958. R. B. Goldschmidt. *Experientia* 14:307–8.

———. 1967. R. B. Goldschmidt, 1878–1958. *Biographical Memoirs of the National Academy of Sciences (U.S.)* 39:141–92.

Stern, Fritz. 1961. *The Politics of Cultural Despair: A Study in the Rise of the Germanic Ideology*. Berkeley: University of California Press.

———. 1987. Fritz Haber: The scientist in power and exile. In *Dreams and Delusions: The Drama of German History* (London: Weidenfeld and Nicolson), 51–76.

Stieve, Hermann. 1923. Neuzeitliche Ansichten über die Bedeutung der Chromosomen unter besonderer Berücksichtigung der Drosophila-Versuche. *Ergebnisse der Anatomie und Entwicklungsgeschichte* 24:491–587.

———. 1939. Rudolf Fick. *Anatomischer Anzeiger* 89:98–127.

Straub, Joseph. 1986. Aus der Geschichte des KWI/MPIs für Züchtungsforschung. *Berichte und Mitteilungen der Max-Planck Gesellschaft*, Nr. 2, 11–36.

Struve, Walter. 1973. *Elites against Democracy: Leadership Ideals in Bourgeois Political thought in Germany, 1890–1933*. Princeton: Princeton University Press.

Stubbe, Hans. 1934a. Die Bedeutung der Mutationen für die theoretische und angewandte Genetik. *Die Naturwissenschaften* 22:781–87.

———. 1934b. Erwin Baur. *ZIAV* 66:v–ix.

———. 1938. Genmutation. *Handbuch der Vererbungswissenschaft* IIF.

———. 1951. Nachruf auf Fritz von Wettstein. *Jahrbuch der deutschen Akademie der Wissenschaften zu Berlin, 1950/51*, 1–12.

———. 1955. Karl Pirschle, 1900–1945. *Berichte der Deutschen Botanischen Gesellschaft* 68a:105–8.

———. 1958. Reinhold von Sengbusch zum 60. Geburtstag. *Der Züchter* 28:1–3.

———. 1959a. Gedächtnisrede auf Erwin Baur. *Der Züchter* 29:1–6.

———. 1959b. Hans Stubbe. *Nova Acta Leopoldina* 21:298–300.

———. 1962. Nachruf auf Otto Renner. *Jahrbuch der Deutschen Akademie der Wissenschaften zu Berlin, 1961*, 882–86.

———. 1972. Elisabeth Schiemann. *Mitteilungen aus der Max-Planck-Gesellschaft*, 3–8.

———. 1980. Arbeit für eine gute Zukunft. In "*. . . einer neuen Zeit Beginn*": *Erinnerungen an die Anfänge unserer Kulturrevolution, 1945–1949* (Berlin/Weimar: Aufbau), 475–79.

———. 1985. Erinnerungen an Rudolf Schick. In Akademie der Landwirtschaftswissenschaften der DDR (ed.), *Rudolf Schick—80 Jahre* (Gross Lüsewitz: Institut für Kartoffelforschung), 3–8.

Stubbe, Hans, and F. von Wettstein. 1941. Über die Bedeutung von Klein- und Grossmutationen in der Evolution. *Biologisches Zentralblatt* 61:265–97.

Stubbe, W. 1976. R. E. Cleland, 1892–1971. *Berichte der Deutschen Botanischen Gesellschaft* 89:91–96.

———. 1987. Peter Michaelis, 1900–1975. *Berichte der Deutschen Botanischen Gesellschaft* 100:69–79.

Sturtevant, A. H. 1915. The behavior of chromosomes as studied through linkage. *ZIAV* 13:234–87.

———. 1932. The use of mosaics in the study of the developmental effects of genes. *Proceedings of the Sixth International Congress of Genetics* 1:304–7.

———. 1947. T. H. Morgan. *Genetics* 32.

———. 1965a. *A History of Genetics*. New York: Harper and Row.

———. 1965b. The early Mendelians. *Proceedings of the American Philosophical Society* 109:199–204.

Sturtevant, A. H., and G. Beadle. 1939. *Introduction to Genetics*. Philadelphia: W. B. Saunders.

Sund, Christian. 1984(?). Zur historischen Entwicklung der Biowissenschaften: Keine sukzessive Verzweigung. Typescript, n.d. Technische Universität Berlin.

Swanson, C. P. 1976. Karl Sax (1892–1973). *Genetics* 83:1–4.

Tembrock, Günter. 1958–59. Zur Geschichte der Zoologie in Berlin. *Wissenschaftliche Zeitschrift der Humboldt-Universität zu Berlin/Math.-Naturw. Reihe* 8:185–95.

Temkin, Owsei. 1977. *The Double Face of Janus*. Baltimore: Johns Hopkins University Press.

Tertulian, Nicolas. 1988. Heidegger—oder: Die Bestätigung der Politik durch Seinsgeschichte. *Frankfurter Rundschau,* 29.1.88.

Thomas, R. H. 1976–77. The uses of "Bildung." *German Life and Letters* 30:177–86.

———. 1985. Nietzsche in Weimar Germany—and the case of Ludwig Klages. In Anthony Phelan (ed.), *The Weimar Dilemma: Intellectuals in the Weimar Republic* (Manchester: Manchester University Press), 71–91.

Thwing, C. F. 1928. *The American and the German University*. New York: Macmillan.

Timoféeff-Ressovsky, N. W. 1934. Verknüpfung von Gen und Aussenmerkmal: Phänomenologie der Genmanifestierung. In W. Kolle (ed.), *Wissenschaftliche Woche zu Frankfurt,* vol. 1: *Erbbiologie* (Leipzig: G. Thieme), 92–115.

———. 1939. Genetik und Evolution (Bericht eines Zoologen). *ZIAV* 76:158–219.

Timoféeff-Ressovsky, N. W., and H. A. Timoféeff-Ressovsky. 1927. Genetische Analyse einer freilebenden D. melanogaster Population. *Archiv für Entwicklungsmechanik der Organismen* 109:70–109.

Titze, Hartmut, et al. 1987. *Datenhandbuch zur deutschen Bildungsgeschichte,* vol. 1, part 1: *Das Hochschulstudium in Preussen und Deutschland, 1820–1944.* Göttingen: Vandenhoeck und Ruprecht.

Tompert, H. 1969. *Lebensformen und Denkweisen der akademischen Welt Heidelbergs im Wilhelminischen Zeitalter*. Lübeck/Hamburg: Matthiesen.

Töpner, Kurt. 1970. *Gelehrte Politiker und politisierende Gelehrte: Die Revolution von 1918 im Urteil deutscher Hochschullehrer.* Göttingen: Musterschmidt.
Toulmin, Stephen. 1961. *Foresight and Understanding.* London: Hutchinson.
Trienes, Rudie. 1989. Type concept revisited: A study of German idealistic morphology in the first half of the twentieth century. *History and Philosophy of the Life Sciences* 11:23–42.
Trommler, Frank. 1985. The rise and fall of Americanism in Germany. In Frank Trommler and Joseph McVeigh (eds.), *America and the Germans: An Assessment of a Three-Hundred-Year History* (Philadelphia: University of Pennsylvania Press), 332–42.
Tschermak-Seysenegg, Erich. 1933. E. Baur. *Wiener Landwirtschaftliche Zeitung.*
Turner, R. Steven. 1971. The growth of professorial research in Prussia, 1818–1848: Causes and context. *Historical Studies in the Physical Sciences* 3:137–82.
von Ubisch, Gerta. 1956. Aus dem Leben einer Hochschuldozentin. *Mädchenbildung und Frauenschaffen* 6, no. 10 (413–22 and 498–507) and no. 11 (35–45).
von Uexküll, Jakob. 1913. Der heutige Stand der Biologie in Amerika. *Die Naturwissenschaften* 1:801–5.
———. 1928. *Theoretische Biologie,* 2d ed. Frankfurt: Suhrkamp, 1973.
———. 1933. *Staatsbiologie: Anatomie, Physiologie, Pathologie des Staates,* 2d ed. Hamburg: Hanseatische Verlagsanstalt.
Ulrich, W. 1950. Hans Nachtsheim. In Hans Grüneberg and W. Ulrich (eds.), *Moderne Biologie: Festschrift zum 60. Geburtstag Hans Nachtsheims* (Berlin: F. W. Peters), 7–14.
Ulrich, Werner. 1960. Karl Heider. In H. Leussink, E. Neumann, and G. Kotowski (eds.), *Studium Berolinense: Aufsätze und Beiträge zur Problemen der Wissenschaft und zur Geschichte der Friedrich-Wilhelms-Universität zu Berlin,* vol. 2 (Berlin: deGruyter), 868–97.
Uschmann, Georg. 1953. Valentin Haecker. *Neue Deutsche Biographie* 7:427–28.
———. 1959. *Geschichte der Zoologie und der zoologischen Anstalten in Jena, 1779–1919.* Jena: G. Fischer.
Veysey, Laurence. 1965. *The Emergence of the American University.* Chicago: University of Chicago Press.
Vierhaus, Rudolf. 1972. Bildung. In O. Bruner, W. Conze, and R. Koselleck (eds.), *Geschichtliche Grundbegriffe,* vol. 1 (Stuttgart: Klett-Cotta), 508–51.
de Vilmorin, P. (ed.). 1913. *IVe Conférence Internationale de Génétique, Paris 1911.* Paris: Masson.
Vogel, Friedrich. 1980. H. Nachtsheim. *Berichte und Mitteilungen der Max-Planck-Gesellschaft,* 25–29.
Vogt, W. Paul, 1982. Identifying scholarly and intellectual communities: A note on French philosophy, 1900–1939. *History and Theory* 21:267–78.

Vondung, Klaus. 1976. Zur Lage der Gebildeten in der wilhelminischen Zeit. In K. Vondung (ed.), *Das wilhelminische Bildungsbürgertum: Zur Sozialgeschichte seiner Ideen* (Göttingen: Vandenhoeck und Ruprecht), 20–33.

Wach, Joachim. 1930. Hochschule und Weltanschauung. In M. Doeberl and Otto Scheel et al. (eds.), *Das Akademische Deutschland*, vol. 3 (Berlin: Weller), 198–205.

Waddington, Conrad. 1975. *The Evolution of an Evolutionist*. Edinburgh: Edinburgh University Press.

Wagenitz, G. 1988. *Göttinger Biologen, 1737–1945: Eine biographisch-bibliographische Liste*. Göttingen: Vandenhoeck und Ruprecht.

Wagner, R. P. 1970. T. S. Painter. *Genetics* 64:87–88.

Weart, Spencer. 1979. The physics business in America, 1919–1940. A statistical reconnaissance. In N. Reingold (ed.), *The Sciences in the American Context* (Washington, D.C.: Smithsonian Institution Press), 295–358.

Weber, Max. 1973. The power of the state and the dignity of the academic calling in imperial Germany, ed. Edward Shils. *Minerva* 11:571–632.

Wehler, H.-U. 1973. *Das Deutsche Kaiserreich 1871–1918*. Göttingen: Vandenhoeck und Ruprecht.

———. 1975. *Modernisierungstheorie und Geschichte*. Göttingen: Vandenhoeck und Ruprecht.

Weidenreich, Franz. 1930a. Bemerkungen zu dem Aufsatz von C. Stern. *Natur und Museum* 60:47.

———. 1930b. Vererbungsexperiment und vergleichende Morphologie. *Paläontologische Zeitschrift* 11:275–86.

Weindling, P. 1981. Theories of the cell-state in imperial Germany. In Charles Webster (ed.), *Biology, Medicine, and Society, 1840–1940*. Cambridge: Cambridge University Press.

———. 1989. *Health, Race, and German Politics between National Unification and Nazism, 1870–1945*. Cambridge: Cambridge University Press.

———. 1991. *Darwinism and Social Darwinism in Imperial Germany: The Contribution of the Cell Biologist Oscar Hertwig (1849–1922)*. Stuttgart: G. Fischer.

Weingart, Peter, Jürgen Kroll, and Kurt Bayertz. 1988. *Rasse, Blut, und Gene: Geschichte der Eugenik und Rassenhygiene in Deutschland*. Frankfurt: Suhrkamp.

Weismann, August. 1892. *Das Keimplasma: Eine Theorie der Vererbung*. Jena: G. Fischer.

Weiss, Sheila Faith. 1987. *Race Hygiene and National Efficiency: The Eugenics of Wilhelm Schallmayer*. Berkeley: University of California Press.

Wendel, Günter. 1975. *Die Kaiser-Wilhelm Gesellschaft, 1911–1914: Zur Anatomie einer imperialistischen Forschungsgesellschaft*. Berlin: Akademie-Verlag.

Werdinger, Jeffrey. 1980. *Embryology at Woods Hole: The Emergence of a New American Biology*. Ph.D. diss., Indiana University.

von Wettstein, Fritz. 1924. Morphologie und Physiologie des Formwechsels der Moose auf genetischer Grundlage. *ZIAV* 33:178– .

———. 1927a. Über plasmatische Vererbung, sowie Plasma- und Genwirkung. *Nachrichten der Gesellschaft der Wissenschaften zu Göttingen, Math.-Phys. Klasse*, 250–81.

———. 1927b. Wie entstehen neue vererbbare Eigenschaften? *Züchtungskunde* 2:241–59.

———. 1928a. Morphologie une Physiologie des Formwechsels der Moose auf genetischer Grundlage II. *Bibliotheca Genetica* 10:1–216.

———. 1928b. Über plasmatische Vererbung und über das Zusammenwirken von Genen und Plasma. *Berichte der deutschen Botanischen Gesellschaft* 46: 32–49.

———. 1930a. Über plasmatische Vererbung, sowie Plasma- und Genwirkung II. *Nachrichten der Gesellschaft der Wissenschaften zu Göttingen, Math.-Phys. Klasse*, 109–18.

———. 1930b. Biologie. In Gustav Abb (ed.), *Aus 50 Jahren deutscher Wissenschaft: Die Entwicklung ihrer Fachgebiete in Einzeldarstellungen* (Berlin: deGruyter), 400–407.

———. 1934a. Über plasmatische Vererbung und das Zusammenwirken von Genen und Plasma. In W. Kolle (ed.), *Wissenschaftliche Woche zu Frankfurt*, vol. 1: *Erbbiologie*, 29–36.

———. 1934b. Die erbbiologischen Grundlagen der Rassenhygiene. In Ernst Rüdin (ed.), *Erblehre und Rassenhygiene im völkischen Staat* (Munich: J. F. Lehmanns), 22–33.

———. 1937. Die genetische und entwicklungs-physiologische Bedeutung des Cytoplasmas. *ZIAV* 73:349–66.

———. 1938. C. E. Correns. *Berichte der Deutschen Botanischen Gesellschaft* 56:140–60.

———. 1939a. Botanik, Paläobotanik, Vererbungsforschung, und Abstammungslehre. *Palaeobiologica* 7:154–68.

———. 1939b. C. E. Correns zum Gedächtnis. *ZIAV* 76:1–10.

———. 1941. Die natürliche Formenmannigfältigkeit. In Th. Roemer and W. Rudorf (eds.), *Handbuch der Pflanzenzüchtung*, vol. 1, 8–45.

———. 1943. Warum hat der diploide Zustand bei den Organismen den grösseren Selektionswert? *Die Naturwissenschaften* 31:574–77.

von Wettstein, Richard. 1928. Das Problem der Evolution und die moderne Vererbungslehre. *ZIAV* (supp.) 1:370–80.

Whaley, W. G. 1983. E. W. Sinnott. *Biographical Memoirs of the National Academy of Sciences (U.S.)* 54:351–72.

Wheeler, W. M. 1921. The organization of research. *Science* 53:53–67.

———. 1923. The dry-rot of our academic biology. *Science* 57:61–71.

Whitley, Richard. 1984. *The Intellectual and Social Organisation of the Sciences*. New York: Oxford University Press.

Wien, Wilhelm. 1918. Theodor Boveri: Erinnerungen an seine Persönlichkeit. In W. C. Röntgen (ed.), *Erinnerungen an Theodor Boveri* (Tübingen: Mohr), 126–60.

Willey, Thomas E. 1977. Liberal historians and the German professoriate: A consideration of some recent books on German thought. *Central European History* 9:184–97.

Wilson, E. B. 1901. Aims and methods of study in natural history. *Science* 13:14–23.

Winkler, Hans. 1924. Über die Rolle von Kern und Protoplasma bei der Vererbung. *ZIAV* 33:238–53.

———. 1930. *Die Konversion der Gene*. Jena: G. Fischer.

Witschi, Emil. 1959. Richard Goldschmidt. *Biologisches Zentralblatt* 78:209–13.

Witkin, H. A., et al. 1962. *Psychological Differentiation*. New York: Wiley.

Wohl, Robert. 1980. *The Generation of 1914*. London: Weidenfeld.

Woltereck, Richard. 1924. Über Reaktionskonstanten und Artänderung. *ZIAV* 33:297–301.

———. 1928. Bemerkungen über die Begriffe "Reaktionsnorm" und "Klon." *Biologisches Zentralblatt* 48:167–72.

———. 1931a. Beobachtung und Versuche zum Fragenkomplex der Artbildung: 1. Wie entsteht eine endemische Rasse oder Art? *Biologisches Zentralblatt* 51:231–53.

———. 1931b. Vererbung und Erbänderung. In H. Driesch and H. Woltereck (eds.), *Das Lebensproblem: Im Lichte der modernen Forschung* (Leipzig: Quelle und Meyer), 225–310.

———. 1932. *Grundzüge einer allgemeinen Biologie: Die Organismen als Gefüge/Getriebe, als Normen, und als erlebende Subjekte*. Stuttgart: F. Enke.

———. 1934. Artdifferenzierung (insbesondere Gestaltänderung) bei Cladoceren. *ZIAV* 67:173–96.

Woodger, J. H. 1930. The "concept of organism" and the relation between embryology and genetics. *Quarterly Review of Biology* 5:1–22.

Worster, Donald. 1979. *Nature's Economy: The Roots of Ecology*. Garden City, N.Y.: Anchor Press.

Wright, Sewall. 1940. Breeding structure of populations in relation to speciation. *American Naturalist* 74:232–48.

———. 1964. Biology and the philosophy of science. *Monist* 48:265–90.

Young, Robert M. 1985a. Darwin's metaphor: Does nature select? In *Darwin's Metaphor: Nature's Place in Victorian Culture* (Cambridge), 79–125.

———. 1985b. Malthus and the evolutionists: The common context of biological and social theory. In *Darwin's Metaphor*, 23–55.

Ziegler, H. 1980. Julius Schwemmle. *Jahrbuch der Bayerischen Akademie der Wissenschaften 1980*, 1–5.

Zierold, Kurt. 1968. *Forschungsförderung in Drei Epochen: Deutsche Forschungsgemeinschaft: Geschichte, Arbeitsweise, Kommentar*. Wiesbaden: Franz Steiner.

Zimmer, K. G. 1982. N. W. Timofeeff-Ressovsky, 1900–1981. *Mutation Research* 106:191–93.

Zirnstein, Gottfried. 1972. Zur gesellschaftlichen Stellung der wissenschaftlichen Pflanzenzüchtung in Deutschland während der zwanziger und dreissiger Jahre des 20. Jahrhunderts. *Naturwissenschaft-Technik-Medizin* 9:60–69.
———. 1987. Aus dem Leben und Wirken des Leipziger Zoologen R. Woltereck (1877–1944). *Naturwissenschaft-Technik-Medizin* 24:113–20.
Znaniecki, F. 1968. *The Social Role of the Man of Knowledge*. New York: Harper and Row.
Zneimer, Richard. 1978. The Nazis and the professors: Social origin, professional mobility, and political involvement of the Frankfurt University faculty, 1933–1939. *Journal of Social History* 12:147–58.

Index

Abhandlungen zur theoretischen Biologie, 28–29
Academics, social status of, in Germany and United States, 189–90. *See also* Education, higher
Acetabularia, 317
Adams, Mark, 137
Advances in Genetics, 83
Agriculture, U.S. Department of, 158, 159, 161
Allen, Garland, 22n.37, 20, 52, 36, 97–98, 178–79
Amerikanisierung, 280
Amerikanismus, 280
Analogies (models), of plasmon theorists
 egalitarian, 331–32, 334
 enzyme-substrate, 332–34
 hierarchical, 331–32, 334
 mechanical, 325, 326, 333, 334
 role of, in scientific thought, 329–30
 social: English-speaking, 335–36, 338, 349; German, 330–31, 333–37; political derivation of, 347–49
Anti-Intellectualism in American Life (Hofstadter), 190
Antirrhinum, 202–3, 232
Anton Dohrn und die Zoologie seiner Zeit (Kühn), 240
Archiv für Rassen- und Gesellschafts-Biologie, 237
Ash, Mitchell, 141–42
Assistenten, 168–69
Atomistic concepts (theories), 315
Ausserordentlicher Professor, 169

Bateson, William, 19, 24, 36
 as a comprehensive, 355–56, 359
 friendship of, with Baur, 244
 views of: on chromosome theory, 42; on *Entwicklungsmechaniker,* 22–23; on evolution, 99–100
Bauer, Hans, 309–10
Baur, Erwin, 8, 36, 43, 62, 148–49, 170, 238–39. *See also* Baur's school (group)
 agricultural affiliations of, 236–37
 biographical sketch, 230–31

 breadth of knowledge, 231–33, 244–45
 conflict with Darré, 219–22, 261–62
 editorial activities of, 35, 135, 233
 friendship of, with Bateson, 244
 microcosm theory applied to, 346
 and pragmatic geneticists, 195–96
 research of, 63, 64–65
 views of: on eugenics, 237–38; on evolution, 101n.6; on Mendelian chromosome theory, 39, 41; on natural selection, 110–11, 114, 129; on specialization, 236
Baur's school (group), 7, 101–3, 211–12
 agricultural affiliations of, 210, 214–18
 compared to Kühn's, 225–26
 development of, after 1933, 218–24, 225
 funding patterns for, 214, 215 table 6.2, 216 table 6.3, 216–18
 organization of, 205 table 6.1
 research programs of, 200–202, 204
 social climate of, 207–8
 social profile of, 290 table 8.2, 292–94
 ties to academic institutions, 209–10, 212–13
 women in, 200–202, 206, 293–94, 312
 working conditions of, 204, 206
Beadle, George, 160, 360
 and Ephrussi, 85–87, 90, 93
 and Kühn, 6, 90, 91, 95–96, 358
 Nobel Prize winner, 6, 92, 361
 research of, 91–93, 159
 views of: on cytoplasmic inheritance, 82, 83; on evolution, 101, 102–3
Becker, Erich, 89–90, 92, 199
Becker, Gustav, 200
Belar, Karl, 39, 41, 151–52
Ben-David, Joseph, 177
Benson, Keith, 22n.37
Berlin, University of, 174–75
Berlin Agricultural College, 35, 36, 143, 144, 157, 210
Bertalanffy, Ludwig, 42
Beyond the Gene (Sapp), 178n
Bildung, 8, 278–79, 283, 284, 286, 307–8

415

Bildung (continued)
  as key to comprehensive style of thought, 276–77
Bildungsbürger, 274, 277
Biochemistry, 34, 140–41
"Biogenetic law," 18–19
Biology, 17–18
Biology, Institute of, 195, 200
Biometrician-Mendelian controversy, 99
Blakeslee, Alfred, 159
Bloor, David, 16
Bohr, Niels, 360–61
Botanical Society, German, 136
Botany, 32–33
Boveri, Theodor, 20, 21, 35, 50, 359
  and KWI for Biology, 176–77
  research of, 62–63
Breadth of knowledge, 135–38. *See also* Comparative analysis
  attitude toward: German, 7, 171–74, 176–78, 181–84; United States, 184, 190
  of comprehensives, 246, 248–49, 251–52, 255–58
  of individuals: Baur, 231–33, 244–45; Dobzhansky, 362; Huxley, 362; Kühn, 232–34, 240–44, 344; Renner, 341; Rensch, 362; Wright, 362
  of pragmatics, 246, 249–51, 253–54, 258
Bridges, Calvin, 50n.2
Brooks, W. K., 22n.37, 167
Brouwer, L. E. J., 353
Bruch, Rüdiger vom, 301–2, 313
Brush, Stephen, 13–14
Butenandt, Adolf, 89–90

California Institute of Technology, 49, 85–87, 92
Caneva, Kenneth, 15, 16
Cannon, Walter B., 349
Career trajectories, 310–12, 311 table 8.3
Carnegie Institute of Washington's Station for the Experimental Study of Evolution, 142
Caspari, Ernst, 32, 52n.7, 61, 83, 92
  research of, 87, 89, 199
Castle, W. E., 23, 35
  Genetics and Eugenics, 99
Caullery, Maurice, 163
Chemotherapy, Institute for Experimental, (Georg-Speyer-Haus), 196–97

Chicago, University of, 25, 144–45
Chromosome theory, Mendelian, 33–34, 36, 39, 41–43
Chun, Carl, 26n
Clark University, 162–63
"Class proposes, institution disposes," 307–10, 312
Cleland, Ralph, 79
Cole, L. J., quoted, 135
Coleman, William, 15
Columbia University, 36, 49, 82
Comparative analysis, 3, 5, 12–14
  sources of: correspondence, 185–88; diary, 188–89, 191; obituaries, 181–84, 183 table 5.1
Comprehensives, 195, 225–26, 306–12. *See also* Breadth of knowledge; Styles of thought, comprehensive
  defined, 189
  individuals as: Bateson, 355–56; 359; Boveri, 359; Dobzhansky, 362; Kühn, 8; Wright, 362
  as mandarins, 299, 351–52
  political outlook of, 266–69
  social profile of, 288–89, 289 table 8.1, 291–92
Conant, James Bryant, 163
Conklin, E. G., 22n.36, 61
Correns, Carl, 34, 36, 41, 43, 50, 135, 170
  rediscovered Mendel, 35
  research of, 62–65, 72–73
  views of: on evolution, 116–17; on plasmon theory, 321–23, 325–28
Coser, Lewis, 363
"Crisis of learning" *(Krise der Wissenschaft),* 280–83
Cross-national studies, 2–4
Culture, Prussian Ministry of, 214
Cytoplasm, *Grundstock* in, 61, 72–73, 106
Cytoplasmic inheritance, 8, 61, 106, 115–16, 120–21, 337. *See also* Plasmon theory
  analogies of: egalitarian, 331–32, 334; enzyme-substrate, 332–34; hierarchical, 331–32, 334; mechanical, 325, 326, 333, 334; social, 330–31, 333–38, 347–49
  in animal species, 73, 75
  difficulties with demonstrating, 62–63
  implications of, for evolutionary theory, 114–19
  Kernmonopolists' views on, 317–20

research on, 65–67, 69, 75–77, 79–80
response to, in United States, 82–84

Danziger, Kurt, 141
Darlington, Cyril, 335
Darré, Walther
  conflict with Baur, 219–22, 261–62
  Neuadel aus Blut und Boden, 222
Darwin, Charles, 18, 22
  Origin of Species, The, 44
Darwinism, 100
Dauermodifications
  controversy over, 119–26
  and cytoplasmic inheritance, 126–29, 317, 320
Dauermodifikationen, translated, 119n.54
Davenport, C. B., 23
Decline of the West (Spengler), 282
Descriptive morphology, 31
Developmental genetics, 86, 94
  funding for, research, 85, 86, 152–54, 172–73, 214, 235
  interest in: German, 50–52, 54–57, 60–61, 138–39; United States, 49, 50, 84–85, 96–97
DeVries, Hugo, 22, 99
  rediscovered Mendel, 35
Dobzhansky, Theodosius, 101–3, 110, 134, 362
  Genetics and the Origin of Species, 44, 102, 115, 117, 117n.51, 118n.53
  research of, 86, 87, 89, 96
Dohrn, Anton, 234, 240
Döring, Herbert, 299–300, 302–3, 313
Douglas, Mary, 15–17
Driesch, Hans, 22n.37, 29, 30, 61
  quoted, 21
Drosophila, 34, 85–87, 90, 91
Dualist theories of evolution, 35, 104–10, 116, 127, 129
Duhem, Pierre, 1, 3
Dunn, L. C., 1, 51, 103, 159, 335
  and developmental genetics, 82, 96
  proponent of Jollos's work, 125–26
Dürken, Bernhard, 106–7

East, E. M., 82, 163
Ecology, 23, 24
Education, higher
  German: effect of economy on, 150–55; fees system in, 164–65; funding of, 145–49; and Kaiser-Wilhelm Society, 173–77; outside influences on, 160–62; power of professor in, 167–73; role of faculty in policy-making, 163–65
  United States: effect of economy on, 154–55; fees system in, 166; outside influences on, 57–62; power of professor in, 166–67; role of faculty in policy-making, 162–63, 165–66; university presidents' power in, 162–63, 165
Education Board, International, 51
Education reform, 307
  in secondary schools, 283–87
  in universities, 287–88
Ehlers, Ernst, 31
Ehrlich, Paul, 196–97
Eimer, Theodor, 52
Einzelelemente, 70
Emergency Council for German Science, *(Notgemeinschaft der Deutschen Wissenschaft),* 152, 172, 214, 236
Emerson, R. A., 35, 96
Emerson, Stirling, 86
Emigrés, German, 179–80, 265
Entfaltungsmechanismus, 64
Entwicklung, 28
Entwicklungsgeschichte, 19, 54, 234
Entwicklungsmechanik, 19–23, 22n.37, 30–31
Entwicklungsmechaniker, 21–23, 31
Entwicklungsphysiologie, 19n.26
Entwicklungsphysiologische Genetik. See Developmental genetics
Ephestia, research on, 57, 58 fig. 2.3, 60, 87, 89–90, 94–96, 199
Ephrussi, Boris, 95–96
  and Beadle, 85–87, 90, 93
  Nucleo-Cytoplasmic Relations in Microorganisms, 96
Epilobium
  research on, 66, 80, 318–19, 324; by Michaelis, 75–77, 79, 128, 324, 328
Erbstock, 107
Eugenics (racial hygiene), 211–12, 237–38
Evolution, 23. *See also* Dualist theories of evolution; Macroevolution
  and cytoplasmic inheritance, 114–19
  neglect of, in United States, 100–103, 132–33
  views on: of Bateson, 99–100; of Baur, 101n.6; of Baur's school, 101–3; of Correns, 116–17; of Mayr, 129–30; of Sturtevant, 101–3; of von Wettstein, 117–18

Experimental biologists, 31
Experimentelle Morphologie, 19n.26
"Expressivity," 55
Extraordinarius, 169

Federley, Harry, 120–21, 135
Fick, Rudolf, 41, 106
Fischer, Emil, 354–55
Fisher, R. A., 100–101
Fleck, Ludwik, 15
Flexner, Abraham, 148
Flour moth *(Ephistia),* research on, 57, 58 fig. 2.3, 60, 87, 89–90, 94–96, 199
Food and Agriculture, German Ministry of, 214
Forced alignment *(Gleichschaltung),* 219, 340
Ford, E. B., 137
Foster, Michael, quoted, 23–24
Franz, Viktor, 29, 110
Frisch, Karl von, 63nn.63–64, 167, 232, 233
Fruton, Joseph, 354

Garrod, Archibald, 86
Genes (Mendelian "factors"), 35
Geneticists
  and morphologists, 101
  reserach interests of, compared, 140–42 table 4.1
Genetics, 4, 24, 44. See also Developmental genetics
  agricultural influences: on German, 160–62, 236–37; on United States, 157–60, 161–62, 178–79
  German, 6, 34–36, 191–92
  United States, 5, 33–34, 191–92
Genetics, 35, 111n.34
Genetics, International Congress of, 36, 72, 109, 135–36, 237
Genetics and Eugenics (Castle), 99
Genetics and the Origin of Species (Dobzhansky), 44, 102, 115, 117, 117n.51, 118n.53
Genetics Society, German, 65, 119, 160, 331
  established, 135, 142, 239
Genetics Society of America, 85, 160
German Society for Racial Hygiene, 237
Gesellschaft deutscher Naturforscher und Ärtze, 26, 29–30
Gestalt school, German, 139–40
Gleichschaltung (forced alignment), 219, 340

Gluecksohn-Waelsch, Salome, 96
Glum, Friedrich, 206
Goldschmidt, Richard, 31, 32, 35, 41, 44, 84–85
  features of work, 51–52
  Material Basis of Evolution, The, 131, 361–62
  phenocopy method named by, 60
  Physiological Genetics, 85
  Physiologische Theorie der Vererbung, 51, 52n.7, 55
  reaction to *Genetics and the Origin of Species,* 103
  research of, 52, 101n.6, 111–12, 123–24
  unorthodox views of, 34, 51
  views of: on chromosome theory, 42, 131; on cytoplasmic inheritance, 317–18; on developmental genetics, 49–50; on natural selection, 129, 131
Graevenitz, Luise von, 200
Gramsci, Antonio, 285
Grégoire, Victor, quoted, 73
Grid-group theory, 15–17
Grundstock hypothesis, 105–6, 108, 116–17, 120
  rejection of, 115–16, 118–19, 127
  support for, 109–10
Grundzüge einer allgemeinen Biologie (Woltereck), 109n.29
Grüneberg, Hans, 171–72

Haeckel, Ernst, 18–19, 22
Haecker, Valentin, 29, 35, 42, 131, 135
  research of, 52, 54–55
Haldane, J. B. S., 86, 100–101
Hall, G. Stanley, 162–63
Halle, University of, 35, 54
Hamburger, Viktor, 95
Hämmerling, Joachim, 317
Harder, Richard, 66, 75
Harnack, Adolf von, 212
Harper, William Rainey, 144–45
Harrison, Ross, 21, 95, 110
  quoted, 93
Hartmann, Max, 44–45
Hausdorff, Felix, 353
Heberer, Gerhard, 110
Heider, Karl, 22n.37, 26, 172, 175
Heilbron, John, 360
Helmholtz, Hermann, 26
Henke, Karl, 57, 60, 89, 199

Hertwig, Oscar, 20, 22n.37, 175
Hertwig, Paula, 116, 201, 328
Hertwig, Richard, 26
Hilbert, David, 353
Historians of science, 2, 3–4, 13–17
Hoffmann, August, 27n.48
Hofmeister, Franz, 354–55
Hofstadter, Richard, *Anti-Intellectualism in American Life*, 190
Holistic concepts (theories), 315
Holtfreter, Johannes, 359
Holton, Gerald, 14–15
Horowitz, Norman, 94–95
Hughes, Stuart, 363
Humanistisches Gymnasium, 285–87
Humboldt, Wilhelm von, 276–77
Husfeld, Bernhard, 222–23
Huxley, Julian, 31, 101, 137, 362

Instrumentalist philosophy, 352–53, 360–61
Ives, P. T., 124

Jennings, H. S., 125–26
Jensen, Paul, 29
Johannsen, Wilhelm, 20, 35, 105–6
John Innes Horticultural Institution, 86
Johns Hopkins University, 20–21, 166–67
Johnson, Jeffrey, 354
Jollos, Victor, 131, 151
  research of, 121, 122–26, 127
Joravsky, David, 219
Jordan, Pascual, 242
Journal for the Inductive Study of Evolution and Heredity, 135
Just, E. E., 335, 338

Kaiser-Wilhelm Institute (KWI)
  for Anthropology, Human Genetics, and Eugenics, 237
  for Biology, 60, 67, 69, 142, 196; committees of, 175–76, 197, 209, 212–13, 219–20; conditions at, 149–50; established, 30–31, 36; funding of, 215 table 6.2; organization of, 173–77, 205 table 6.1
  for Brain Research, 55
  for Breeding Research, 75, 209, 219; established, 142–43; funding of, 215 table 6.2, 216 table 6.3; organization of, 173–74, 205 table 6.1; resemblance of, to industrial laboratory, 196, 197, 218, 222–23, 225
  for Experimental Therapy, 197
Kaiser-Wilhelm Society, 149–50, 173–77
Kalmus, Hans, 32
Kay, Lily, 92
Kernmonopol (nuclear monopoly), 65, 315, 316, 318–19, 331–32
Kernmonopolists, 317–20, 328, 340, 347–48
Kinsey, Alfred, 110
Klein, Felix, 353
Knapp, Edgar, 328, 345–46
Kohler, Robert, 34, 142
Köhler, Wolfgang, 140
Korschelt, Eugen, 31–32
Krise der Wissenschaft ("crisis of learning"), 280–83
Kuckuck, Hermann, 202, 204, 223–24
Kühn, Alfred, 7–8, 28, 32n.63, 83, 85. *See also* Kühn's school (group)
  as ally of plasmon theorists, 73, 75
  Anton Dohrn und die Zoologie seiner Zeit, 240
  and Beadle, 6, 90, 91, 95–96, 358
  biographical sketch, 229–30
  breadth of knowledge, 232–34, 240–44, 344
  and comprehensive geneticists, 195–96
  microcosm theory applied to, 344–45
  research of, 57, 60, 89–90, 94–96, 199, 231–32
  "Über den biologischen Wert" (Kühn), 117n.51
  views of: on mechanism, 241–42; on specialization, 234–36; on teamwork, 235–36; on vitalism (mystics), 242
Kühn's school (group), 211–12
  compared to Baur's 225–26
  development of, after 1933, 218–19, 224–25
  funding patterns for, 214, 215 table 6.2, 218
  organization of, 205 table 6.1
  research programs of, 197, 199–200, 204
  social climate of, 207–8
  social profile of, 290 table 8.2, 294
  ties to academic institutions, 208–11
  working conditions of, 204, 206–7
Kükenthal, Willy, 172
Kultur, 279–83, 288, 297

420   Index

Kuratorium, 175–76, 197, 209, 219–20
Küster, Ernst, 29
KWI. *See* Kaiser-Wilhelm Institute

Lamarck, Jean-Baptiste-Pierre-Antoine de Monet de, 18
Lamarckian theories, 28
Landauer, Walter, 142
Land-grant universities, 158–60
Lehmann, Ernst, 66, 318–19
Lehmann, Wolfgang, 113n.39
Lindegren, Carl, 116
Lochow, F. von, 214, 217
Lowell, James, 163
Ludwig, Wilhelm, 110, 113–14
Lüers, Herbert, 113n.39

McClintock, Barbara, 159
MacDougal, D. T., 23
Mach, Ernst, 26, 353
MacKenzie, Donald, 15, 355
Macroevolution, 100, 105, 136–37
Maienschein, Jane, 22n.37
Mandarins, 8, 276
  as comprehensives, 299, 351–52
  modernist, 296–97, 300–301
  orthodox, 296–97, 300–301
  and outsiders, 314, 363
  in the United States, 359–60
Mannheim, Karl, 9–11, 15, 17, 313, 352
  quoted, 179
Marine Biological Laboratory, 22n.37
Martin, H. Newell, 166–67
Material Basis of Evolution, The (Goldschmidt), 131, 361–62
Matrix (species plasma), 108–9
Mayr, Ernst, 6, 100–101, 103, 129–30, 136–37
  quoted, 134
Melchers, Georg, 200
Mendel, Gregor, 34–35
Mandelian-biometrician controversy, 99
Mendelian chromosome theory, 33–34, 36, 39, 41–43
Mendelian "factors," 35
Medelians, 28, 62, 99
Mendelism, 23, 50, 100
  German interest in, 6–7, 33–36, 38–39, 41–45
Merogones, 62–63, 66
Merriam, C. Hart, 22, 23, 24
Merz, John Theodore, 1
Methodology, problems of, 227–29
Meyer-Abich, Adolf, 42, 242

Michaelis, Peter, 83, 117, 128–29, 312.
  See also Plasmon theory, revisionist conceptions of
  career of, 75
  microcosm theory applied to, 344
  research of, 75–77, 79
  social analogy developed by, 333–34
Microcosm theory, 8–9, 352
  applied to: audiences of cytoplasmic debate, 344–47; critics of plasmon theory, 340; orthodox plasmon theorists, 338–40; revisionist plasmon theorists, 340–44
  described, 338–39
  summarized, 347–50
Microevolution, 100, 105
Misfits, 311–12
Modernization, 274–76
  effects of: in Britain, 355; in Germany, 351–55, 358–59; in United States, 356–58
Morgan, T. H., 24n.42, 35, 49, 101, 102.
  See also Morgan's school (group)
  chromosome theory developed by, 33–34, 36, 39, 41–43
  Nobel Prize winner, 34, 92, 361
  Physical Basis of Heredity, 38–39
  Theory of the Gene, 39
  views of: on cytoplasmic inheritance, 70, 82; on developmental genetics, 50, 96; on systematists, 133–34
Morgan's school (group), 36, 38, 39, 85–86, 96–98
Morphologists, 101
Morphology, 17–23
Muller, H. J., 83, 84, 119, 123
  Nobel Prize winner, 361
Mutationism, 99
Mystics *(Schwärmgeister),* 242

Nachtsheim, Hans, 38, 39, 160, 161, 202
Nanney, D. L., quoted, 335–36
Naples Zoological Station, 232, 234, 240
Nationalist Socialist German Workers' Party (NSDAP), 218, 222–23, 225, 259–60
Naturalists, American Society of, 25, 30, 85, 99
Natural selection, 18, 28
  debate over, 104–14
  endorsed, 129–30
  and plasmon theory, 118–19
  skepticism about, 100, 131–32
Naturwissenschaften, Die, 28, 41

Neo-Lamarckism, 39, 99, 118–19
Neuadel aus Blut and Boden (Darré), 222
Neurospora, 91–93, 94
Nilsson-Ehle, Hermann, 217
Noack, Konrad L., 322–23
Nobel Prize, 6, 34, 92, 361
Notgemeinschaft der Deutschen Wissenschaft (Emergency Council for German Science), 152, 172, 214, 236
NSDAP (Nationalist Socialist German Workers' Party), 218, 222–23, 225, 249–60
Nuclear monopoly *(Kernmonopol)*, 65, 315, 316, 318–19, 331–32
Nucleo-Cytoplasmic Relations in Microorganisms (Ephrussi), 96

Oberrealschule, 286–87
Obituaries, comparative analysis of, 181–84, 183 table 5.1
Oehlkers, Friedrich, 50, 69, 82, 83, 117
research of, 79–80
views of, on plasmon theory, 321, 322, 324, 326
Oenothera, 66–67, 69, 319, 322, 324–26
"One gene—one enzyme," 85
Onslow, Muriel Wheldale, 86
Oppenheimer, Jane, quoted, 233, 257
Origin of Species (Darwin), 44
Orthogenetic sequences, 122
Osborn, H. F., 19, 24
Outsiders, 308
and mandarins, 314, 363
as pragmatics, 351–52
within professoriate, 312–14
sympathetic to modernization, 298 figure 8.1, 303–6, 305 figure 8.2
in United States, 359–60

Paleontological Society, German, 119
Pätau, Klaus, 113
quoted, 259–60
Paulsen, Friedrich, quoted, 281
Pearl, Raymond, 35, 159
Pearson, Karl, 355
"Penetrance," 55
Petzoldt, Joseph, 241
Phänogenetik, 54–55
Physical Basis of Heredity (Morgan), 38–39
Physiological genetics. *See* Developmental genetics
Physiological Genetics (Goldschmidt), 85

Physiologische Theorie der Vererbung (Goldschmidt), 51, 52n.7, 55
Piepho, Hans, 199
Plagge, Ernst, 89–90, 92, 199
Planck, Max, 29, 219–21, 223, 353
Plant Breeding and Hybridization, International Congress for, 158
Plasmagenes, 65
Plasmon, 70, 72–73, 126–29, 325–29
effects of, 322–24
function of, 321–22
genetic structure of, 324–25
Plasmon theory, 104, 116–17
critics of, 316–20, 340
debate, 315–16, 316 table 9.1; political outlooks reflected in, 338–44, 352
and Kernmonopolists, 347–48
and natural selection, 118–19
research on, 69–70, 72–73, 75–77, 79–80, 82
revisionist conceptions of: Michaelis, 321–23, 325, 328; Renner, 322–27, Schwemmle, 321–22, 324–26
Plate, Ludwig, 35, 107–8
Plough, H. H., 124
Poincaré, Henri, 353
Political outlook. *See also* Professoriate, political outlook of, analyzed
of groups: Baur's, 258, 260–61; comprehensives, 266–69; Kernmonopolists, 340; Kühn's, 258, 263; pragmatics, 266, 268–69; von Wettstein's, 264–65
of individuals: Baur, 261–63; Kühn, 263–64; von Wettstein, 265–66
reflected in plasmon theory debate, 338–44, 352
Pontecorvo, Guido, 95
Population genetics, 100–01, 111–14
Pragmatics, 195–96, 225–26, 306–12. *See also* Styles of thought, pragmatic
breadth of knowledge, 246, 249–51, 253–54, 258
defined, 189
political outlook of, 266, 268–69
social profile of, 288, 289 table 8.1, 291–92
as unmandarin, 299–301
Privatdozenten, 167–69, 209
Professoriate, political outlook of analysis of, 295–96, 298 figure 8.1,

422   Index

Professoriate (*continued*)
303–6, 305 figure 8.2; vom Bruch's, 301–2, 313; Döring's, 299–300, 302–3, 313; Ringer's, 296–97, 300–302, 304, 312–13; Schwabe's, 297–99, 302, 312–13; Töpner's, 298–99, 312–13
Provine, William, 100
Przibram, Hans, 29

Racial hygiene (eugenics), 211–12, 237–38
Radical critics, 300–301
Realgymnasium, 286–87
Realist philosophy, 352–53, 360–61
Reche, Otto, 113n.39
Reinig, W. F., 111
Renner, Otto, 28, 69, 75, 117, 134n.87. *See also* Plasmon theory, revisionist conceptions of
microcosm theory applied to, 341–42
research of, 66–67, 69, 80, 82
Rensch, Bernhard, 101, 110, 122, 362
Rhoades, Marcus, 159
Riddle, Oscar, 96
Ringer, Fritz, 7, 279, 281, 296–97, 300–302, 304, 312–13
Rockefeller Foundation, 85, 86, 152–54, 172, 214, 235
Rosenberg, Charles, 1–2
Roux, Wilhelm, 19–23, 31, 32, 61
Rubner, Max, 26n
Rudorf, Wilhelm, 223–24
Rudwick, Martin, 16

Sapp, Jan, 83–84, 130–31
Beyond the Gene, 178n
Schaxel, Julius, 22n.36, 27–30
Scheler, Max, 277
Schick, Rudolf, 204, 223–24
Schiemann, Elisabeth, 200–201
Schlösser, L. A., 200
Schulze, F. E., 32
Schwabe, Klaus, 297–99, 302, 312–13
Schwantiz, Franz, 113n.39
Schwärmgeister (mystics), 242
Schwartz, Viktor, 199
Schwemmle, Joachim, 66, 69, 312. *See also* Plasmon theory, revisionist conceptions of
microcosm theory applied to, 342–43
research of, 80, 82, 319
Science, historians of, 2–4, 13–17
"Science as a Vocation" (Weber), 303

Sciences, Prussian Academy of, 32, 172, 174–75
Seiler, Jakob, 32n.63, 171
Selectionists, German, 110
Sengbusch, Reinhold von, 217–18, 224
Serum Testing and Research, Institute for, 196
Shull, George, 23, 35, 36, 133
Simpson, G. G., 101
Sinnott, Edmund, 82, 96, 359
Sonneborn, Tracy, 82–84, 116, 125–26
Specialization, response to
German, 26–33
United States, 23–25
Species plasma (matrix), 108–9
Spemann, Hans, 21, 28, 30, 233, 243–44, 359–60
Spencer, Warren, 134
Spengler, Oswald, *Decline of the West,* 282
Stebbins, Ledyard, 103, 110
Stein, Emmy, 201–2
Steinmann, Gustav, 135
Stern, Curt, 39, 85, 91, 308–9
Stieve, Hermann, 41, 106
Straub, Joseph, 118n.52
Stresemann, Erwin, 101
Strohl, Jean, 235–36
Stubbe, Hans, 204, 223–24
Sturtevant, A. H., 35, 38, 101–2, 160
research of, 86, 87, 89
views of: on cytoplasmic inheritance, 82, 83; on evolution, 101–3
Style, concept of, described, 9–12
Styles of science, 13–17
Styles of thought, 9, 14–17, 228 table 7.1, 272–73
basis for determining, 11–13
coherence of, 269–72
compared, 247 table 7.2, 248 table 7.3
comprehensive, 352–55; and *Bildung,* 275–77, 283; relative success of, 361–62; in United States, 356–58
defined, 245
national, 359–60
pragmatic, 283, 352–55; relative success of, 361–62; in United States, 356–48
psychological, 10, 11n.14
and school choice, 291–92
and social class, 289, 291
Süffert, Fritz, 110
Sumner, F. B., 101, 111n.34
Sutton, Walter, 35
Svalöf, 217

Synthesizers, 136–37
Systematics, 24, 31, 32

Tatum, Edward, 85, 87, 90–93
Theory of the Gene (Morgan), 39
Timoféeff-Ressovsky, Helena, 55–56, 111
Timoféeff-Ressovsky, N. W., 55–56, 110, 111, 114–15, 122, 129
Töpner, Kurt, 298–99, 312–13
Transmission genetics, 51, 138–39
Tschermak-Seysenegg, Erich, rediscovered Mendel, 35

"Über den biologischen Wert" (Kühn), 117n.51
Ubisch, Gerda von, 200
Unmandarin, pragmatics as, 299–301

Vererbung, 28
Vererbungslehre, 27
Vererbungswissenschaft, 35
Verwaltungsausschuss, 209, 219
Vienna Circle, 284–85, 361
Vogt, Oskar, 55–56n.14
Volksgemeinschaft, 284
Von Wettstein's school (group). *See also* Kühn's school (group)
Social profile of, 290 table 8.2, 294

Waddington, Conrad, 95, 359
Weaver, Warren, 85
Weber, Max, 275, 301, 302
"Science as a Vocation," 303
Weidel, W., 90
Weidenreich, Franz, 121
Weimar Circle, 302–3
Weinberg, Wilhelm, 34
Weismann, August, 20, 22, 52, 61
Weldon, W. F. R., 23
Weltanschauung, 282, 339
Wettstein, Fritz von, 44, 50, 70, 72, 80, 110. *See also* Von Wettstein's school (group)
career, 69
and comprehensive geneticists, 195–96
Grundstock hypothesis rejected by, 116

Nachtsheim's analogy challenged by, 335
research of, 69–70, 72, 76, 77
views of: on evolution, 117–18; on dauermodifications, 127; on departmental structure in U.S. universities, 167; on plasmon theory, 321, 322, 325–28
Wettstein, Richard von, 109–10, 135
Wheeler, William Morton, 24, 163
Whitman, C. O., 25
Wilson, E. B., 20–22, 22n.37
Winkler, Hans, 35, 50, 65–66, 70, 72, 76, 106
Wissenschaftlicher Beirat, 175–76, 209, 212
Witschi, Emil, 31, 50n.2
Woltereck, Richard, 108–9
Grundzüge einer allgemeinen Biologie, 109n.29
microcosm theory applied to, 347
Women, in Baur's school, 200–202, 206, 293–94, 312
Woodger, Joseph, 61–62
Wright, Sewall, 96, 100–101, 114, 134, 159, 362

Zarapkin, S. R., 111
Zeitschrift für induktive Abstammungs- und Vererbungslehre (ZIAV), 21, 35, 38, 54, 135
Zeitschrift für vergleichende Physiologie, 232
Zentralblatt für allgemeine und experimentelle Biologie, 19
ZIAV (*Zeitschrift für induktive Abstammungs- and Vererbungslehre*), 21, 36, 38, 54, 135
Zimmer, K. G., 55–56n.14
Zimmermann, Klaus, 111
Zimmermann, W., 110
Zivilisation, 279–83, 288, 297
Znaniecki, Florian, 363
Zoological Society, German, 32, 136
Zoologists, American Society of, 2, 24
Zoologists, skeptical of cytoplasmic inheritance, 319–20
Zoology, Institute of, 56, 195, 197, 200